Our Daily Poison

Also by Marie-Monique Robin

*The World According to Monsanto: Pollution, Corruption,
and the Control of Our Food Supply*

OUR DAILY POISON

From Pesticides to Packaging, How Chemicals Have Contaminated the Food Chain and Are Making Us Sick

MARIE-MONIQUE ROBIN

Translated by Allison Schein and Lara Vergnaud

THE NEW PRESS

NEW YORK
LONDON

Cet ouvrage publié dans le cadre du programme d'aide à la publication bénéficie du soutien du Ministère des Affaires Etrangères et du Service Culturel de l'Ambassade de France représénté aux Etats-Unis. This work received support from the French Ministry of Foreign Affairs and the Cultural Services of the French Embassy in the United States through their publishing assistance program.

The New Press gratefully acknowledges the Florence Gould Foundation for supporting publication of this book.

Published in the United States by The New Press, New York, 2014
Distributed by Perseus Distribution

LIBRARY OF CONGRESS CATALOGING-IN-PUBLICATION DATA
Robin, Marie-Monique.
  [Notre poison quotidien. English]
  Our daily poison : from pesticides to packaging, how chemicals have contaminated the food chain and are making us sick / Marie-Monique Robin ; translated by Allison Schein and Lara Vergnaud.
     pages cm
  Originally published in French as: Notre poison quotidien (Paris : Dicouverte, 2011).
  Includes bibliographical references and index.
  ISBN 978-1-59558-909-5 (hardcover : alk. paper)—ISBN 978-1-59558-930-9 (e-book)
  1. Environmental toxicology.   2. Chemical industry—Environmental aspects.
  3. Chronic Disease—Environmental aspects.   4. Chemicals—Physiological
  effect.   5. Pesticides—Physiological effect.   6. Food contamination.   I. Title.
  RA1226.R6213 2014
  363.738'498—dc23

                                                           2014027768

The New Press publishes books that promote and enrich public discussion and understanding of the issues vital to our democracy and to a more equitable world. These books are made possible by the enthusiasm of our readers; the support of a committed group of donors, large and small; the collaboration of our many partners in the independent media and the not-for-profit sector; booksellers, who often hand-sell New Press books; librarians; and above all by our authors.

www.thenewpress.com

Composition by Westchester Publishing Services
This book was set in Fairfield Light Italic

Printed in the United States of America

2 4 6 8 10 9 7 5 3

Contents

Our Daily Poison

INTRODUCTION

Knowledge Is Power

Will this book be the sequel to *The World According to Monsanto?*[1] I have constantly been asked this question since 2008 when I announced at a lecture or a debate that I was working on a new project. Yes and no: this book is and is not a "sequel to *Monsanto*," even though the material is obviously related to the earlier investigation. Indeed, books and films—for me the two are closely connected—are like pearls on a necklace or pieces of a jigsaw puzzle: they follow one another and fit together without my realizing it. They arise indirectly from and are nourished by the questions growing out of the work that went before. And they end up taking their place as links in a single chain. In every case the process at work is the same: the wish to understand in order to then communicate to the widest audience the knowledge gained.

Three Questions About the Role of the Chemical Industry

So, *Our Daily Poison* is the outcome of a long process that began in 2004. At the time, I was worried about the threats weighing on biodiversity: in two documentaries broadcast on Arte on the patenting of live organisms and the history of wheat,[2] I had described how multinationals secure unwarranted patents on plants and know-how of countries of the global South. At the same time I was shooting a documentary in Argentina detailing the disastrous consequences of the cultivation of transgenic soy, the notorious Roundup-ready soy from Monsanto.[3] For these three films, I had traveled to the four corners of the planet, calling into question the agro-industrial model established after World War II whose avowed purpose was to "feed the world." I had observed that it brought about an expansion of monocultures at the expense of family food-producing agriculture, leading to a drastic reduction in biodiversity. In the long term, this poses a threat to the food security and sovereignty of the peoples of the world. I also found that the celebrated green revolution went along with an impoverishment of natural resources (soil and water quality) and widespread pollution of the environment because of the massive use of chemical products (pesticides and artificial fertilizers).

This trilogy quite naturally led me to develop an interest in the American company Monsanto, one of the major promoters and beneficiaries of the green revolution: first because it was (and continues to be) one of the principal manufacturers of pesticides in the twentieth- and twenty-first centuries; and because it has become the largest seed producer and is trying to take control of the food chain by means of patented transgenic seeds (genetically modified organisms [GMOs]). I can never adequately express my surprise at discovering the many lies, manipulations, and dirty tricks the Saint Louis firm is capable of to keep highly toxic chemical products on the market, whatever the environmental, health, and human costs.

And as I advanced in this "thriller of modern times," to quote the sociologist Louise Vandelac, who wrote the preface for the Canadian edition of *Le Monde selon Monsanto*, three questions constantly plagued me. Was Monsanto an exception in industrial history, or, on the contrary, does its criminal conduct (I choose my words carefully) characterize the majority of chemical product manufacturers? One question led to another, and I also wondered how the approximately one hundred thousand synthetic chemical molecules that have invaded our environment and our dinner plates for a half century are evaluated and regulated. Finally, is there a link between exposure to these chemical substances and the spectacular increase in cancers, neurodegenerative diseases, reproductive disorders, diabetes, and obesity that have been recorded in developed countries, to such a degree that the World Health Organization (WHO) speaks of an "epidemic"?

To answer these questions, I decided in this new investigation to focus solely on the chemical substances that come into contact with the food chain, from the farmer's field (pesticides) to the consumer's plate (food additives and food grade plastics). Thus this book will not address electromagnetic waves, mobile phones, or nuclear pollution, but solely the synthetic molecules we are exposed to in our environment and our food—our "daily bread" that has largely become our "daily poison." Knowing that the subject is highly controversial—which is not surprising considering the importance of the economic stakes involved—I have chosen to proceed methodically, starting from the simplest example that is least open to question, namely, acute and chronic poisoning of farm workers directly exposed to pesticides, and moving gradually to the most complex issue, the effects at low doses of the residues of chemical products that we all have in our bodies.

Gathering the Pieces of the Puzzle

Our Daily Poison is the product of a long investigation that mobilized three kinds of resources. I first consulted about a hundred books, written by historians, sociologists, and scientists—the majority from North America. My study thus owes a good deal to the invaluable research carried out by highly talented academics like Paul Blanc, professor of occupational and environmental medicine at the University of California, his historian colleagues Gerald Markowitz and David Rosner, and David Michaels, an epidemiologist appointed head of the U.S. Occupational Safety and Health Administration (OSHA), the agency in charge of workplace safety. Their thoroughly documented books provided access to a mass of unpublished archives and helped me locate the subject of my investigation in a broader context of industrial history.

So, I went back to the origins of the industrial revolution, which preceded the green revolution, two faces of the same insatiable monster: progress, supposed to bring us universal happiness and well-being; but all indications are that, like a modern-day Saturn, progress threatens to devour its own children. Without this historical view, it is indeed impossible to understand how the regulatory system for chemical products was invented and still operates today—a system rooted in the persistent contempt of manufacturers and public authorities for the factory workers who paid a heavy tribute to the chemical madness of so-called developed societies.

This book also relies on many archival documents that I was able to procure from lawyers, nongovernmental organizations, experts, and particularly stubborn individuals, all of whom have accomplished considerable work to document the misdeeds of the chemical industry. One example is the amazing Betty Martini of Atlanta, whom I salute for her perseverance in gathering evidence against the highly suspect artificial sweetener, aspartame. I have, of course, kept copies of all the documents I cite in these pages. All these documents helped me to reconstruct the puzzle of which this book aims to present a clear, if not definitive, picture.

But the task would have been incomplete if it had not also been informed by the fifty personal interviews that I conducted in the ten countries where my investigation led me: France, Germany, Switzerland, Italy, Great Britain,

Denmark, the United States, Canada, India, and Chile. Among the major witnesses I questioned were seventeen representatives of agencies that evaluate chemical products, such as the European Food Safety Authority (EFSA), the U.S. Food and Drug Administration (FDA), and the International Agency for Research on Cancer (IARC), under the authority of the WHO, as well as the Joint Meeting on Pesticide Residues (JMPR), a joint committee of the WHO and the UN Food and Agriculture Organization (FAO) charged with evaluating the toxicity of pesticides. I also interviewed thirty-one scientists, primarily European and American, to whom I would also like to pay tribute, because they are continuing to fight to maintain their independence and to defend a conception of science at the service of the common good, not private interests. These long conversations were all filmed, and they are also a part of my film *Our Daily Poison*[4] that goes along with this book.

The Devil Is in the Details

Our Daily Poison is, finally, the product of a conviction I would like to be shared: we have to retake control of what is on our plates and gain an understanding of what we eat so that we are no longer fed small doses of poisons that provide no benefits. As Erik Millstone, a British academic, explained to me, in the current system "it's the consumers who take the risks and the companies that get the profits." But to be able to criticize the many failings of the system and demand that it be reformed from top to bottom, we have to understand how it operates.

I must admit that it was not easy to decipher the mechanisms that control the establishment of the norms governing exposure to what the euphemistic jargon of the experts calls "chemical risks." It was extremely difficult, for example, to trace the origin of the "acceptable daily intake" (ADI) for poisons to which we are all exposed. I even suspect that the complexity of the system of evaluation and regulation of chemical poisons, which always operates behind closed doors and with the greatest secrecy, is also a way of guaranteeing its permanence. Who would stick his nose into the history of the ADI, or the "maximum residue limits"? And if, by chance an overly curious journalist or consumer dares to ask questions, the regulatory agencies generally answer: "It works more or less. Besides, you know, it's very complicated; trust us, we know what we're doing."

The problem is that there cannot be any more or less when it comes to toxicological data, when what is at stake is the health of consumers, including future generations. This is why, convinced rather that the devil is in the details, I decided to take the opposite tack. I hope readers will forgive me for what they might sometimes consider an excessive concern for precision or explanation, and the proliferation of notes and references. But my aim is for everyone to have available the rigorous arguments enabling them to act within the limits of their resources, and even to influence the rules of the game governing our health, because knowledge is power.

PART I

Pesticides Are Poisons

1

The Ruffec Appeal and the Battle
of Paul François

Humanitarianism consists in never sacrificing a human being to a purpose.
—Albert Schweitzer

It was a beautiful winter day, cold and sunny. And the date, Sunday, January 17, 2010, will remain forever stamped on my memory, and also on the history of French agriculture. Thirty farmers, suffering from serious illnesses—cancer, leukemia, or Parkinson's disease—had agreed to meet at the initiative of the Movement for Law and Respect for Future Generations (Mouvement pour le droit et le respect des générations futures, MDRGF),[1] an association that has been fighting for fifteen years against the ravages of pesticides. Planned far in advance, this first meeting of its kind in the world had been organized in Ruffec, a town of 3,500 in Charente. I had left Paris the day before on a TGV with Guillaume Marin, cameraman, and Marc Duployer, sound engineer, my two unfailing associates who have traveled with me to the four corners of the earth to film the investigation that is the source of this book.

As soon as I was settled in the train, I had opened my laptop, thinking I would use the two and a half hours of the trip to work. But as the countryside rolled past the misted-up window, I was unable to write a line. Overwhelmed with memories, I explained to my two companions why this trip had a special meaning for me, blending a professional search by an investigative journalist with a more personal quest of a daughter of farmers, born just fifty years ago on a farm in Deux-Sèvres, located in a town in Gâtine a hundred kilometers from Ruffec.

The Tremendous Promises of the Green Revolution

When I was born in 1960, the green revolution was in its infancy. A few years earlier, more precisely on April 1, 1952, the first Renault tractor had replaced the team of oxen on my family's farm, soon followed by the first tanks of pesticides, including the deadly atrazine—a herbicide that I will discuss at length. Very involved with the Catholic Agricultural Youth (Jeunesse agricole catholique, JAC), a breeding ground for political and union leaders in the rural world, my father had welcomed these "tools from America" as a "new opportunity."[2] They would, he thought, relieve farmers from the heaviest labor while at the same time guaranteeing France's food independence. No more shortages or famines: industrial agriculture would be able to "feed the world" by providing cheap, abundant food.

Proud to have "the greatest profession on earth," because all human activity depends on it, my father was a committed participant in the inexorable process of the transformation of agricultural production that was radically changing the countryside, as the baby boom generation was experiencing the euphoria of postwar prosperity. Mechanization, the massive use of "inputs"—fertilizer and chemical pesticides—replacement of mixed farming with grain monoculture, consolidation, expansion of planted areas, indebtedness to the unavoidable agricultural bank: the farm of my forebears became a laboratory for the green revolution, breaking away from the family-farming model that had prevailed for generations. Inspired by the teachings of the JAC and subsequently the Christians in the Rural World (Chrétiens dans le monde rural, CMR)—who wanted to "change the world" even before May 1968—my parents established one of the first collective farming groups (Groupement agricole d'exploitation en commun, GAEC). Based on pooling the means of production and equal shares of income, this agricultural community, which included three associates and three paid employees, made it possible to go on vacation, a rare privilege among farming families.

Unusual in this very conservative region, the experiment caused a lot of talk, to the point that at the village school I was called the "girl from the kolkhoz." From those years, I recall a happy childhood amid a swarm of kids, where I was taught to stand up proudly for my peasant origins, because the emancipation of the rural world would come through the unselfconscious assertion of one's identity. Thanks to the green revolution, supposed to be a step in the irresistible march of humanity toward universal progress and well-being, people sometimes called rubes or hicks were standing up and embarking on the "Adventure," a little-known song that Jacques Brel wrote in 1958 at the request of the JAC.

"It was a wonderful time," my father told me recently. "How could we imagine that this new agricultural model was going to sow the seeds of destruction and death?" After a troubled silence, he went on: "How could we imagine that the pesticides the agricultural cooperative sold us were highly toxic products that would pollute the environment and make farmers ill?" It would indeed be unjust to cast stones only at farmers, who performed amazing feats to fit into a technological and chemical agricultural model promoted as a panacea by the National Federation of Agricultural Holders' Unions (Federation nationale des syndicats d'exploitants agricoles, FNSEA)—the largest farmers' organization—and the Ministry of Agriculture, at the cost of a rural exodus as massive as it was painful and countless suicides.[3]

It was not until I produced the film and book, *The World According to Monsanto*[4] in 2008 that all of a sudden hitherto private questions could be spoken aloud in my family: suppose illnesses and premature deaths were due to pesticides. Were they the cause of the Parkinson's disease that struck one of my father's cousins before he was fifty? Of the prostate cancer of one of my uncles, a former associate in the GAEC? Of the liver cancer of another associate, who died before he was sixty? Of the amyotrophic lateral sclerosis of a neighbor, former activist in the CMR, recently deceased? And the list is far from exhaustive.

The Ruffec Appeal

"Why is this meeting being held today? We have been working on chemical pollution for fifteen years, particularly pollution related to pesticides, and for fifteen years in rural France we have seen farmers who are ill or who tell us they have colleagues who are ill. This day is intended to allow you to express yourselves and to find some answers to questions you have been asking yourselves about toxicology, both medical and legal questions, because we have experts here at your disposal." With these words, François Veillerette, president and founder of the MDRGF, opened the special meeting on January 17, 2010, which closed with the "Ruffec Appeal." Having lived for twenty-five years in Oise—a region of intensive agriculture where he developed his ecological convictions—this teacher who headed Greenpeace France from 2003 to 2006 before being elected vice president of the Picardy region on the Europe Écologie ticket is one of the best French specialists on the issue of pesticides. His book, *Pesticides, le piège se referme* (Pesticides: The Trap Closes),[5] is a treasure trove of scientific references which I went through exhaustively before embarking on my investigation.

Among the experts he had invited to Ruffec was André Picot, a chemist who worked for the pharmaceutical giant Roussel-Uclaf before joining the National Scientific Research Center (Centre national de la recherche scientifique, CNRS). Renowned for his courageous independence, in a milieu where complicity with industry is frequent, he quit the French Food Safety Agency (Agence Française de sécurité sanitaire des aliments, AFSSA)[6] in 2002, because he dissented from the institution's manner of dealing with sensitive issues. Also present was Genon Jensen, executive director of the Health and Environmental Alliance (HEAL), a nongovernmental organization based in Brussels that coordinates a network of sixty-five European associations, including the MDRGF; in November 2008 it launched a campaign titled Pesticides and Cancer, backed by the European Union. Also in attendance were Maître Stéphane Cottineau, the MDRGF's lawyer, and Maître François Lafforgue, an adviser to the National Association for the Defense of Asbestos Victims (Association nationale de défense des victimes de l'amiante, ANDEVA), as well as to the Association of Veterans of Nuclear Tests, and the association of the victims of the catastrophe at the AZF factory in Toulouse.

Lafforgue also represents Paul François, a farmer suffering from serious chronic ailments caused by an accidental acute poisoning in 2004, who has become the emblem of the Network for the Defense of Victims of Pesticides established in June 2009 by the MDRGF.[7] Operating a farm in Bernac, a few kilometers from Ruffec, it was he who had suggested organizing the meeting on his land, because his story has become a symbol of the tragedy tearing apart many farming families everywhere in France. François Veillerette asked him to open the session of personal testimony as a reverent silence fell over the conference room of the Escargot Hotel amid the corn fields on the outskirts of Ruffec.

Sitting in a circle like a support group, some of the farmers and their wives had traveled several hundred kilometers to come to the little Charente town despite their debilitating illness. Among them was Jean-Marie Desdion, from the Centre region, suffering from myeloma, a bone cancer; Dominique Marshall, from the Vosges, being treated for myeloproliferative disorder, a leukemia-like disease; Gilbert Vendé, a farmer from Cher suffering from Parkinson's disease; and Jean-Marie Bony, who worked in an agricultural cooperative in

Languedoc-Roussillon until he was diagnosed with non-Hodgkin's lymphoma. As we shall see, some of their ailments had been recognized as occupational diseases by the agricultural social mutual fund after a long battle, and others were in the process of being recognized (see Chapter 3).

Aware of the reticence of these men and women, hard-working and not inclined to complain outside the family circle, I had no difficulty recognizing the effort they had to make to participate in the Ruffec Appeal, addressed to the public authorities to have them withdraw from the market as quickly as possible pesticides dangerous to the health, and to farmers so that they might stop experiencing their diseases as their fate, and eventually take their cases to court.

"I'm glad you came," said Paul François, visibly moved, "because I know it's not easy. Diseases caused by pesticides are a taboo subject. But it's time we broke the silence. It's true that we share responsibility for the pollution contaminating the water, air, and food, but we must not forget that we are using products approved by the authorities and that we are also the first victims."

Victim of Acute Poisoning by Monsanto's Lasso Herbicide

This wasn't the first time I'd met Paul François. In April 2008, I had participated in a showing of my film *The World According to Monsanto*, at the request of an association in Ruffec headed by Yves Manguy, a former member of the JAC who had known my father well and was the first spokesman of the small farmers' confederation (Confédération paysanne) when it was established in 1987.[8] More than five hundred people had packed the village hall and the evening had concluded with a book-signing session. A man approached and asked to speak to me. He was Paul François, forty-four at the time, and amid the crowd he began to tell me his story. Encouraged by Yves Manguy, who had led me to understand that his case was serious, I invited the farmer to visit me in my home near Paris whenever he came to the capital. He arrived a few weeks later, with a huge file under his arm and we spent the day dissecting it together.

Operating a six-hundred-acre farm, where he grew wheat, corn, and rapeseed, Paul François acknowledged with a contrite smile that he had been a "prototype of the conventional farmer." He meant a practitioner of chemical agriculture who had no qualms about using the many molecules—herbicides, insecticides, and fungicides—recommended by his cooperative for the treatment of grains. Until the sunny day in April 2004 when his "life was turned upside down,"[9] after a serious accident due to what toxicologists call "acute poisoning," caused by the inhalation of a large quantity of pesticide.

The farmer had just sprayed his corn fields with Lasso, a herbicide manufactured by the American multinational Monsanto. In the firm's television advertisement praising the qualities of the herbicide one can see a forty-year-old farmer, a cap jammed on his head, who, after enumerating the weeds "polluting" his fields, concludes, staring into the camera: "My answer is chemical weed control. When properly used, nobody gets hurt, only the weeds." This kind of spot was commonplace in the United States in the 1970s, when chemical manufacturers had no hesitation in using the TV screen to persuade farmers, and consumers as well, of the usefulness of their products for the good of all.

After spraying, Paul François went about other business and came back a few hours later to verify that the sprayer tank had been thoroughly rinsed by the automatic cleaning system. Contrary to what he thought, the tank was not empty but contained residues of Lasso, in particular of monochlorobenzene (also known as chlorobenzene), the compound's principal solvent. The heat of the sun had turned it into a gas whose vapors the farmer inhaled. "I was taken with violent nausea and hot flashes," he told me. "I immediately told my wife, who is a nurse, and she took me to the emergency room in Ruffec, being careful to bring the Lasso label. I lost consciousness when I got to the hospital, where I stayed for four days, spitting blood, with terrible headaches, memory loss, inability to speak, and loss of balance."

The first strange anomaly (we shall see that Paul François's file is full of them) was that, when contacted by the Ruffec emergency physician, who had been informed of the product inhaled, the Bordeaux poison center twice advised against taking blood and urine samples, which would have made it possible to measure the level of poisoning by detecting traces of Lasso's active ingredient,[10] alachlor, as well as of chlorophenol, the major metabolite—that is, the product of its degradation by the organism—of chlorobenzene. The lack of these samples was felt severely when the farmer sued the St. Louis multinational. But I'm getting ahead of myself.

After his hospitalization, Paul François was on sick leave for five weeks, during which he suffered from stammering and spells of amnesia of varying lengths. Then, despite profound fatigue, he decided to go back to work. In early November 2004, more than six months after his accident, he had a momentary lapse: while driving his combine, he abruptly left the field he was harvesting and crossed a road. "I was completely unconscious," he says today.

"I might very well have run into a tree or landed in a ditch." Thinking it was an aftereffect of the April poisoning, his treating doctor contacted the Angers poison center, which, like its counterpart in Bordeaux, refused to examine him or to take blood and urine samples.

In 2007, when Paul François's lawyer François Lafforgue asked Professor Jean-François Narbonne, director of the biochemical toxicology group at the University of Bordeaux and a qualified expert for such institutions as the AFSSA, to prepare a report, the professor did not mince words: "I must insist here on the aberrant conduct of French poison centers that, against all scientific logic, several times advised against conducting procedures to measure biomarkers for exposure, despite repeated requests from Paul François's family," he wrote on January 20, 2008. "These astonishing lapses are incomprehensible for a toxicologist and leave the door open to all kinds of hypotheses, ranging from serious incompetence to a deliberate desire not to provide evidence that might implicate a commercial product and ultimately the manufacturing company. . . . This serious error warrants judicial proceedings."

If they had done their work, respecting their public health mission, the toxicologists in the poison centers of Bordeaux and Angers could easily have consulted the technical specifications of Lasso; Monsanto first received authorization to market the pesticide on December 1, 1968. They would have been able to note that the herbicide contains an active ingredient, alachlor, in the proportion of 43 percent, and several additives or inert ingredients, including chlorobenzene (used as a solvent), making up 50 percent of the product. This substance was declared by Monsanto when it asked for Lasso's authorization, but it is not listed on the labels of tanks sold to farmers. And if one adds together the percentages attributed to alachlor and chlorobenzene, something is still missing: the remaining 7 percent is protected by a "trade secret" and, as we shall see, does not appear in the herbicide's technical specifications.

Had they reviewed the specifications for chlorobenzene developed by the National Institute of Research and Safety for the Prevention of Work Accidents and Occupational Diseases (Institut national de recherche et de sécurité pour la prévention des accidents du travail et des maladies professionnelles, INRS), poison center officials could in any case have read that this "organic synthesis intermediate" used in the "manufacture of coloring agents and pesticides" is "harmful by inhalation" and "produces harmful long-term effects." Further, it "concentrates in the liver, kidneys, lungs, and especially in fatty tissue. . . . Inhalation of vapors produces irritation of the eyes and the respiratory tract with exposure on the order of 200 ppm (930 mg/m³). At high doses, there can be neurological damage, creating drowsiness, lack of coordination, and depression of the central nervous system, followed by a lowering of consciousness." Finally, the experts at the INRS recommend "measuring 4-chlorocatechol and 4-chlorophenol [the two metabolites of chlorobenzene] in urine for the biological monitoring of exposed subjects." This is precisely what the two poison centers consulted had refused to do. Finally, it should be noted that the solvent is included in the document's table 9, which lists occupational diseases covered by social security, because it may cause acute neurological accidents.

As for alachlor, the active ingredient in Lasso that confers its function as a herbicide, a 1996 document from the World Health Organization (WHO) and the Food and Agriculture Organization (FAO) notes that in "rats exposed to lethal amounts" death is preceded "by salivation, tremors, collapse, and coma."[11] With regard to labeling, the UN organizations recommend specifying that the product is a "possible human carcinogen" and that clean protective clothing, including gloves, and face mask must be worn when handling alachlor. Finally, they specify that, although there have been "no reported cases," "symptoms [of acute poisoning] would probably include headache, nausea, vomiting, and dizziness. Severe poisoning may induce convulsions and coma." For all these reasons, Canada banned the use of Lasso as of December 31, 1985, followed by the European Union in 2007.[12]

In early 2007, a document issued by the French Ministry of Agriculture announced that the "definitive withdrawal" of the herbicide was scheduled for April 23, 2007, but that a "distribution deadline" had been granted until December 31, and the "use deadline" had been set for June 18, 2008. This would allow Monsanto and the agricultural cooperatives to quietly sell off their stocks, as evidenced by an article on April 19, 2007, in the weekly *Le Syndicat agricole* which announced several "scheduled withdrawals" of pesticides, including alachlor-based pesticides, such as Lasso, Indiana, and Arizona. "However," the paper explained, "as European directive 91/414 provides, member states may enjoy a grace period enabling them to destroy, sell, and use existing stocks."[13]

It is interesting to note that the article at no point explains why the European Union decided to "suspend marketing authorizations," in clear terms, banning Monsanto herbicides whose active ingredient had been shown to be carcinogenic in rodent studies. It was as though agronomic concerns prevailed over health concerns, whereas it hardly needs repeating that if herbicides are withdrawn from sale, this is because they endanger the health of their users, in this case the readers of *Le Syndicat agricole.*

Paul François's Battle

For Paul François, his work accident turned into a nightmare. On November 20, 2004, he abruptly went into a coma at home; his two daughters, then nine and thirteen, raised the alarm. He was hospitalized in the Poitiers teaching hospital for several weeks. In a diagnosis of January 25, 2005, the emergency service doctor described a "deeply altered state of consciousness"; the patient "does not respond to simple commands"; "the electroencephalogram . . . shows acute, slow, proleptic activity suggesting epilepsy." The same day, a neurologist noted: "Slurred speech (dysarthria) and amnesia are continuing."

There followed seven months of intermittent hospitalization, including sixty-three days in La Pitié-Salpêtrière in Paris, transfers from one hospital to another, and repeated comas. Oddly, the various specialists consulted stubbornly and unanimously persisted in disregarding the origin of the farmer's illness: his poisoning by Lasso. Depression, mental illness, epilepsy, various hypotheses were examined in turn, with plenty of tests. Paul François was subjected to scans and encephalograms and even had a psychiatric evaluation, but in the end all hypotheses were dismissed.

Worn out by these prevarications and encouraged by his wife, Paul François contacted the Toxicology and Chemistry Association (Association Toxicologie-Chimie, ATC) headed by Professor André Picot, one of the experts at the Ruffec meeting. Picot advised him to have Lasso analyzed to determine the precise composition of the herbicide and in particular the ingredients not appearing in the technical specifications. The analysis by a specialized laboratory revealed that the herbicide contains 0.2 percent of acetic acid chloromethyl ester, an additive derived from an extremely toxic product, methyl chloroacetate, which can produce cellular asphyxia from inhalation or skin contact.[14]

Wanting to understand the origin of his neurological disorders so he might seek better treatment, Paul François asked the assistant director of the cooperative that had supplied the Lasso to contact Monsanto. The assistant director told him that he had already reported the accident to the multinational's French subsidiary, located in a suburb of Lyon, but the company had not followed up. "I was very naïve," François now says. I thought Monsanto would cooperate to help me find a solution to my health problems. But that didn't happen." Finally, thanks to the tenacity of the cooperative's representative, there was a telephone conversation between François's wife Sylvie and Dr. John Jackson, a former Monsanto employee who had become a consultant to the firm in Europe. "My wife was shocked," François says, "because, after asserting he knew of no previous poisonings by Lasso, he offered financial compensation, in exchange for an agreement to give up any claims against the firm." These are their usual tactics, which I described at length in *The World According to Monsanto*. Faced with Sylvie François's insistence, Jackson agreed to set up a telephone conference with Dr. Daniel Goldstein, head of the toxicology department at the firm's St. Louis headquarters. Not speaking English, François asked a friend, the head of a company, to conduct the conversation. Like his colleague in Europe, Goldstein started by offering financial compensation. "We really had the impression that my health problems were no concern of his," says François. "He even went so far as to deny the presence of acetic acid chloromethyl ester in the formulation of Lasso. But when we offered to send him the results of the analyses of two samples of Lasso with a two-year interval between dates of manufacture, he changed his strategy and said that the molecule's presence must be due to a process of degradation of the herbicide. If that's the case, it's odd that the level is exactly the same in each sample." Putting it plainly, for the Monsanto representative, acetic acid chloromethyl ester is the result of an accidental chemical reaction caused by the aging of the herbicide. "This is bad faith," says André Picot, who believes "'chloroacetate was used for its energizing power to intensify the weed-killer's action'."[15]

"Monsanto's Bêtes Noires"

This was how Paul François became "one of Monsanto's bêtes noires," as *La Charente libre* put it, a characteristic I certainly share with him. But he soon also became "a textbook case of controversy among scientists and toxicologists."[16] In fact, observing a deterioration in the farmer's neurological condition, La Pitié-Salpêtrière hospital decided to take the urine samples the poison centers had not thought worth recommending. Carried out on February 23, 2005, ten months after the initial accident, the tests revealed, against expectations, a peak in the excretion of chlorophenol, the principal metabolite of chlorobenzene, along with products of the degradation of alachlor. All indications were that a portion of the herbicide had been stored in Paul François's body, in particular in his fatty tissue, and that its gradual release into the bloodstream was the source of the comas and serious neurological disorders that regularly afflicted him.

But instead of facing facts and acting accordingly, "specialists," with poison center toxicologists in the lead, maintained that it was impossible. To justify their denial, they put forward the fact that chlorophenol or monochlorobenzene could not last longer than three days in the body and that in no instance could one find a trace of those molecules beyond that period. This is an entirely theoretical explanation based on the toxicological data provided by the manufacturers, which, as we shall see, are often open to question (see Chapter 5).

If we take the example of the technical specifications established by the INRS for chlorobenzene, obviously based on studies provided by manufacturers, one sees that the data concerning the organism's elimination of the substance, after *oral* administration of a relatively high dose (500 mg/kg of body weight, twice daily for four days), were derived from an experiment on a rabbit. This rodent is, to be sure, a mammal with which we share a certain number of characteristics, but to conclude from that, eyes closed, that excretory mechanisms observed in the animal can be extrapolated to humans, is a step too hastily taken. Especially when this argument is used to deny the link between acute human poisoning by *inhalation* and its long-term neurological effects.

The only available data concerning humans involves samples taken at the end of a shift from workers in factories manufacturing chlorobenzene (or using it—the data do not specify). According to the INRS experts, "in humans 4-chlorocatechol and 4-chlorophenol appear in urine soon after the start of exposure, with a peak in elimination reached at the end of exposure (around eight hours). Elimination in urine is biphasic: the half-lives of 4-chlorocatechol are 2.2 and 17.3 hours for each phase respectively, and 3 and 12.2 hours for 4-chlorophenol. Excretion of 4-chlorocatechol is approximately three times more abundant than of 4-chlorophenol." It must be acknowledged that the specifications are laconic: they do not indicate the workers' level of exposure, but it is reasonable to suspect that it was lower than the "gassing," to adopt the term used by Professor André Picot, experienced by Paul François, otherwise they would have ended up in hospital. Nor do the data say whether the excretion mechanism concerned all or some of the metabolites, which, the INRS specifies, tend to "concentrate in fatty tissue."

All that would amount to a rather tedious battle of specialists, were it not for the shameful conclusion (I choose my words carefully) drawn by the brilliant toxicologists of three French poison centers: if metabolites of chlorobenzene were found in Paul François's urine and even in his hair in February and again in May 2005, this was because he had inhaled Lasso a few days earlier.

"The first time I heard that argument, I got pretty annoyed," François says. "It came from Dr. Daniel Poisot, chief medical officer of the Bordeaux poison center. Putting it plainly, he was accusing me of mainlining Lasso. When I pointed out that the first urine sample had been taken in the middle of a long hospitalization at La Pitié-Salpêtrière, where it was hard to be in contact with the herbicide, he answered that nothing was stopping me from hiding a vial in my hospital room. I was so astounded that I made a crack about the ties between some toxicologists and the chemical industry. He laughed and said that that was a fiction and that in any case the firms existed to create healthy products, not to put the planet, much less people, in danger."

The notion of Paul François's alleged drug addiction was also brought up by Dr. Patrick Henry, head of the Angers poison center, in a telephone conversation with Sylvie François, as stated in her prepared testimony before the Angoulême Social Security Court (Tribunal des affaires de sécurité sociale, TASS). "He bluntly stated that the test results could only be explained by the voluntary inhalation of the product."

As for Dr. Robert Garnier, chief medical officer of the Paris poison center, he did not openly put forth the possibility of "voluntary inhalation," preferring a psychiatric explanation for Francois's problems. "Monochlorobenzene can account for the initial accident and the disorders observed in the following hours or even days, but it is not the direct source of the disorders that appeared in subsequent weeks and months," he wrote in a letter to Dr. Annette Le Toux on June 1, 2005. "His acute poisoning sufficiently alarmed the farmer for him to fear having been permanently poisoned; the repeated episodes of illness could be the somatization of this anxiety." In her answer, two weeks later, the Agricultural Social Mutual Fund (Mutualité sociale agricole, MSA) doctor pointed out that the "disorders" were "complete loss of consciousness" and that medical examination "excluded the psychiatric origin of the problems observed." Then, obviously a little ill at ease, she added that there was no "central thread" in the case.

And for good reason: the toxicologists consulted had stubbornly denied the chronic effects of Lasso and its ingredients to put Monsanto's poison in the clear. Why? We shall see later that some toxicologists and chemists have very close ties to the chemical industry, even those (and that's the real problem) who hold positions in public institutions, such as in this instance the poison centers. Sometimes there are real conflicts of interest that the parties involved are careful not to make public; sometimes what is involved is simply an "incestuous relationship" due to the fact that scientists specializing in chemistry or toxicology "come from the same family," in the words of Ned Groth, an environmental expert I met in the United States (see Chapters 12 and 13).

These intimate connections are clearly illustrated by the example of Robert Garnier, head of the Paris poison center. When he came to my house, Paul François showed me a document he had printed from Medichem's website, and I kept a copy.[17] This "international scientific association," which is concerned exclusively with "occupational and environmental health in the production and use of chemicals," was established in 1972 by Dr. Alfred Thiess, former medical director of the German chemical firm BASF. Among its backers are some of the largest global chemical companies, most of which have a past—and a present—as admitted polluters.

Medichem organizes an international conference every year. In 2004 it was held in Paris under the chairmanship of Robert Garnier, who was then on the board of the association, along with, for example, Dr. Michael Nasterlack, a BASF executive and secretary of the association. The list of conference participants included Daniel Goldstein, Monsanto's chief toxicologist, the man who had proposed a financial transaction to Paul François in exchange for surrendering any claims. In a meeting with Garnier, the farmer from Ruffec asked him if he knew his colleague from Monsanto, which Garnier denied. In any event, as I write this, I have not found the document François gave me on the Web, because it has simply disappeared.

Legal Proceedings Against the MSA and Monsanto

"To tell you the truth, my case made me lose my naïveté," says François, "and so, for the first time in my life, I found myself in court." Confronted with a refusal by the MSA and the AAEXA—the insurance arm of the MSA responsible for work accidents—to recognize his serious health problems as an occupational disease, François decided to bring a case in the Angoulême TASS.

On November 3, 2008, the TASS found in his favor, declaring "that his declared relapse on November 29, 2004, is directly related to the work accident he suffered on April 27, 2004, and that it must be addressed in accordance with occupational legislation." In its decision, the court referred to the report by Professor Jean-François Narbonne cited above, which noted that the disorders are due to the "massive accumulation of substances in fatty tissue and/or [to the] persistent blockage of metabolic activity." In other words: with the extremely high level of poisoning, metabolization of toxic substances was blocked, bringing about an accumulation of those substances in the body. "Although unusual, this hypothesis is entirely plausible," stated André Picot, an opinion shared by Professor Gérard Lachâtre, expert in the pharmacology and toxicology department of the Limoges teaching hospital, the only specialist who considered a link between Paul François's recurrent neurological disorders and his "gassing" with Lasso.

The decision of the Angoulême TASS was a first victory for the Ruffec farmer. But he didn't stop there: he filed suit against Monsanto in the High Court (Tribunal de grande instance, TGI) of Lyon, on the grounds that the firm had "failed in its obligation to inform [users] concerning the composition of the product." "On the packaging provided with Lasso, only the presence of alachlor is mentioned as entering into the composition of the weed killer; the presence of monochlorobenzene is not noted," wrote the lawyer François Lafforgue in the proposed conclusions he submitted to the court on July 21, 2009. "The risk of inhaling monochlorobenzene, a very volatile substance, the precautions to be taken in handling the product, and the secondary effects of an accidental inhalation are not mentioned."

On the other side, with incredible cynicism, Monsanto's proposed conclusions exploited the absence of blood and urine samples, which the Bordeaux poison center refused to take just after the accident: "M. Paul François has never established that the product he is alleged to have inhaled on April 27, 2004, was Lasso," argued the multinational's lawyers. "In fact, there is no medical document reporting an inhalation of Lasso on April 27, 2004. . . . This evidence, that M. Paul François attempts to explain by negligence on the part of hospital services, is clear." And they coolly concluded: "It ensues from the elements cited above that no causal link can be established (or even presumed) between the April 27, 2004, accident and M. Paul François's state of health."

To back up its blunt conclusions, Monsanto attached two documents as appendices. The first was from Dr. Pierre-Gérard Pontal, who had conducted a "scientific medical evaluation of the case of poisoning of M. Paul François." A Web search easily turns up the curriculum vitae of Pontal, which he himself put online. He had worked at the Paris poison center, then for five years as chief medical officer in a Rhône-Poulenc Agrochimie factory, before moving on to head the human risk assessment team at Aventis CropScience. His ties to the chemical industry are obvious. Generally speaking, his report serves up all the clichés of institutional toxicology, invoking "scientifically established facts," such as the inviolable principle of Paracelsus, "only the dose makes the poison," which I will consider at length (see Chapter 8).

But to sum up the biased nature of his evaluation, it suffices to quote his criticism of the report by Jean-François Narbonne, which, he alleges, "neglects to ask the question of determining the dose to which M. François was exposed." This is preposterous, considering that Professor Narbonne clearly denounced the negligence of the poison centers that refused to take samples, which would precisely have enabled the measurement of the level of poisoning experienced by Paul François.

Drafted by Daniel Goldstein, head of the Monsanto Product Safety Center in St. Louis, the second document cited by the company's lawyers amounts to special pleading in favor of the poison centers, which at least has the virtue of clarity: "Considering that what is involved is identified exposure to a substance that is in theory swiftly excreted and should not have chronic toxicity, the fact of obtaining concentrations in blood or urine offers little or no interest for the patient," Goldstein notes. Then he hammers the point home by ostentatiously supporting the men whom his remarks elevate to the rank of accomplices in what strongly resembles an organized denial: "We confirm the statements from the French poison center that the conduct of analyses shortly after exposure would not have provided useful information, and that M. François should have recovered from the brief exposure by inhalation without difficulty." No comment is necessary.[18]

2

Chemical Weapons Recycled
for Agriculture

The world will not be destroyed by those who do evil, but by those who watch
them without doing anything.

—Albert Einstein

Paul François's story is exemplary because it points to evidence that the euphemistic language of the chemical industry as well as of the public authorities has sought to obscure: pesticides are poisons. As Dr. Geneviève Barbier
and the writer Armand Farrachi show in their book *La Société cancérigène*
(Carcinogenic Society), "their use has become so commonplace that we forget they were designed to kill."[1] They go on to say: "It is fruitless to look at
the packaging of these products for the kinds of warnings found on cigarette
packs intended to alert smokers: 'Spraying weeds kills' or 'Spraying mosquitoes or cockroaches causes cancer.'"[2]

From "Killers of Plagues" to "Phytopharmaceutical Products"

Pesticides are even "unique in being the only chemicals designed and deliberately released into the environment by humans, to kill or damage other
living organisms," said the Pesticide Action Network (PAN), an international
network against pesticides, in a brochure published in 2007 with financial
assistance from the European Union.[3] The broad family of pesticides is identifiable through their common suffix "cide" (from the Latin *caedere* "kill" or
"cut down"), for, according to the word's etymology, pesticides are killers of
pests (harmful animals, insects, or plants; the word "pest" is itself derived from

the Latin *pestis* meaning plagues or contagious diseases): weeds (herbicides), insects (insecticides), fungi (fungicides), snails and slugs (molluscicides), worms (nematicides), rodents (rodenticides), or crows (corvicides).

In the 1960s, when atrazine appeared on my ancestral farm, the promoters of chemical agriculture had no hesitation in pointing out the highly toxic, even fatal, nature of pesticides to justify their prevention programs. For example, in American audiovisual archives I found a television spot from 1964 showing a man in a white coat—the distinctive sign of a scientist—standing behind a table full of cans of chemical products, reciting in learned tones: "Always remember pesticides are poisons. Their safe use depends on you. Use pesticides safely!"[4]

A half century later, it is futile to look for such an explicit warning in the advertising of the major companies in the sector, as can be verified in France, for example, by visiting the website of the Crop Protection Industry Association (Union des industries de la protection des plantes, UIPP), now bringing together the "nineteen companies that market phytopharmaceutical products and agricultural services." The words chosen to present the professional organization, which includes in its membership the French subsidiaries of the six global giants in the business—BASF Agro SAS, Bayer CropScience, Dow AgroSciences, DuPont, Monsanto, and Syngenta—are telling on the process of euphemization that gradually took hold from the 1970s on. In the small powerful world of industrial agriculture, people carefully avoid speaking of "pesticides," preferring the term "phytosanitary products," recently replaced by the no doubt even more reassuring "phytopharmaceutical products." This is the definition provided on the UIPP website: "*Phytopharmaceutical* products play the role of *protecting* agricultural products against multiple attacks that may present obstacles *to the proper development* of plants: harmful insects, diseases (fungi), weeds. . . . *They foster regular harvests of sufficient quality and quantity.*"[5]

The shift from "pesticide" to "phytosanitary" or "phytopharmaceutical product" represents more than semantic mumbo jumbo: it is directly aimed at deceiving farmers—and consequently consumers—by passing off "products designed to kill" as medicines intended to protect the health of plants and hence the quality of food: a well-designed obfuscation that might be considered anodyne, typical of companies' advertising manipulation, if it were not echoed by government organizations at the highest level.

The home page of the Ministry of Agriculture's website[6] is very enlightening in this regard: the word "pesticide" does not appear at all. By way of contrast, it contains a section titled "Health and Protection of Plants," where we learn that the ministry "carries out many activities for the prevention and management of health and phytosanitary risks inherent in plant production." This is highly skilled evasiveness. Reading ministerial prose, one has the impression that the fact of producing plants *in itself* leads to "health and phytosanitary risks," whereas it is obviously the poisons used—never mentioned—that are the source of those risks. And what follows provides no further clarification: "The services charged with the protection of plants have three goals: health and phytosanitary monitoring; oversight of the conditions of plant production; and the production of farming practices more respectful of health and the environment."

The same orthodoxy is found on the Agricultural Social Mutual Fund (Mutualité sociale agricole, MSA) website, even though it is entrusted with the health of farmers. In an April 2010 article full of good intentions and presenting the "Phyt'attitude plan, a specific monitoring program for the risks connected with the use of phytosanitary products,"[7] obfuscation is so well integrated into the text that the authors fall on their faces: *Phytosanitary products, also known as pesticides* . . . are preparations intended to: protect plants or plant products from all harmful organisms or to prevent the action of those organisms; affect their vital processes, ensure their conservation; destroy undesirable plants or some of their parts."[8] The reader will have noted the surprising inversion, because in reality it is pesticides that are also known as phytosanitary products and not vice versa. The term imposed by the chemical industry to mask the harmfulness of its products has overridden the original term, now denounced by apostles of chemical agriculture as the sign of a retrograde obsession of ecologists and hippies.

But the message was long ago thoroughly assimilated in the countryside: in the village where I grew up I never heard "pesticides" mentioned, only "phyto products" that you got from the "phyto store," like a drug from the medicine chest.

From Arsenic to Mustard Gas

As the biologist Julie Marc points out in a doctoral thesis on Monsanto's Roundup, the most widely sold herbicide in the world, "the use of pesticides goes back to Antiquity,"[9] but until the early twentieth century the "killers of plagues" were derivatives of mineral compounds or plants, of natural origin (lead, sulfur, tobacco, or neem leaves in India).[10] The fact that they were natural did not mean some of them were not extremely dangerous, such as arsenic, recommended by Pliny the Elder in his monumental *Natural History*. Used in China and Europe as an insecticide as early as the sixteenth century, the well-known poison—more precisely its byproduct arsenite of soda—was banned in vineyards in 2001.[11]

Previously of limited use, pesticides went through a first surge with the advent of mineral chemistry in the nineteenth century. The symbol of this development is the well-known Bordeaux mixture, a blend of copper sulfate and slaked lime used on vines, starting in 1885, to counteract mildew, and later as a herbicide. In the same period, copper arsenite, better known as "Paris green" because it was used to kill rats in Paris sewers, had huge success in the United States, where it was used as an insecticide in orchards.[12] A little later, it was discovered that, spread on grain fields, copper sulfate destroyed weeds without harming the grain.

But it was World War I that laid the foundations for the massive production of pesticides, which profited from the development of synthetic organic chemistry and research on battlefield gases. Indeed, the history of most "phytosanitary products" in wide use today is intimately connected to the history of chemical warfare, whose paternity can be traced to the German Fritz Haber. Born in 1868, this chemist first achieved fame by inventing a procedure for manufacturing ammonia by synthesizing hydrogen with atmospheric nitrogen, which earned him the 1918 Nobel Prize for Chemistry. His work on the process of fixation of atmospheric nitrogen was used for the production of chemical nitrogen fertilizers (which replaced Chilean and Peruvian guano[13] and went along with the development of industrial agriculture), as well as in the production of explosives. When the Great War broke out, he was the head of the prestigious Kaiser Wilhelm Institute in Berlin, and his laboratory was asked to participate in the war effort. Heading a group of 150 scientists and 1,300 technicians, his mission was to develop irritant gases, intended to drive Allied soldiers out of their trenches, even though chemical weapons had been banned by the 1899 Hague Declaration.

The lab work was assigned to Ferdinand Flury, charged with testing the toxicological effects and mechanisms of all kinds of toxic gases on mice, rats, monkeys, and even horses. But only one really stood out from the others: chlorine gas. At the time, the industrial use of chlorine, abundant in nature, combined with other elements—such as, for example, sodium in salt form (sodium chloride)—was in its infancy. Since the well-received presentation in 1785 by the chemist Claude-Louis Berthollet, who had described the whitening property of Javel water (bleach)—a solution of chlorine and potassium invented in a factory in the Javel neighborhood of Paris—the element had dazzling success as a whitening agent (in the textile and then the paper industry) and later as a disinfectant. But its use had remained limited, because as an uncompounded element chlorine is a yellow-green gas—its name derives from *chloros*, meaning pale green in Greek—and is extremely toxic, with a very unpleasant suffocating odor that violently attacks the respiratory tract. Added to the fact that it is heavier than air and has a tendency to concentrate close to the ground—which is very useful in trench warfare—it was precisely the toxic properties of chlorine gas (also known as dichlorine) that interested Fritz Haber.

On April 22, 1915, the German army released 146 tons of gas at Ypres, at the scientist's direction; he had no hesitation in going to the front to supervise chemical attacks. He was the one who organized the secret installation of five thousand barrels of chlorine over a distance of six kilometers and ordered that the valves be opened at five in the morning. Driven by the breeze, the gas drifted over the Allied trenches. Taken by surprise, the French (mainly Algerian), British, and Canadian troops fell like flies, while trying to protect themselves with urine-soaked handkerchiefs. "I shall never forget the horrible agony of surprise in the eyes of the men who got that first dose," a Canadian survivor recalled. "It was the look of a dog being suddenly beaten for a thing it hadn't done. . . . They began gulping and coughing, and then fell down with their faces in their hands. . . . I had rolled and writhed, with the agony of the gas in my lungs, in a pool of slush in the bottom of the trench."[14]

Fritz Haber, who had the rank of captain, paid a heavy price for this first victory: a few days after the Ypres trenches were gassed, his wife Clara Immerwahr, also a chemist, committed suicide by shooting herself in the heart with her husband's service revolver. It was reported that she strenuously opposed his work on poison gas.

But Haber did not give up his work. Learning that the Allies had equipped their troops with gas masks, which made chlorine ineffective, he perfected phosgene, a mixture of two extremely toxic gases, dichlorine and carbon monoxide. Less irritating to the eyes, nose, and throat than chlorine gas alone, it was nonetheless the deadliest of the chemical weapons concocted in the laboratories of Berlin, because it violently attacked the lungs, filling them with hydrochloric acid. The few infantrymen who survived attacks died of the aftereffects in the years following the great slaughter. It is noteworthy that phosgene is still used today as a chemical ingredient in the pesticide industry. It is one of the ingredients in Sevin (carbaryl), the insecticide that produced the December 1984 disaster in Bhopal (see Chapter 3).

Toward the end of the war, after tens of thousands had been gassed, the German army released Haber's latest find: mustard gas, also known as yperite because, like chlorine gas, it was used first in the Ypres trenches. Its effects are terrible: it produces huge blisters on the skin, burns the cornea, causing blindness, and attacks bone marrow, causing leukemia. Few soldiers survived mustard gas attacks.

While poison gas was unquestionably first used by the Germans, finally all the belligerents mobilized their chemical industries to produce and use it. The Great War was in general a boon for industrialists, who took advantage of the war effort to lay the foundations for veritable empires, whose heirs today are multinationals specializing in the production of pesticides or transgenic seeds. For example, Hoechst (which merged with the French company Rhône-Poulenc in 1999 to produce the biotechnology giant Aventis) supplied the German Army with explosives and mustard gas. In the same period the American company DuPont (now one of the world's largest seed producers) supplied the Allies with gunpowder and explosives. Likewise, Monsanto (the world leader in genetically modified organisms [GMOs]) that had been established at the beginning of the century to produce saccharine, centupled its profits by selling chemical products used to manufacture explosives or poison gas, including sulfuric acid and the deadly phenol.

Haber's Law and Zyklon B

"During peace time a scientist belongs to the world, but during war time he belongs to his country." A zealous patriot, it was in these terms that Fritz Haber justified his work on poison gas, which, after the armistice, earned him a place on the roster of war criminals whose extradition the Allies demanded. He took refuge in Switzerland until the demand was officially withdrawn in 1919. A year later in Stockholm, he received the Nobel Prize in Chemistry for his work on the industrial process for the synthesis of ammonia. His nomination caused an uproar in the international scientific community, and the French, English, and American laureates of previous prizes boycotted the prestigious ceremony. In their eyes, Haber embodied precisely what Alfred Nobel, the fabulously wealthy inventor of dynamite, had denounced in his will: the alliance between science and war.

But although his role as father of chemical warfare has been lost in the annals of science, his name is well known to toxicologists, who still use Haber's law as a reference for assessing the toxicity of chemical products contaminating our environment, in particular pesticides. "Fritz Haber was not a toxicologist but a physical chemist," notes Professor Hanspeter Witschi of University of California, Davis, in *Inhalation Toxicology*, "but he profoundly influenced the science of toxicology."[15] In fact, as he was developing fearsome chemical weapons, he simultaneously applied himself to comparing the toxicity of gases to derive a law making it possible to assess their "effectiveness," that is, their lethal power. "Haber's law" expresses the relationship between the concentration of a gas and the exposure time necessary to cause the death of an individual. This is the definition Haber provided: "For each war gas, the amount (c) present in one cubic meter, of air is expressed in milligrams and multiplied by the time (t) in minutes necessary for the experimental animal inhaling this air to obtain a lethal effect. The smaller this product (c × t) is, the greater the toxicity of the war gas."[16]

While carrying out the observations that led him to formulate his terrible law, Haber also noted that exposure to a weak concentration of poison gas over a long period often had the same fatal effect as exposure to a high dose for a short period. Curiously, as we shall see, the regulatory agencies that make ample use of Haber's teachings to assess the toxicity of pesticides seem to have forgotten this part of his conclusions. Indeed, although they have little trouble acknowledging that phytosanitary products may have severe, even fatal, effects in a case of acute poisoning, they often deny the long-term effects caused by chronic exposure to weak concentrations.

In the meanwhile, one thing is certain: Haber's law "is often used in setting exposure guidelines for toxic substances," as David Gaylor, a U.S. Food and Drug Administration (FDA) toxicologist, acknowledges.[17] Indeed, it directly inspired the creation of one of the basic tools for the assessment and management of chemical risks: "lethal dose, 50%" or LD_{50}. Officially invented by the Briton John William Trevan in 1927, this indicator of toxicity measures the dose of a chemical substance necessary to kill half the animals—usually mice and rats—exposed to it, generally by inhalation, but also by ingestion, or cutaneous application. It is expressed as units of mass of the substance per unit of body mass of the test subject (mg/kg). An example: if a pesticide has an LD_{50} of 40 mg/kg, then 3,200 mg (3.2 g) is calculated to kill half the humans who weigh 80 kg.

According to a World Health Organization (WHO) document, it is estimated that a chemical product with an LD_{50} lower than 5 mg/kg of body weight (solids) or lower than 20 mg/kg (liquids) can be considered "extremely hazardous." It is "slightly hazardous" if its LD_{50} is over 500 and 2,000 mg/kg respectively.[18] As examples, the LD_{50} of vitamin C is 11,900 mg/kg, of table salt 3,000 mg/kg, cyanide from 0.5 to 3 mg/kg, and dioxin 0.02 mg/kg (0.001 for a dog).

What about Zyklon B? It is 1 mg/kg.[19] It is a tragic irony of history that Fritz Haber, who was of Jewish ancestry, was also the inventor of the deadly Zyklon B, used by the Nazis to exterminate the Jews in the gas chambers of the death camps. In the 1920s he was contacted by the German pesticide company Degesch, which asked him to resume his work on hydrocyanic acid to develop an application as an insecticide. Haber was familiar with the gas: according to the criteria of Haber's law, it is so toxic that it is extremely hazardous to handle, which explains the decision not to use it as a chemical weapon. Regardless, Haber developed a formulation enabling it to be safely transported and sprayed on crops. It is noteworthy that Zyklon B was authorized in France in 1958 for the treatment of cereal seeds and the protection of stored grain. Marketed by the Eden Vert company, it was banned in 1988.[20] The French subsidiary of Degesch continued to use a product derived from Zyklon B as a disinfection agent for storage sites until 1997.[21]

Meanwhile, the life of this zealous patriot, a Protestant convert for pragmatic reasons, came to a sad end. After Hitler came to power in 1933, the National Socialist Party asked him to fire all his Jewish associates. Seeing that it was impossible to resist, Haber decided to resign. "You cannot expect that a man of 65 years will change the way he thinks, a way that guided him so well during the past 39 years in his academic life, and you will understand that the pride with which he served Germany, his country, during his entire life, now requires him to ask to be relieved from his duties."[22]

Suffering from chronic angina, Haber went into exile in Switzerland, thinking he would restore his health before going to Palestine, at the urging of his friend Chaim Weizmann. But the journey never came. He died on January 29, 1934, unaware that members of his family would be asphyxiated by Zyklon B in the death camps.

DDT and the Beginning of the Industrial Age

"Can anyone believe it is possible to lay down such a barrage of poisons on the surface of the earth without making it unfit for all life?" Rachel Carson posed this question in *Silent Spring*, published in 1962, considered the founding work of the ecological movement. "They should not be called 'insecticides' but 'biocides.'" She went on: "This industry is a child of the Second World War. In the course of developing agents of chemical warfare, some of the chemicals developed in the laboratory were found to be lethal to insects. The discovery did not come by chance: insects were widely used to test chemicals as agents of death for men."[23]

Fritz Haber's work on chlorinated gases did indeed open the way to the industrial production of synthetic insecticides, the most well-known of which is DDT (dichlorodiphenyltrichloroethane), one of the large family of organochlorines. An organochlorine is an organic compound in which one or more hydrogen atoms have been replaced by chlorine atoms, forming an extremely stable chemical structure that is therefore resistant to environmental degradation. Some are considered "persistent organic pollutants" (POPs), because they accumulate in animal and human fatty tissue and because their extreme volatility enables them to move through the atmosphere to contaminate the remotest areas of the planet. I will return to the damaging effects of POPs, several of which—known as the "dirty dozen" (from the 1967 Robert Aldrich film)[24]—were banned by the Stockholm Convention adopted on May 22, 2001, by the United Nations Environment Programme (UNEP), but still pollute the environment and even mothers' milk. Among them are Monsanto's polychlorinated biphenyls (PCBs),[25] along with nine pesticides, including DDT, the "miracle insecticide" that began its brilliant career during World War II, bringing in its wake many molecules developed between the wars.

Synthesized by the Austrian chemist Othmar Zeidler in 1874, DDT was left in a laboratory drawer until 1939, when the Swiss chemist Paul Müller, who was working for the Geigy company,[26] identified its properties as an insecticide. His discovery had such great success that, only nine years later (record time) he won the Nobel Prize in Medicine. Appearing in solid form, insoluble in water—to be used it has to be dissolved in an oil—DDT was first used by the U.S. Army in Naples in 1943, to contain a typhus epidemic; the disease, transmitted by lice, was decimating Allied troops. The massive operation was repeated in the South Pacific to eradicate the anopheles mosquito, the carrier of malaria, and later as an antiseptic for death camp survivors, Korean prisoners, and the German civilian population when the defeated country was occupied.

Yet the organochlorine pesticide was never used for military purposes during World War II, because it seems all high commands had learned the lessons of the Great War. In any event, this is what Major William Buckingham suggested in a book published in 1982 by the U.S. Office of Air Force History, where he notes that "the Allies and Axis in World War II abstained from using the weapon either because of legal restrictions, or to avoid retaliation in kind."[27] But in the aftermath of the war, DDT was universally celebrated as a "miracle insecticide" able to defeat any harmful insect. I have been able to consult some hallucinatory audiovisual archives in which one can see entire cities in the United States treated with DDT in the 1950s. Sprayers go up and down the streets spewing huge white clouds, while housewives are asked to disinfect their cupboards with sponges soaked in the insecticide. Authorized in agriculture in 1945, DDT was later used massively in the treatment of crops, forests, and rivers, in an impressive expenditure of resources.

In 1955, the WHO launched a vast campaign against malaria in many parts of the world—Europe, Asia, Central America, and North Africa. But initial successes, sometimes accomplishing complete eradication of the disease, were followed by disillusionment, because the mosquitoes carrying the parasite that causes the disease very rapidly developed resistance to DDT, resulting, in particular in India and Central America, in a spectacular resurgence of the scourge.[28] But for the chemical industry, with Monsanto and Dow Chemical in the lead, it was a jackpot: from 1950 to 1980 more than forty thousand tons of DDT were sprayed around the world every year, with production reaching a record of 82,000 tons in 1963 (making for a total of 1.8 million tons between the early 1940s and 2010). In the United States alone, some 675,000 tons were sprayed before the agricultural use of DDT was banned in 1972.[29]

As Rachel Carson pointed out in *Silent Spring*, "the myth of the harmlessness of DDT rests on the fact that one of its first uses was the wartime dusting of many thousands of soldiers, refugees, and prisoners, to combat lice."[30] In addition, there is its low acute toxicity in mammals: classified as "moderately hazardous" by the WHO, its LD_{50} is only 113 mg/kg (for rats). On the other hand—I will come back to this in Chapters 16 and 17—its long-term effects are terrible: acting as an endocrine disruptor, it leads to cancer, birth defects, and reproductive disorders, in particular for those subject to prenatal exposure.[31]

Boosted by the success of DDT and other organochlorine pesticides, a second category of insecticides appeared in the wake of World War II. These were the organophosphates,[32] whose development was directly connected to research on new poison gases, but which, for the same reasons as for DDT, were never used for military purposes. As the official site of the French Observatory for Pesticide Residues (Observatoire des résidus de pesticides, ORP), established by the French government in 2003, soberly states: "not having been used during hostilities, they were used against insects."[33] Designed to attack the nervous system of insects, these molecules have a much more elevated acute toxicity than organochlorines, but they degrade more quickly. In this family are highly hazardous insecticides like parathion (LD_{50}: 15 mg/kg), used as early as 1944, malathion, dichlorvos, and chlorpyrifos, as well as carbaryl (responsible for the Bhopal disaster), and sarin (LD_{50}: 0.5 mg/kg), a highly toxic gas developed in 1939 in the IG Farben laboratories and now considered a "weapon of mass destruction" by the United Nations.[34]

The Precursors of Agent Orange

Launched at top speed thanks to synthetic insecticides, the green revolution also involved the marketing of chemical herbicides developed in British and American laboratories during World War II.[35] In the early 1940s, researchers succeeded in isolating the hormone that controls plant growth, and synthetically reproduced the molecule. They observed that, injected in small doses, the artificial hormone strongly stimulated plant growth, while, in contrast, high doses caused the death of plants. This led to the creation of two highly effective weed killers that initiated a veritable "agricultural revolution and laid the corner stone of present-day weed science," in the words of the American botanist James Troyer.[36] The two herbicides were 2,4-dichlorophenoxyacetic acid (2,4-D) and 2,4,5-trichlorophenoxyacetic acid (2,4,5-T), two chemical molecules in the chlorophenol family.[37]

Researchers soon recognized the wartime potential of these extremely powerful weed killers, because they made it possible to destroy crops and thereby starve enemy armies and populations. In 1943, the UK Agricultural Research Council launched a secret testing program that bore fruit in Maylasia in the 1950s where, for the first time in history, the British army used herbicides to destroy the communist insurgents' harvests. At the same time in the United States the Fort Detrick, Maryland, Biological Warfare Center was testing Dinoxol and Trinoxol, mixtures of 2,4-D and 2,4,5-T, the ancestor of Agent Orange, the defoliant used massively by the U.S. Army during the Vietnam War.

Indeed, although the Allies had renounced the use of chemical weapons, fearing above all an escalation that would have produced a terrible backlash, the emergence of the Cold War lifted this circumstantial taboo; for the White House any means were justified to combat the communist threat. So, from January 13, 1962, the launch date of Operation Ranch Hand, to 1971, some 80 million liters of defoliants were dumped on Vietnam, contaminating for decades more than 8 million acres and three thousand villages; 60 percent of the products used were Agent Orange, which is still causing birth defects thirty-five years after the end of the war.

The extreme toxicity of this chemical weapon is principally due to 2,4,5-T, a dreadful poison that is characteristically polluted by very small quantities of dioxin or TCDD.[38] Considered the most toxic substance ever created by man—a by-product of industrial processes, it does not exist in nature—the molecule was isolated in a Hamburg laboratory in 1957.[39] It is now known that its LD_{50} is 0.02 mg/kg (for rats) and that, according to a Columbia University study published in 2003, dissolving 80 grams of dioxin in a drinking water system could eliminate a city of 8 million people.[40] And estimates agree that in Vietnam 400 kilograms of pure dioxin were dumped in the southern part of the country.[41]

For the general public, TCDD emerged from the secrecy of laboratories on July 16, 1976, with a serious industrial accident known as the Seveso disaster. On that day, a reactor explosion in an Italian 2,4,5-T factory owned by the multinational Hoffmann-La Roche caused the release of an extremely toxic cloud in the Seveso region of Lombardy. Cattle died en masse, and officially 183 people contracted chloracne, an extremely serious condition resulting from dioxin poisoning, which manifests itself by an eruption of pustules all over the body, lasting several years and sometimes permanently.[42]

The characteristics of this human-created disease had been widely discussed in the medical literature beginning in the late 1930s, after the entry onto the market of pentachlorophenol, a cousin of 2,4,5-T, made by Monsanto and Dow Chemical and used as a fungicide in the treatment of wood as well as in the whitening of paper pulp. For his 2007 book, *How Everyday Products Make People Sick*, Paul Blanc, professor of occupational and environmental medicine at the University of California, consulted the archives of the *Journal of the American Medical Association* (*JAMA*),[43] where he found many letters from doctors asking for advice on the treatment of patients suffering from this dreadful skin disease, which was then unknown. "Nowhere in the literature have I found any case of caustic or chemical burn which lasted over several years unless the patient was in constant contact with the agent," reported a baffled Dr. Karl Stingily of Mississippi in a paper presented at a conference of the Southern Medical Association.[44] At the same conference, where this "new epidemic" was discussed at length, Dr. M. Toulmin Gaines of Alabama reported the case of a patient who worked in a lumber factory, a father of two young children: "He had acne . . . with comedones [medical term designating the specific lesions of acne] all over his face and back and shoulders and arms and thighs. His two children were a girl about five years old and a little boy about three. They had comedones all over their faces. They had a typical acne on the face. The boy had an indurated acne on the back of his neck such as you would see on a man about thirty years old. . . . I diagnosed it as chlorine acne and the children got it from the patient's clothing. He said that when he came home with his overalls on, the children would grab him around the legs and hug him and he would take them up in his lap."[45]

The same symptoms were secretly observed by Monsanto after an explosion in a 2,4,5-T factory in Nitro, West Virginia, on March 8, 1949. Victims of dioxin poisoning, the workers present for the accident or called on to clean up the site, experienced nausea, vomiting, and persistent headaches, and developed a severe form of chloracne. On November 17, 1953, a similar accident occurred in a BASF factory producing the herbicide that was then flooding the fields of Europe and America. Followed just as secretly at the firm's request by Dr. Karl Schultz, the exposed workers developed the same skin disease, which the Hamburg physician named chloracne. Throughout the 1950s many cases of this extremely disfiguring disease were recorded in the four corners of the United States, while an "amazing rain of death" fell upon the surface of the earth.[46]

3

"Elixirs of Death"

There would be no future peace for me if I kept silent.

—Rachel Carson

"*Silent Spring* is now a noisy summer," wrote the *New York Times* on July 22, 1962, after the *New Yorker* published the series of articles that became the book. On its publication in book form in September, it became an immediate bestseller (six hundred thousand copies sold in one month). It is indeed unusual for a scientific work dealing with theoretically difficult questions such as the effects of pollution on the environment to have such popular success and to create a months-long controversy in the scientific community, the press, industry, and even the White House.

Silent Spring, or Rachel Carson's Battle

The upheaval unleashed by Rachel Carson's book has been compared to the one provoked by Darwin's *Origin of Species* in its time. On its publication in France in 1963, the preface, written by the well-respected Roger Heim, at the time director of the French National Museum of Natural History and president of the Academy of Sciences, caused quite a stir: "We arrest gangsters, we shoot at hold-up men, we guillotine assassins, we execute despots—or alleged despots—but who will jail the public poisoners who distribute every day the products that synthetic chemistry provides for their profit and their recklessness?" he asked.[1] Fifty years later, *Silent Spring* remains a benchmark, because it is unique. At a time when chemical agriculture was conquering the world, for the first time a scientist dared to question the model of industrial agriculture that was supposed to produce universal abundance and well-being; Carson methodically exposed the damage caused by the "elixirs of death"[2] to both wildlife and people.

There was nothing to suggest that Rachel Carson would become the author of a bestseller that contributed to the creation of the ecological movement, the establishment of the Environmental Protection Agency (EPA), and the ban of the agricultural use of DDT. Born in 1907 in the small town of Springdale, Pennsylvania, not far from Pittsburgh, the polluted capital of iron and steel, young Rachel discovered nature in the company of her mother, from whom she learned how to observe birds in the course of long walks on the banks of the Allegheny River. From a modest background, she won a scholarship to study marine biology at Johns Hopkins, where women students were rare. According to her biographer Linda Lear, "In postwar America, science was god, and science was male."[3] With a passion for writing as well as for the sea, Carson was hired by the U.S. Bureau of Fisheries, where she worked as a laboratory assistant while writing her first articles for the *Baltimore Sun*. In them she campaigned for the regulation of industrial waste dumped into Chesapeake Bay, which was polluting the oyster habitat. In order to be taken seriously, she posed as a man, signing E.L. Carson.

In 1936, she was appointed to a full-time position in the Bureau of Fisheries, merged into the U.S. Fish and Wildlife Service in 1940, where she became chief editor of publications in 1949. In 1941, she published *Under the Sea Wind*, followed by *The Sea Around Us* (1951), and *The Edge of the Sea* (1955), a trilogy on the sea that had great success and established her as the most prominent scientific writer of her time. The recipient of many awards and elected to the American Academy of Arts and Letters, she was working on her next book when an event transformed her life.

In 1957, the U.S. Department of Agriculture (USDA) launched a highly publicized campaign to eradicate fire ants, an insect from Latin America that had entered the United States in the 1930s through the port of Mobile, Alabama. The red ant that had conquered the southern states suddenly became the bête noire of the USDA's newly established Plant Pest Control Division, responsible for aerial spraying of pesticides. Carson wrote in *Silent Spring*, "The fire ant suddenly became the target of a barrage of government releases, motion pictures, and government-inspired stories portraying it as a despoiler of southern agriculture and a killer of birds, livestock, and man" (162). However, she pointed out, the "president of the Entomological Society of America states that his department [at the Alabama Polytechnic Institute] 'has not received a single report of damage to plants in the last five years'" (163). Likewise, although feared because of its painful sting, "the Alabama State Health Officer declare[d] that 'there has never been a human death recorded from the bites of imported fire ants'" (164).

The eradication program planned to "treat" 20 million acres by spraying DDT, along with dieldrin and heptachlor; it began in 1958 and lasted until 1961. Relying on reports from many scientists—biologists, entomologists, and zoologists—as well as from elected officials and local associations, Carson assessed the results of this "rain of death." In the first year a large portion of the wildlife was exterminated. Everywhere were corpses of birds, beavers, possums, and armadillos. Domestic animals were not spared: chickens, livestock, cats, and dogs paid the price for this incredible ant hunt.

"Never has any pesticide program been so thoroughly and so deservedly damned by practically everyone but the beneficiaries of this 'sales bonanza,'" Carson wrote. "It is an outstanding example of an ill-conceived, badly executed, and thoroughly detrimental experiment in the mass control of insects, an experiment so costly in dollars, in loss of animal life, and in public confidence in the Agriculture Department that it is inconceivable that any funds could still be devoted to it" (162). And this was happening even after the operation had become a complete fiasco. In 1962 the director of entomology research at Louisiana State Agricultural Experiment Station summarized the dismal state of the program: "'The imported fire ant "eradication" program which has been conducted by state and federal agencies is thus far a failure. There are more infested acres in Louisiana now than when the program began'" (172).

"Chains of Poisonings"

"Who has made the decision that sets in motion these chains of poisonings, this ever-widening wave of death that spreads out, like ripples when a pebble is dropped into a still pond? . . . Who has decided—who has the *right* to decide—for the countless legions of people who were not consulted that the supreme value is a world without insects?" (127). This question haunted Rachel Carson throughout the research leading to *Silent Spring*. After her battle against the fire ant eradication program, she conducted an enormous amount of research on the environmental damage caused by the pesticide mania. She consulted countless reports and university studies and obtained confidential information thanks to her relations with many scientists in government agencies, such as the National Cancer Institute, accumulating data on what she called "a chain of poisoning and death" (6). And she asked ironically, "How could intelligent beings seek to control a few unwanted species by a method that contaminated the entire environment and brought the threat of disease and death even to their own kind?" (8).

A half century later we have to reread *Silent Spring* to grasp the magnitude of the madness that seized humanity in the aftermath of World War II. With supporting data, Carson tells, for example, the story of Clear Lake, California. Located one hundred miles north of San Francisco, this lake was prized by recreational fishermen. But, unfortunately for them, it was also an ideal habitat for the small gnat *Chaoborus astictipus*, which "although closely related to mosquitoes," "is not a bloodsucker." However, local residents found it annoying. The solution was simple: chemical insecticides would resolve the problem, in this case DDD, an insecticide related to DDT and "apparently offering fewer threats to fish life" (46).

Unfortunately, after a first application with "great dilution" the insects were still there. It was decided to increase the concentration to 50 ppm (a dilution factor of 1 mg per liter of liquid). The effects were terrible: dozens of western grebes, an aquatic bird species that feeds on fish, began to die on the lake. After the third application—the gnats continued to resist—the slaughter was so great that not a living grebe remained at Clear Lake. Intrigued, scientists carried out autopsies of the dead birds and found that their fatty tissue contained extremely high concentrations of DDD—up to 1,600 ppm— even though the concentration of the insecticide had never exceeded 50 ppm.

It was by analyzing the fish in the lake that the biologists understood the phenomenon at work: bioaccumulation, a process "in which the large carnivores had eaten the smaller carnivores, that had eaten the herbivores, that had eaten the plankton, that had absorbed the poison from the water" (48). The DDD used to exterminate the gnats had contaminated the plankton in the lake and had accumulated in organisms at every step in the food chain, reaching record levels in the western grebes, who added to the fish they ate all the intermediary species. We will see later that it is this process of bioaccumulation that explains why humans, the final predators in the food chain, are under particular threat from persistent organic pollutants (POPs), because our dinner plates are the receptacles of all the pollutants accumulated by the lower predators that contribute to our food.

If we add to bioaccumulation the phenomenon of bioconcentration— which designates the capacity of a living organism to accumulate the poison ingested in its fatty tissue—we can understand why birds were the first victims of this planned assault on what Carson called the "ecological web of life" (75).

The Silence of the Birds

An expert ornithologist ever since her long walks on the banks of the river of her childhood, Rachel Carson had thought of titling her book *The Silence of the Birds*, because the fate of those innocent creatures seemed to her emblematic of the process of destruction at work. In her research she had consulted hundreds of letters to government agencies and universities, such as a letter from a housewife of Hinsdale, Illinois, found in the archives of Robert Cushman Murphy, a renowned ornithologist at the American Museum of Natural History: "When we moved here six years ago, there was a wealth of bird life," she wrote. "After several years of DDT spray, the town is almost devoid of robins and starlings; chickadees have not been on my shelf for two years, and this year the cardinals are gone too; the nesting population in the neighborhood seems to consist of one dove pair and perhaps one catbird family. It is hard to explain to the children that the birds have been killed off, when they have learned in school that a Federal law protects the birds from killing or capture" (103).

These individual observations—chemical industry skeptics called them "anecdotal"—were confirmed all through the 1950s in reports from such official organizations as the U.S. Fish and Wildlife Service (which employed Carson). It noted the striking phenomenon of "blank spots weirdly empty of virtually *all* bird life" (104). The same thing occurred in Europe, as shown by "the deluge of reports of dead birds [that] reached . . . the Royal Society for the Protection of Birds"; the cause was the treatment of seeds with fungicides and insecticides before planting, which indirectly led to the death of 1,300 foxes between November 1959 and April 1960 (123). The foxes died because they ate the poisoned birds, who had filled up on earthworms, themselves stuffed with the poison covering the seeds.

To fully understand the twofold phenomenon of bioaccumulation and bioconcentration—I repeat this because it is of the greatest importance—it is necessary to refer to the long study conducted by Professor George Wallace, an ornithologist at the University of Michigan, following DDT spraying of the campus and the surrounding area in 1954. The purpose of the "program" was to exterminate the bark beetles thought to be the carriers of Dutch elm disease. The following spring everything seemed normal: robins returned to the leafy campus to build their nests. Then suddenly, the campus turned into a "graveyard." According to Wallace, "'in spite of the assurances of the insecticide people that their sprays were "harmless to birds," the robins were really dying of insecticidal poisoning; they exhibited the well-known symptoms of loss of balance, followed by tremors, convulsions, and death'" (107).

Perplexed, the ornithologist contacted Dr. Roy Barker, a member of an Illinois research center, whose work had "traced the intricate cycle of events by which the robins' fate is linked to the elm trees by way of the earthworms" (108). DDT forms a "tenacious film" over leaves and bark, killing, along with the targeted bark beetles, beneficial insects, predators invaluable for ecological balance and plant protection. In the autumn the worms swallow the insecticide deposited on the dead leaves and in the earth through poisoned insects, and accumulate it in their fatty tissue without being directly affected. Pesticides are like Russian roulette: their effects vary from species to species and in this case earthworms are not sensitive to DDT (by contrast, however, Monsanto's Roundup is fatal to them). The following spring, the heedless robins sign their death warrant by eating the earthworms. According to Barker, a fatal dose takes only eleven worms.

But that was not the end of the story. In the years after the campus spraying, Wallace observed that the robins that had survived had lost the ability to produce offspring. The numbers are eloquent: in 1953 the adult bird population was 370; five years later it had fallen to "two or three dozen." This drastic population reduction was coupled with a disturbing phenomenon: Wallace had "'records of robins and other birds building nests but laying no eggs, and others laying eggs and incubating them but not hatching them. We have one record of a robin that sat on its eggs faithfully for 21 days and they did not hatch'" (108).

Although not all robins have been exterminated, the survivors live under what Carson called the "shadow of sterility." At the time, no one was yet able to explain the process at the origin of this dysfunction threatening the survival of the species. As we shall see in Chapters 16 and 17, it is now known that DDT acts as an endocrine disrupter, which affects the development of exposed organisms in the fetal phase. In a 1960 congressional hearing Wallace reported finding extremely high levels of DDT in bird ovaries and testicles. In her chapter on the collapse of bird populations, Carson cites "important studies" showing that "insecticidal poison affects a generation once removed from initial contact with it. Storage of poison in the egg, in the yolk material that nourishes the developing embryo . . . explains why so many . . . birds died in the egg or a few days after hatching" (123).[4]

American Industry's Arrogance and Denial

"The major claims of Miss Rachel Carson's book, *Silent Spring*, are gross distortions of the actual facts, completely unsupported by scientific, experimental evidence, and general practical experience in the field. Her suggestion that pesticides are in fact biocides destroying all life is obviously absurd in the light of the fact that without selective biologicals these compounds would be completely useless." In transcribing these words from Robert White-Stevens, a biochemist working for American Cyanamid (one of the major pesticide manufacturers at the time), I wondered if the CBS correspondent interviewing him on April 3, 1963, had pointed out how counterproductive and even ridiculous his argument was.[5] The designated spokesman for the chemical industry, a man with a low voice and a mechanical delivery, was one of Carson's most vitriolic critics; he described her as an obscurantist opposed to sacrosanct "progress": "If man were to follow the teachings of Miss Carson, we would return to the Dark Ages, and the insects and diseases and vermin would once again inherit the earth."[6]

This apocalyptic vision of a world without pesticides was the theme of a parody published by Monsanto only a month after *Silent Spring*, titled "The Desolate Year." It's hard to find a copy today; the insipid work seems to have fallen through history's trapdoor. The firm uses science fiction to describe the horrors that would afflict the United States if DDT were banned. Here is a typical example of the painful prose: "Without pesticides, the pest control firms had automatically gone out of business. Of a sudden, some of the starkness of the time dawned on other people. No more protection against moths in clothing, furniture, carpets; no weapon but a flyswatter against rampant bedbugs, silverfish, fleas, slithering cockroaches, and spreading ants. More people shuddered, then, and still the desolate year was young."[7]

Taken by surprise—this was the first time the usefulness of their "miracle products" had been called into question—the pesticide manufacturers reacted violently and with the full force of their arrogance. This was nothing like the subtle disinformation campaigns of the 2000s carefully orchestrated by public relations agencies working in the shadows; in the early 1960s, chemical manufacturers were untouchable gods, arousing respect and gratitude because they were considered guarantors of the progress and abundance that were supposed to characterize civilized society. Certain of his position, in his letter sent along with "The Desolate Year" to the country's decision makers, the CEO of Monsanto was not afraid to resort to sexist insults, calling "Miss Rachel Carson" a "hysterical woman," "a bird and bunny lover," and a "member of the cult of the balance of nature."

The critics of *Silent Spring* also received support from the press which had adopted the reigning orthodoxy, such as *Time*, which in September 1962, denounced the "emotional and inaccurate outburst" of a book "full of over-simplifications and downright errors."[8] This did not prevent the same magazine thirty-seven years later from classifying Rachel Carson as one of the hundred most influential people of the twentieth century, correctly recalling the "huge counterattack organized and led by Monsanto, Velsicol, American Cyanamid—indeed, the whole chemical industry—duly supported by the Agriculture Department as well as the more cautious in the media."[9] In a letter to former president Dwight D. Eisenhower, the former secretary of agriculture, Ezra Taft Benson,[10] who actively encouraged the development of chemical agriculture in the 1950s, wondered "why a spinster with no children was so worried about genetics." His explanation was that she was "probably a Communist."[11]

But outrageous denials by supporters of pesticides did not succeed in stifling the incredible response to *Silent Spring*, even in the White House. In a press conference on August 29, 1962, a reporter questioned President John F. Kennedy "as to the possibility of dangerous long-range side effects from the widespread use of DDT and other pesticides. Have you considered asking the Department of Agriculture or the Public Health Service to take a closer look at this?" The president replied: "Yes, and I know that they already are. I think particularly, of course, since Miss Carson's book, but they are examining the matter."

Indeed, in the days following its serial publication in the *New Yorker*, Kennedy asked his science adviser Jerome Wiesner to set up a committee to study the "use of pesticides." The committee presented its report on May 15, 1963.[12] Its conclusions "add[ed] up to a fairly thorough-going vindication of Rachel Carson's *Silent Spring* thesis," according to an article in *Science*, because it recommended as a goal the "elimination of persistent toxic pesticides."[13] In their introduction, the authors acknowledge that "until the publication of *Silent Spring*, people were generally unaware of the toxicity of pesticides."

Following the publication of the report the Senate held a series of hearings on environmental risks, including testimony from Rachel Carson. Her work contributed to the establishment of the EPA on December 3, 1970, the first such agency in the world. Two years later, despite industry delaying tactics, the new agency banned the agricultural use of DDT, because it "posed unacceptable risks to the environment and potential harm to human health."[14] This was a notable posthumous victory for Rachel Carson, dead prematurely from cancer on April 14, 1964, at the age of fifty-six. When they voted to ratify the establishment of the EPA, no doubt some American congressmen recalled her words: "The question is whether any civilization can wage relentless war on life without destroying itself, and without losing the right to be called civilized."[15]

From Bhopal to Pakistan and Sri Lanka: Pesticides, "Poisons of the Third World"

"Birds started to drop from the sky. Streets and fields were littered with the corpses of water buffalo, cows, and dogs, swollen after a few hours in the heat of Central Asia. And everywhere were people dead of suffocation—curled up, foam at the mouth, hands desperately gripping the ground." These words are not from a story of the Great War, but from a report on the Bhopal disaster published in *Der Spiegel* in December 1984. Horrified by this "historically unprecedented industrial apocalypse," the German weekly put it on the front page with an unambiguous title: "The Fatal Gas of the Poison Factory."[16]

The tragedy happened during the night of December 2–3, 1984, in the Indian factory of the American firm Union Carbide established four years earlier in Bhopal, Madhya Pradesh, for the purpose of manufacturing annually five thousand tons of Sevin (carbaryl), a chemical insecticide used in agriculture. It is made of two gases: phosgene—invented by Fritz Haber (see Chapter 2)—and monomethylamine. When mixed, the two molecules produce methyl isocyanate (MIC), an extremely toxic substance that breaks down under the effect of heat into hydrocyanic acid, which is just as fatal. On that fateful night, technical failures produced the explosion of a tank containing forty-two tons of MIC and the release of a gas cloud that "descended like a shroud on a densely populated sixty-five square kilometers."[17] The toll was at least twenty thousand dead and between 250,000 and 500,000 injured.

I visited Bhopal in December 2004, on the occasion of the twentieth anniversary of the disaster, with Vandana Shiva, winner of the Alternative Nobel Prize, and a figure in the fight against genetically modified organisms (GMOs). At the time I was preparing a documentary on the abusive patents filed by multinationals on plants around the world. For instance, in September 1994, W.R. Grace, an American manufacturer of pesticides among other products, had obtained a patent on neem leaves, known in India as the medicine tree. Its medicinal properties had been described in ayurvedic medical treatises for at least three thousand years. Among them were the insecticidal properties of the leaves, whose "active principle" W.R. Grace had deciphered, thereby justifying its patent application.[18]

"It was the Bhopal disaster and the act of piracy by W.R. Grace that brought about my battle against any form of appropriation of living things," Shiva explained from the podium. "We have to reject chemical pesticides and use our plants, which are much more effective and threaten neither the environment nor human health." I was deeply moved when a delegation of blind women then spoke, demanding that Union Carbide executives finally be put on trial, the victims compensated, and the surroundings of the industrial site decontaminated.

While the Bhopal disaster reminded the world that pesticides are deadly poisons, few people know that every year around 220,000 people die as a result of acute poisoning by these products. This figure comes from a World Health Organization (WHO) study published in 1990,[19] according to which every year there are between 1 and 2 million cases of involuntary poisoning that occur in accidents related to spraying (the cause of 7 percent of the total number of deaths). In addition, there are 2 million suicide attempts (the cause of 91 percent of the fatalities), carried out primarily in the countries of the Global South.[20] The remaining 2 percent are linked to food poisoning. In addition, 500 million people, essentially peasants and farm workers, are victims of "less severe" poisoning.

A 1982 study in Sri Lanka by Dr. Jerry Jeyaratnam shows that between 1975 and 1980 an average of fifteen thousand people annually were admitted to hospital as a result of pesticide poisoning, 75 percent of whom were suicide attempts (out of a total population of 15 million); about one thousand people died. The pesticides responsible were generally organophosphates, but also included paraquat.[21] The same highly troubling situation can be found in Indonesia, Thailand, and Malaysia, where the average rate of occupational poisoning is thirteen per hundred thousand, so that Jeyaratnam judges that "pesticides illness is the new Third World disease."[22]

Poisonings were sometimes massive. This was the case during a malaria eradication program in Pakistan in 1976, when 2,800 agricultural workers recruited to spray malathion were severely poisoned, some fatally.[23] The WHO document also reveals that in Sichuan province in China 10 million farm workers (12 percent of the population) are in contact with pesticides; on average, 1 percent of them, or one hundred thousand individuals, annually suffer acute poisoning. To remedy this dramatic situation the WHO advocates training programs at all levels, from pesticide users to health workers.

To that end, in 2006 WHO experts prepared a 332-page manual intended to promote the prevention of acute and chronic poisoning due to pesticides.[24] You have to read this document to recognize the extent to which even an international organization like the WHO, whose mission is to protect human health, has things backward. The preparation of a prevention manual is, of course, a laudable effort, but in the face of the horrors it describes, it would be reasonable to expect a more radical stance, calling for an indefinite ban of all poisons that endanger agricultural workers (and, as we shall see, consumers). Instead, the venerable institution applies itself to managing as well as it can—necessarily badly—the terrible harm that can be caused by poisons "designed to kill," whose use in food production therefore never should have been authorized.

Page after page, pesticide by pesticide, the UN experts describe the clinical symptoms of acute poisoning and ways of treating it, when it is not too late. Module 6, for example ("First Aid for Pesticide Poisoning"), tells us that in a case of poisoning by organophosphorus insecticides, "the person begins to sweat and salivate (water at the mouth); he or she may vomit, have diarrhoea and complain of stomach cramps; the pupils become very small, and the person may mention blurred vision; the muscles twitch, and the hands shake; breathing becomes bubbly, and the person may have a fit and become unconscious" (214).

With regard to Roundup, the Monsanto herbicide which the firm has always claimed is as inoffensive as table salt—some agricultural cooperatives even go so far as to tell their members that it can be drunk without danger—the WHO explains that it "can be acutely toxic to humans and animals" (224). "The clinical manifestations after ingestion of glyphosate [the active ingredient in Roundup] vary according to the severity of poisoning. Mild: stomach cramps, nausea and vomiting, diarrhoea, mouth and throat pain, hypersalivation. . . . Severe: respiratory failure, renal failure, respiratory pneumonitis, secondary organ dysfunction, seizure, coma, death" (271). As for by 2,4-D, the Agent Orange ingredient still in wide use today,[25] acute poisoning causes "tachycardia, muscle weakness and muscle spasms . . . [which] may progress to profound muscle weakness and coma . . . resulting in death within 24h" (225).

A final example is paraquat, one of the most widely sold herbicides in the world. Departing from their usual discretion, the WHO experts write: "If it is swallowed . . . the effects are catastrophic, with very high mortality. . . . in severe cases rapid death from pulmonary oedema and acute oliguric renal failure; in less severe cases, signs of renal failure and liver damage; anxiety, ataxia[26] and convulsions may occur" (270).

The Poisoned of Chile

"If I were to offer you an apple with residues of the insecticide chlorpyrifos (see Chapter 13) and other pesticides, would you eat it?" The question obviously surprised Dr. Clelia Vallebuona, head of the toxicovigilance program in the Chilean Ministry of Health, whom I was interviewing in Santiago on November 11, 2009. After a long silence she said "No."

She said nothing further, evidence that in Chile, as elsewhere, the subject of pesticides is extremely sensitive. Yet Vallebuona can be proud of her work: in 1992, along with "particularly motivated" colleagues, she decided to apply the recommendations of the 1990 WHO report literally by proposing to establish a national network for the epidemiological surveillance of pesticides (Red de vigilancia epidemiológica de plaguicidas, REVEP), because the country was at the time confronting serious health problems. "Ten years earlier, the government had decided to develop export crops," she explained, "and suddenly thousands of farm workers found themselves exposed to highly toxic substances with no protection. Something absolutely had to be done because we knew that there were numerous cases of poisoning even though we had no official data."

From 1997 to 2006, 6,233 cases of poisoning, more than thirty of which were fatal, were reported to REVEP, an annual average of six hundred cases. "There was no legal obligation to report cases of acute poisoning until 2004, but we are convinced that the real figure has to be at least five times as much," said Vallebuona, who showed me the statistics gathered year by year. The pesticides most often incriminated were the insecticides chlorpyrifos, methamidophos, dimethoate, and cypermethrin, and the herbicide glyphosate. Thirty-four percent were organophosphates, 12 percent carbamates, and 28 percent pyrethroids.

"I imagine it wasn't easy to set up this pesticide surveillance network. Did you come under any pressure?"

"People are always challenging our statistics," she answered, visibly choosing her words.

"Who are these people?"

"Industry," she said wearily.

"And how does the Ministry of Agriculture behave?"

"It's not always easy. Sometimes we manage to cooperate, but their logic is completely different from ours."

"As an official in the Ministry of Health, do you think pesticides represent a real public health problem?"

"Yes. When you know the quantity of pesticides used in this country and the seriousness of their potential effects, it's a huge health problem that spares no one, from children to the elderly."

Indeed, from 1985 to 2009 the annual consumption of pesticides in Chile quintupled, rising from 5,500 to 30,000 tons. Chemical poisons are used principally in the Central Valley, which begins south of the capital, where intensive farming of crops for the European and American markets is concentrated. Along with Marc Duployer and Guillaume Martin, I traveled through this magnificent region bordered by the Andes in early November 2009. We were accompanied by Patricia Bravo and Maria Elena Rosas, two officials in the Chilean section of RAP-AL, the Latin American branch of the Pesticide Action Network (PAN), an international network of nongovernmental organizations promoting sustainable alternatives to the use of pesticides.

On the road south we stopped at the renowned San Pedro vineyard, where a farm worker wearing no protection was spraying dimethoate (an organophosphate insecticide whose LD_{50} for rats is 255 mg/kg). "Unfortunately," Patricia Bravo told me, "many companies still don't provide protective equipment for their workers. This young man may never suffer acute poisoning, but what effect will repeated exposure to small doses of pesticides have on him?"

My colleagues and I decided to go discreetly into the middle of the vines to film the spraying. Posted at the end of perfectly aligned rows of vine stocks, we were able to capture the white cloud constantly spewing out of the sprayer attached to the back of a mini-tractor with no cabin. When we were still at least two hundred meters from the machine we could clearly smell the bitter odor, which irritated the throat and stung the eyes. We resolved not to film like that again without first putting on protective clothing, mask, and goggles.

Edita and Olivia, Two Chilean Seasonal Workers Given Second-Degree Burns by Pesticides

We got back on the road and headed for the Maule region, where there is intensive growing of fruits (kiwis, apples, red fruits, table grapes) and vegetables, some of which pass through the Rungis market near Paris before ending up on French plates. Here, for four months of the year, the harvests provide a livelihood for thousands of seasonal workers (one-third of them women), the first victims of acute poisoning.

We had an appointment with two of them, Edita Araya Fajardo, sixty-three, and Olivia Muñoz Palma, thirty-nine, whose tragic story had received wide publicity five years earlier. The meeting had been scheduled in the home of Jacqueline Hernandez, head of an association for the defense of the rights of *temporeras* (seasonal workers), which had placed her on the large agricultural producers' blacklist. Seated in the living room of the modest cinder block house, Edita and Olivia had agreed to recount their ordeal, "so the world knows," even though their testimony might cause them trouble.

At dawn on October 22, 2004, they were part of a group of twenty-one women hired by a "crew boss" named Alejandro Esparza. He transported them in a cattle truck to a place called El Descanso ("rest") in the Pelarco area. Paid by the day, with no employment contract, their job was to harvest a field of broad beans. "As soon as we got there we smelled a penetrating odor of a chemical product," said Edita in a troubled voice. "The beans were wet. The boss told us they had been sprayed with a pesticide the day before, but that there was no problem. He had given me ten sacks to fill. I had a lot of trouble getting as far as the fifth, because I really felt like vomiting."

"I also felt very ill," said Olivia, taking up the story. "I felt violent itching on my legs, feet, and arms; I had the impression I was being sprayed with boiling water."

In the middle of the morning, three-quarters of the women decided to consult the emergency service in San Clemente, the nearest city. The doctors diagnosed acute dermatitis and severe erythema, the cause of which they did not understand, despite the patients' unanimous story. From the Bordeaux poison center to the hospitals of Chile, we always find the same denial mixed with cowardice and complicity. Even though they still felt sick, all the *temporeras* were told to go home, except for Edita and Olivia, whose condition had worsened.

Chilean television station Canal 13 broadcast a program that same evening, October 22, 2004, about their story and about pesticide poisoning. The program made a big splash: it was the first time television had dealt with this taboo subject, because in Chile it's hard to criticize the agricultural export model, which provides substantial foreign earnings. In scenes shot at the regional hospital of Talca where Edita and Olivia had been transported by ambulance, they can be seen lying in bed, unable to move because a large part of their bodies—abdomen, back, and legs—had second-degree burns. The reporter was particularly virulent that evening, not because poisons were being used in food production, but against the "irresponsible company heads who don't respect labor law or the lives of their workers, transporting them like cattle and exposing them to dangerous products with neither employment contract nor protection. But these are human beings who pick all the fruit we export and are so proud of. All this has to be denounced and the guilty punished."

Not so easy. . . . Of the twenty-one *temporeras* poisoned only Edita and Olivia filed complaints, the others "preferring to keep quiet for fear of reprisals." And for once, thanks to media exposure, the judges were heard. On August 26, 2005, the Supreme Court of Chile imposed fines of 6 million pesos (about $1,500) on Antonio Navarrete Rojas, the owner of the bean field, and 5 million pesos (about $1,200) on Alejandro Esparza, the crew boss. The press later revealed that Esparza never paid the fine, thanks to help from the mayor of San Clemente, Juan Rojas Vergara, known to have a long arm in this leading agribusiness region.

Backed by various associations, including RAP-AL and the National Association of Rural and Indigenous Women (ANAMUR), Edita and Olivia sued for damages so that they could at least pay for their medical expenses, but their civil suit never came to trial. On September 3, 2005, they held a press conference, attended by the deputies Juan Pablo Letelier and Adriana Muñoz, sponsors of a bill to improve the regulation and control of the pesticides used in the country. They reported that 279 cases of acute poisoning had been recorded in 2004 in the Maule region alone. "It is a disgrace that in twenty-first-century Chile a kiwi or an apple is worth more than the workers who pick them," said Muñoz. Edita and Olivia still suffer from hypersensitivity syndrome, manifested primarily by a severe allergic reaction to the sun. Edita said: "As soon as I go out without protection, I get red spots on the face and I feel intense fatigue. In spite of everything I had to resume work picking, because I'm a widow and have no other means of support."

The story of the two Chilean women is unfortunately sadly commonplace. According to a study by the Pan American Health Organization, annually there are four hundred thousand pesticide poisoning victims in the seven countries of Central America. In Brazil, the figure is three hundred thousand. In Argentina, where 40 million acres of transgenic soy are sprayed with Roundup at least twice a year, victims are counted in the thousands. "And acute cases of poisoning are only the tip of the iceberg," says Maria Elena Rosas, director of RAP-AL Chile. "What is not seen are the cases of chronic poisoning with small doses that years later produce cancer, birth defects, or fertility problems."

Impossible Prevention

"The chief difficulty you will have in using phytosanitary products is learning how to perceive what is invisible, that is, finding out that the 'phyto product' you started out with in the tank has gradually found its way into your environment. You understand it's not red paint, *it can't be seen*.[27] It's especially difficult because the spraying equipment is nothing special, the formulations are hard to use, and the products hazardous. Despite all that, you will have to learn how to manage your own prevention."

This bizarre scene took place on February 9, 2010, in the Catholic agricultural lycée in Bonne-Terre de Pézenas, Hérault. Doctor of occupational medicine for the Agricultural Social Mutual Fund (Mutualité sociale agricole, MSA) Gérard Bernadac had come to conduct a session on the "prevention of phytosanitary risks," along with Édith Cathonnet, prevention adviser at the Languedoc MSA, and Dr. Jean-Luc Dupupet, doctor in charge of chemical risk, who had specially come from Paris, where the MSA's headquarters are located. The training session was addressed to thirty students—all boys—in the wine-growing course, sons of winegrowers preparing to join the family business.[28] It was part of a unit that would enable these future farmers to obtain the "certiphyto," a diploma authorizing the professional use of phytopharmaceutical products, that will be required beginning in 2015, following an October 2009 European directive "for the sustainable use of pesticides." Between now and then, the MSA has its work cut out, because the Ministry of Agriculture has given it the task of training users, warehousemen, and traders—about a million people. Before this, anyone could use these poisons with no preliminary training.

Observing the young students sitting quietly in the pretty private school chapel, I couldn't help thinking about the many hazards they would inevitably face in their working lives. Every year 220,000 tons of pesticides are released into the European environment: 108,000 tons of fungicides, 84,000 tons of herbicides, and 21,000 tons of insecticides.[29] If we add the seven thousand tons of "growth regulators"—hormones designed to shorten grain stalks—that makes about one pound of active substances for every European citizen.

France has the lion's share because, with its eighty thousand annual tons, it is the largest European user of pesticides and the fourth largest in the world, after the United States, Brazil, and Japan. Eighty percent of the substances sprayed involve four crops, which, however, represent only 40 percent of cultivated land: grains, corn, rapeseed, and vines, one of the agricultural sectors that uses the greatest amount of phytopharmaceutical products.

The training at the Bonne-Terre lycée began with a session of "Phyto theater," a sketch performed by Bernadac and his colleague from the MSA to sensitize the future farmers to "good practices," enabling them to avoid the worst. In her introduction, incidentally, Édith Cathonnet had made a strange confession: after enumerating all the phases in the process involving risks—opening the can, preparing the mixture, filling or cleaning the tank, the spraying itself, especially if the cabin is not sealed or is soiled—she ended by letting slip a cry from the heart: "The ideal way to protect yourself is to not spray, because you have no contact with the product."

Then as the thoroughly realistic sketch went on—I had seen these gestures a thousand times on the farms of my home town—my discomfort grew. The whole demonstration was based on the use of the space suit farmers are supposed to wear to protect themselves, with the indispensable accessories of gas masks and frogman goggles that make them look like extraterrestrials. Three months earlier, on January 15, 2010, the French Agency for Environmental and Occupational Health Safety (Agence française de sécurité sanitaire de l'environnement et du travail, AFSSET) had published a very disturbing report on the ineffectiveness of these suits.[30] In the study, the experts explained in detail that they had tested ten models of suit: "Only two models of the ten tested according to the standard attain the level of performance declared. For the other suits, the passage of chemical products was almost immediate through the material of three of them, and through the stitching of two others, which constitute serious non-conformities. The final three are to be rejected for at least one substance."

Hammering the point home, they observed that the tests conducted by the manufacturers "are conducted in the laboratory in conditions too far removed from the real conditions of exposure. The essential factors, such as length of exposure, outside temperature, type of activity, length of contact, do not enter into consideration." And their conclusion was implacable: "An inspection for conformity of all the suits for protection against liquid chemical products on the market should be conducted and the non-conforming suits should be withdrawn without delay."

Phyt'attitude: The MSA Campaign in France

Of course, we can be pleased that the MSA, which long underestimated and even denied the risks inherent in pesticides, abandoned its inactivity and launched a vast program of prevention. In 1991, the MSA set up a toxicovigilance network, similar to the Chilean REVEP, called "Phyt'attitude." Data are centralized at the National Institute of Agricultural Medicine (Institut national de médecine agricole, INMA) in Tours.

In 1999, an internal study showed that "one user of phytopharmaceutical products out of five has experienced problems (skin irritation, respiratory problems, vomiting, headaches) at least once in the past year." To encourage victims to break their silence, the MSA set up a toll-free number (0800 887 887), where they "can report their symptoms with no charge and anonymously," as the MSA website states.

"Why anonymously? Are farmers ashamed of being poisoning victims?" I posed the question to Jean-Luc Dupupet, who supervises the Phyt'attitude program, in an interview he granted me after the day of training sessions in the agricultural lycée. It caused him not a moment's hesitation: "Of course. The potential toxic effects of phytopharmaceutical products is still a taboo subject and, for some users, poisoning indicates a handling error or even professional misconduct, especially shameful because it proves that those who claim agriculture is a source of contamination of food and the environment are right."

"How many reports did you receive in 2009?"

"Two hundred seventy-one. The complaints observed affected primarily the mucous membranes and the skin, with irritation, burns, itching, or eczema (40 percent of cases studied), the digestive system (34 percent), the respiratory system (20 percent), then the rest of the organism, including attacks on the neurological system, including headaches (24 percent); 13 percent of those reporting mentioned hospitalization following the poisoning, and 27 percent took sick leave. According to our estimates, every year around one hundred thousand farm workers complain of problems after using phytopharmaceutical products, but our network places a priority on cases of acute poisoning."

"What types of products are most often incriminated?"

"In general, headaches, that is, neurological symptoms, are caused by insecticides; with fungicides we observe more skin problems; and with herbicides, the effects are both digestive and on the skin."

"And what are the chronic illnesses that can now be recognized as occupational diseases by the MSA?"

"Well, there are neurodegenerative diseases, such as Parkinson's disease, types of cancer, like blood cancers—leukemia or non-Hodgkin's lymphoma—cancer of the brain, prostate, skin, lung, and pancreas. Indeed, talking about chronic diseases helps get our prevention message to farmers, because if you simply tell them they risk slight eye symptoms, sneezing, a runny nose, or skin irritation that disappears in twenty-four hours, it doesn't do much good. But when we tell them we see more Parkinson's disease, more brain and prostate cancer in farmers than in the rest of the population, that makes them think and our prevention messages get through better."[31]

It doesn't seem like much, but an interview like this, and on film, would have been impossible five years earlier. The frankness of Dupupet and the MSA is a break from the posture of the public authorities and manufacturers, as well as that of the agricultural cooperatives which, as we shall see, continue to deny the long-term health effects of chronic exposure to the poisons used in food production.

4

Ill from Pesticides

The obligation to endure gives us the right to know.

—Jean Rostand

"I'm sorry, but I can't let you film." Rather pleasant-looking in his business suit, Jean-Marc de Cacqueray, director of the Regional Office for Labor, Employment, and Training (Direction régionale du travail, de l'emploi, et de la formation, DRTEFP) of Brittany, looked openly embarrassed. "Why?" I insisted. "Who's against it?" The director glanced desperately at François Boutin, his assistant for professional risk prevention, who, under pressure from his boss, finally said "Coop de France."

"OK," I said, a bit amused, as my cameraman filmed the incredible scene with a hidden camera, "in that case I'd like to speak with a representative of Coop de France."

"Go get Lacombe," Cacqueray said. Boutin followed orders and went into the amphitheater of the Faculty of Trades of Ker Lann, near Rennes, which I'd managed to get into a few minutes earlier, before being escorted out by a very aggressive bodyguard, who I presume was a representative of Coop de France Ouest. But Étienne Lacombe, the organization's official representative, did not deign to come to explain why he wanted to keep me from filming the seminar on "Farmers and Their Health" being held that day, December 1, 2009, by the DRTEFP and the Agricultural Social Mutual Fund (Mutualité sociale agricole, MSA), and open to all the "shippers and sellers of phytosanitary products" in the region of Brittany.

When Agricultural Cooperatives Make Law

This interesting program was being held in connection with setting up the "certiphyto," the diploma that will be obligatory in 2015 for anyone recommending, selling, or using phytosanitary products for professional purposes. "Products that are not inoffensive, because some preparations are classified for their carcinogenic, mutagenic, and reprotoxic (CMR) effects," as the invitation that I have carefully kept explains.

Yet everything had started out well. I had been told about the seminar a few days earlier by Dr. Jean-Luc Dupupet, who was scheduled to give a presentation on the link between pesticide exposure and cancer; he had put me in touch with François Boutin. When I contacted Boutin, he immediately sent me an e-mail with all the "documents related to the seminar" so that I could prepare for filming. On November 26, my answering machine had a slightly embarrassed but cordial message from Boutin. I provide its content here not to discredit him but to show the power of agricultural cooperatives, who are able to lay down the law for a representative of the state whenever they feel their interests threatened.

"This is about the seminar on phytosanitary products," he said. "As a matter of principle, I asked other participants, and the leader of the trading companies is in favor; the Regional Labor Director, my boss, is also in favor of your participation; on the other hand, my counterpart in Coop de France is a little hesitant." Then he read me a convoluted e-mail in which the representative of the agricultural cooperatives asked that we give up filming the seminar, with a very strange argument: "The main reason is the short time between now and December 1, which will not allow us to prepare with Arte the conditions under which this documentary will be produced. We are open to exchanges about propositions we could present jointly, for example organizing visits to and conversations in cooperatives."

Nonetheless, François Boutin seemed rather confident: "I'm in the process of trying to defuse this argument so you can be allowed to come in any case, but I can't betray or be disloyal to my partners in this matter. I'll keep you informed during the day by telephone or e-mail." Indeed, a few hours later, I received an e-mail asking me in the end to give up our trip to Rennes. But the National Audiovisual Institute (Institut national audiovisuel, INA), the film's producer, and I had decided to go, thinking that the obstacle could be removed when we got there. However, despite the intervention of Dupupet, who had tried to persuade the regional director to allow us to film at least his presentation, we returned to Paris empty-handed.

When I got home, I conducted a brief investigation of Coop de France. I discovered that, established in 1966, during the boom in chemical agriculture, the "unified professional organization of agricultural cooperation" brought together "three thousand industrial and commercial enterprises and more than fifteen hundred subsidiaries," which produced a "combined turnover evaluated at more than 80 billion euros in 2008." With "at least 150,000 permanent employees," Coop de France represents a huge business adding up to "40 percent of French food processing" and controls the majority of agricultural production since "three-fourths of the 406,000 farms belong to at least one cooperative." On the other hand, what the website of Coop de France does not say is how much revenue is provided to cooperatives by phytosanitary products, which makes up a significant part of their fabulous earnings.

Incidentally, it is interesting to note what bad press these products seem to have received even on the websites of agricultural cooperatives. One example is the site for Terrena, a large Breton cooperative which advocates an "ecologically intensive agriculture" and has an annual turnover of 3.9 billion euros. It is futile to look for the earnings it derives from phytosanitary products: the information never appears, even in its annual report, which is online. If you look under the heading "Agronomy and Agricultural Supplies," a subdivision of "Animal Production and Large-Scale Crops," you find "some figures": "enrichment and fertilizers" (300,000 tons); "plant health" (3.9 million acres); "seeds" (790,000 acres); "agricultural and rural equipment" (35 million euros); "total turnover" (216 million euros). Chemical poisons are concealed under the term "plant health," but the only indication provided concerns the number of acres treated with products sold by the cooperatives.

The Terrena site also explains that the cooperative has a 43 percent stake in Odalis, whose "profession" is to "connect suppliers to distributors and farmers." The "suppliers" are pesticide manufacturers, whose attractive cans can be seen in a video posted by Odalis to present its know-how.[1] We learn that "26 thousand tons of products are shipped annually," for revenues of 3.6 million euros. But the portion accounted for by pesticides is not specified, because the amount indicated includes "agricultural seeds" as well as "plant health products."

Surfing the Web, I discovered in any case that in January 2009 Coop de France had sponsored a little brochure titled "The Proper Use of Glyphosate in Agriculture," apparently with no financial support from Monsanto.[2] One of its authors was none other than Étienne Lacombe.

Chronic Poisoning of Farmers by Pesticides: An Infernal Trap

"Do you understand why Coop de France kept me from filming the seminar in Rennes?" Three months after the unfortunate Breton incident, I could not resist the wish to record the testimony of Jean-Luc Dupupet, when we met at the agricultural lycée in Pézenas. It was obviously a sensitive question. "Well," the doctor in charge of chemical risks for the MSA mumbled. After a long silence, he said, "There you've got me stuck. It's very hard for me to give you an explanation. Um, you know the chronic effects of phytosanitary products is still a taboo subject and obviously the agricultural cooperatives prefer that it be talked about, let's say, privately, without the media being present."

"Are they afraid their members and employees will turn against them, accusing them of complicity in poisoning or for not assisting a person in danger, as Sylvain Médard did recently?"

"Um."

"You know who Sylvain Médard is?"

"Yes, of course. He was a technician in an agricultural cooperative and developed a rare form of myopathy that was recognized as an occupational disease."

Indeed, it was even a first, which was widely reported and caused a stir in agricultural circles. Sylvain Médard had worked for thirteen years in an agricultural cooperative in Picardy, Capsom (located at Corbie, Somme), when in 1997 doctors diagnosed him with "acquired mitochondrial myopathy," a neuromuscular disease with a gloomy prognosis which causes degeneration of muscle tissue. As its name indicates, unlike other types of myopathy, the one the thirty-three-year-old man is suffering from is not genetic in origin, but caused by a toxic agent in medication or chemicals. The agricultural technician's main work consisted of testing new pesticides on behalf of manufacturers who had filed a request for marketing authorization. In the professional jargon, he was "in charge of tests on samples." For this purpose, companies sent unlabeled cans to the cooperative, each with a number written on it. For years, the technician had handled dozens of poisons, protected only by a cotton suit and a simple paper mask, just enough to protect him from inhaling dust.

Sylvain Médard decided to bring his case to the Social Security Court (Tribunal des affaires de sécurité sociale, TASS) in Amiens. On May 23, 2005, determining that "respiratory protection was insufficient," the judges found the cooperative liable for "inexcusable negligence," on the grounds that it "could not have been unaware at the time of the health risks tied to the toxic products to which its employees were exposed." "This decision gives hope to the victims of occupational diseases in agriculture," according to a press release from Michel Ledoux, Médard's lawyer.[3] Indeed, the case marked a turning point in the way pesticides are seen in France—first of all by the agricultural cooperatives, paralyzed by the prospect of what some were calling the "new asbestos scandal."[4]

"That's a little exaggerated," according to Dupupet, who obviously does not appreciate the comparison. "What I can tell you is that the attitude of the cooperatives is changing: it's true that until recently, they were interested only in the agronomic results of phytosanitary products, but now they're beginning to talk about health risks, warning users as a pharmacist does when a patient buys a medication after a medical consultation."[5] The MSA chief doctor said nothing further, but we must acknowledge his frankness and the efforts he has been making to break the implacable law of silence that surrounds the long-term consequences of repeated exposure to pesticides. Indeed, we have to acknowledge that, although it is still very cautious, the new posture adopted by the MSA, long denounced for its silence on the issue, has clearly broken with the denial that continues to characterize the beneficiaries of this deadly commerce—the merchants, among whom are agricultural cooperatives, and the manufacturers—as well as the public authorities.

It is one thing to acknowledge that pesticides can cause acute poisoning; faced with a farm worker who starts to vomit or suffers second-degree burns after handling phytosanitary products, it is hard to deny the causal link, even though, as we saw with Paul François in Chapter 1, the victims are often confronted with bad faith on the part of their employers or the manufacturers. But it is another thing to venture onto the more unstable, indeed frankly mined, territory of the long-term consequences of chronic poisoning—repeated small doses—by the descendants of poison gas.

Incidentally, in the Paul François case, it's a safe bet that Monsanto would not have persisted in denying his acute poisoning if the farmer from Ruffec had not dug in his heels. What the company did not want to admit is that accidental poisoning can produce serious chronic effects, because that would mean opening a Pandora's box and would lead to a challenge to the toxicologists' dogma that "the dose makes the poison"—I'll come back to this.

The fact that cases of accidental poisoning represent only the "tip of the iceberg" in the words of Maria Elena Rosas, director of RAP-AL Chile (see Chapter 3), had already been glimpsed by Rachel Carson in *Silent Spring*: "We know that even single exposures to these chemicals, if the amount is large enough, can precipitate acute poisoning. But this is not the major problem. The sudden illness or death of farmers, spraymen, pilots, and others exposed to appreciable quantities of pesticides are tragic and should not occur. For the population as a whole, we must be more concerned with the delayed effects of absorbing small amounts of the pesticides that invisibly contaminate our world."[6]

What Carson describes for the "population as a whole" is particularly true for farmers who handle numerous pesticides for many years without ever being victims of acute poisoning, but who are in regular contact with these substances, inhaling them or absorbing them through the skin—especially because, as the French Agency for Environmental and Occupational Health Safety (Agence française de sécurité sanitaire de l'environnement et du travail, AFSSET) report cited in Chapter 3 showed, protective clothing is usually ineffective. The problem is that when they develop a serious illness, such as cancer or Parkinson's disease, it is very hard for them to demonstrate a relationship between their complaints and their occupational activity, precisely because they have been exposed to a multitude of agents that might cause the same effects, which complicates the identification of a causal link with a particular substance. And without an established causal link, there is no official recognition of an occupational disease and hence neither provision of treatment nor indemnification for the harm suffered.

This situation, which has long guaranteed the impunity of the manufacturers of poisons, leads to what the Quebec toxicologist Michel Gérin and his co-authors, in their seminal work *Environnement et santé publique* (Environment and Public Health), call an "under-reporting of environmental diseases," beginning with those linked to chronic exposure to pesticides: "Recognition of the real impact of the environment on health suffers from the difficulty of establishing, on an individual basis, the environmental origin of a disease. The problem is particularly acute in the case of effects linked to the exposure to toxic substances, often medium- or long-term effects whose 'signature' escapes doctors' grasp. Several factors contribute to this underestimate. A major obstacle comes from the often significant latency between exposure and diagnosable effect, which makes the establishment of a causal link problematic. Past exposure or use is forgotten, or there no longer is objective information about exposure. Further, the non-specificity of most effects tied to the environment means that their possible environmental origin goes unnoticed."[7]

Indeed, the situation of farmers is very different from that of workers in Saint-Gobain factories who were exposed to asbestos fibers while manufacturing fiber cement panels. As Fabrice Nicolino and François Veillerette accurately explain, "the inconceivable tragedy of asbestos had, if we dare say it, a considerable advantage over the tragedy of pesticides. This carcinogenic fiber leaves traces, a kind of fingerprint, even a genetic print of the crime, which takes the lively form of a cancer specific to the pleura, correlated so closely to contact with asbestos that everyone, including specialists, calls mesothelioma 'asbestos cancer.'"[8] Nothing of the kind is true for pesticides, which are moreover made up of both an active molecule—such as alachlor for Monsanto's Lasso—and of various highly toxic substances which, as we have seen in the case of Paul François, are not always reported when licensing of the formulation is requested. When a sick farmer knocks on the MSA's door seeking acknowledgment of his occupational disease, he must expect a lengthy obstacle course, often beyond his strength and resources.

The Dominique Marchal Case

Nothing better illustrates this difficult process of acknowledgment than the story of Dominique Marchal, a farmer from Meurthe-et-Moselle who participated in the Ruffec Appeal. In 1978, he established a collective farming group (Groupement agricole d'exploitation en commun, GAEC) with three associates on the 1,300-acre family farm near Lunéville. The work was carefully shared out: his uncle and his cousin took care of the cattle, his brother of sowing, and he of "crop health," meaning the application of phytosanitary products on their fields of wheat, barley, and rapeseed.[9] In January 2002, when he had a knee operation, the doctors noticed he had an abnormally elevated level of blood platelets and, after further tests, they diagnosed him with "myeloproliferative syndrome," a disease of the bone marrow that might develop into leukemia. "Since I was the only one who treated the crops, I immediately thought of the phytosanitary products," Dominique Marchal explained at the Ruffec meeting. "Especially because myeloproliferative syndrome is in the table of agricultural occupational diseases associated with exposure to benzene."

Before continuing the incredible story of the farmer from Lorraine, I have to explain the French "tables of occupational diseases of the Social Security general account and agricultural account," which can be consulted on the National Institute of Research and Safety for the Prevention of Work Accidents and Occupational Diseases (Institut national de recherche et de sécurité pour la prévention des accidents du travail et des maladies professionnelles, INRS), website. They go back to October 1919, when a law officially recognized as occupational diseases a certain number of illnesses linked to the use of lead and mercury in industrial or craft work.[10] This decision came after many clinical observations of workers in factories or workshops using heavy metals like lead, whose toxicity had been known since Antiquity and had been the subject of numerous medical reports beginning in the early twentieth century. At the First National Conference on Industrial Diseases, held in Chicago in 1910, Alice Hamilton, an occupational health doctor, described the ailments affecting painters using white lead-based paints (also known as lead carbonate), now classified as lead poisoning.[11] Even now, the first table of occupational diseases of the general account concerns "ailments due to lead and its compounds," such as anemia, nephropathy, and encephalopathy, listed in the left column of the table. The middle column presents the "treatment delay time," that is, the maximum period between the end of exposure to the risk and the first medical observation of the disease. Finally, the right column indicates the work likely to cause the ailment in question, in this case "the extraction, treatment, preparation, use, and handling of lead, lead ore, its alloys, its compounds, and any product containing it."

Since 1919, the list of occupational diseases in the general account has lengthened considerably—it now includes 114 tables. Established by decree, they have been added as medical knowledge of the effects of occupationally used poisons used occupationally has grown. But the creation of a new table, as we shall see in Chapter 6, is the outcome of a long process often delayed by manufacturers' maneuvering; before a chemical substance and the diseases associated with it get on the list, disease and death continue.[12]

A June 17, 1955, decree created the first seven tables of occupational diseases under the agricultural account, listing infectious diseases such as tetanus, leptospirosis, and brucellosis, but also some illnesses linked to arsenic (the latest revision of table 10 dealing with "arsenic and its mineral compounds" dates from August 22, 2008: additions are skin, lung, urinary tract, and liver cancer). The list now contains fifty-seven tables designating ailments associated with lead, mercury, coal tar, and wood and asbestos dust. But only two tables deal with pesticides: table 11, which concerns certain "organophosphates and anticholinesterase carbamates" ("weeding work and anti-parasite treatments of crops and plant products"), and table 13, related to "nitric derivatives of phenol" and "pentachlorophenol associated with Lindane" (for the "treatment of cut wood and timber"). As I explained earlier, the near absence of agricultural poisons in the list is tied to the difficulty of establishing a causal link between a substance and a given disease, because farmers are exposed to many different pesticides throughout their working lives.

On the other hand, as Dominique Marchal pointed out, table 19 concerns "hemopathies caused by benzene and products containing it" such as "anemia, myeloproliferative syndrome, and leukemia."[13] I will come back to the history of benzene (see Chapter 9) which, like that of lead, perfectly illustrates how the regulation of highly toxic substance can be delayed because of organized denial on the part of manufacturers, with the paid complicity of some scientists—which is also true for pesticides and for any other poison coming into contact with our food. Here, it is enough to know that originally benzene was a byproduct of coal tar, the industrial production of which began in the middle of the eighteenth century, with a growing number of uses (a solvent for the manufacture of glue and synthetic dyes, a detergent to remove grease from metals, a material used in the manufacture of synthetic rubber, plastics, explosives, and pesticides, and a gasoline additive).

Classified as a "new domestic poison" by *The Lancet* in 1862,[14] benzene has been classified since 1981 as "carcinogenic for humans" by the International Agency for Research on Cancer (IARC), which, after years of procrastination, finally took into account the many studies showing that chronic exposure to small doses causes serious bone marrow lesions. Indeed, by the late 1920s, medical reports coming primarily from North America and Europe revealed an epidemic of aplastic anemia and leukemia among workers in contact with benzene. In October 1939, the *Journal of Industrial Hygiene and Toxicology* published a special issue on chronic exposure to benzene in which it listed fifty-four studies showing a link between that substance and bone marrow cancer.[15]

Alone Against Everyone

"I had always heard there was benzene in phytosanitary products," Dominique Marchal said at the Ruffec meeting, "and I thought I wouldn't have any trouble being found to suffer from an occupational disease. That was a major mistake." His wife Catherine nodded in agreement with a knowing air. Indeed, in December 2002, the couple sent a request for acknowledgment to the MSA referring to table 19 of occupational diseases under the agricultural account. The MSA took no action on the grounds that benzene did not appear on the warning labels of the pesticides used by the farmer between 1986 and 2002, the not insignificant quantity of 250 products, the invoices for which he had been careful to preserve. Needless to say, had he been, as he would say, a "slipshod farmer," he would have been "on his own."

As we saw with the Paul François case, the additives in the formulation are not mentioned on the labels of cans, and when they are, it is at best under the vague name of "aromatic solvent" or "derivative of petroleum products." Moreover, to justify its decision the MSA referred to a report prepared by Dr. François Testud, an occupational health doctor and toxicologist at the Lyon poison center, who asserted that "the petroleum hydrocarbons used to dissolve certain active ingredients have been free of benzene since the mid-1970s. Later questioned about his gross "mistake" by *L'Express*, the expert, once again serving the interests of industry, punted: "It was inaccurate," he said. "I should have indicated that benzene was not present in proportions posing a health risk."[16]

Hammering the point home, the MSA pointed out that the occupational activity referred to by Dominique Marchal, the spraying of pesticides, was not on the "indicative list of work likely to cause illness" as provided in the right-hand column of table 19: "Preparation and use of varnish, paint, enamel, putty, glue, ink, cleaning products containing benzene."

Faced with the MSA's refusal, the Marchals decided to file a claim with the Épinal TASS; the court appointed a toxicologist, who was unable to move the case forward because he constantly came up against the same problem: the lack of data on the precise composition of the pesticides used. "I was discouraged and wanted to give it all up," said Marchal. "But my wife didn't want to drop it." Catherine's amazing story riveted the audience in Ruffec.

Convinced that benzene was indeed the cause of her husband's serious illness, she decided to ask for help from Senator Christian Poncelet from Vosges, president of the Senate, who wrote to the National Institute of Agronomic Research (Institut national de la recherche agronomique, INRA). In a letter dated January 28, 2005, its president, Marie Guillou, refused to intervene, arguing that the "complete list of ingredients of phytosanitary products is a trade secret."[17] The president of a public institute, whose ties with pesticide manufacturers are an open secret, refused to come to the assistance of a sick farmer, invoking a "trade secret" that has no justification other than the protection of those manufacturers' private interests.

But Catherine did not give up. Encouraged by the family lawyer, Marie-José Chaumont, she decided to conduct the investigation herself. Armed with the names of the molecules her husband had used and a pair of dishwashing gloves, she went around the neighboring farms to collect samples that she meticulously decanted into jam jars. In this way, she managed to collect sixteen "elixirs of death." They next had to be analyzed. Several laboratories refused to carry out the delicate task, but the Chem Tox company, located in a Strasbourg suburb, agreed.[18] "Half the pesticides analyzed contained benzene," said Catherine Marchal, to the applause of the participants in the Ruffec Appeal. "From then on, we knew the case was won."

Indeed, in its September 18, 2006, verdict the Vosges TASS classified Dominique Marchal's myeloproliferative syndrome as an occupational disease. Following Sylvain Médard, the technician of the Picardy agricultural cooperative, he was the second pesticide user to obtain that result. The courageous decision by the Lorraine TASS opened the way for other farmers suffering from leukemia. According to Jean-Luc Dupupet, four years later four of them had been recognized as suffering from an occupational disease; one was Yannick Chenet, who made the effort to participate in the Ruffec meeting. The testimony of this farmer, who works a farm in Saujon, Charente-Maritime, made up of 148 acres of grains and 16 acres of vines for cognac production, once again stirred the audience. After developing "myeloid leukemia type 4" in October 2002, he underwent a "bone marrow transplant which was not 100 percent compatible," he explained, speaking with great difficulty. "My body reacted against the transplant, and I now suffer from retracted tendons, scleroderma of the skin, dry eyes, and lots of other problems." Recognized as suffering from an occupational disease in 2006, the farmer does receive a disability pension, but he has to keep his farm running and to do that he had to hire a farm worker. "All the savings we'd been able to make before my illness have been put into the business to try to save it, but my wife and I are at the end of our rope. I would like to know what my rights are to be able to get out of this situation."[19]

"The only thing you can do," answered Paul François's lawyer François Lafforgue, "is sue the manufacturers to get financial compensation that will enable you to pay the worker you need. It's not easy and the outcome is uncertain, but the more of you who do it, the more chance you'll have of obtaining reparation for the harm you have suffered. That's what happened with the asbestos victims who, by organizing and systematically suing, were finally compensated."

"Counting the Sick and the Dead in the Morgue"

Sick farmers have not yet reached that point, not even those who went to Ruffec, because some are still fighting to be recognized as suffering from occupational diseases. The stories of Dominique Marchal and Yannick Chenet are exceptions, because their illnesses (myeloproliferative syndrome and leukemia) are found in the tables of occupational diseases appended to the Social Security Code. For all other illnesses, patients have to file what is called a request for recognition "off table," following a usually long and trying procedure that was established in 1993. It provides that individuals considering themselves victims of an occupational disease not listed in the tables can address the Regional Committee for the Recognition of Occupational Diseases (Comité régional de reconnaissance des maladies professionnelles, CRRMP) if they have a permanent partial disability of at least 25 percent or if they are dead (in which case the request is made by the widow or the orphans). This is what Sylvain Médard did; he had the "luck" to have contracted such a rare disease, acquired mitochondrial myopathy, that its chemical origin was not too hard to demonstrate.

The CRRMPs—there is one per region—are composed of three expert doctors: the regional medical officer or his representative, a labor inspector doctor, and a university professor and/or hospital practitioner, whose task is to examine the medical file to determine whether there is a causal link between the disease and the occupational activity of the complainant. And this is where things get difficult, because for much more "banal" diseases than Sylvain Médard's myopathy, on what grounds can the experts base their evaluation?

To be able to state with certainty that a given poison causes a given disease, the ideal thing would be to conduct an experiment in which you expose volunteers to the poison at a *certain* dose, for a *certain* period of time, to observe after a *certain* number of years how many contract the disease in question. Further, to avoid contamination of the human guinea pigs by other substances—which might be used by poison manufacturers to cast doubt on the relevance of the results—it would be appropriate to confine them to an isolated site throughout the length of the experiment while strictly controlling their environment. This is clearly impossible, first of all for obvious ethical reasons. After the horrors perpetrated by Nazi doctors on the victims of the extermination camps, the Nuremberg trials pointed out that this kind of experiment was a crime. And then, assuming morality did not forbid it, to be conclusive the study would have to be repeated several times, varying the profile of the human guinea pigs (age, sex, state of health), doses, length of exposure, and observation of effects (especially because the latency period for chronic diseases is estimated to be at least twenty years). Given that one hundred thousand potentially toxic molecules have been released into the environment since the end of World War II, it is not hard to imagine the magnitude of the task.

Before going any further, I would like to point out that if we have reached this point, namely, considering how *best* to measure the link between a serious illness and exposure to a chemical product, it is precisely because at one moment in their history humans decided that they could, with impunity, dump poisons on their fields, their factories, their houses, the water they drink, the air they breathe, and their food. And by doing this they de facto transformed the inhabitants of our planet into guinea pigs, because fifty years later we are reduced to "counting the sick and the dead in the morgue," in the words of the American epidemiologist David Michaels, who correctly points out that it is a "very simplistic method" and "remarkable in this day and age."[20]

And we have reached this point also because politicians have allowed manufacturers to lay down the law, which consists of "demanding that one prove the toxicity of their products before any regulation, which amounts to applying the principles of criminal law to substances, presuming them innocent until proven guilty," as Geneviève Barbier and Armand Farrachi explain in their book *La Société cancérigène* (Carcinogenic Society). "But if the ecosystem as a whole is contaminated, it becomes impossible to isolate the responsibility attributable to one of them."[21]

In the meantime, what morality forbids being practiced on laboratory humans is authorized on animals, who have paid a heavy price for the frenzied industrialization imposed by humans. Indeed, as we shall see in Chapter 9, for about thirty years manufacturers have been required to conduct toxicological studies to obtain marketing authorization for their products. Conducted on animals, usually rodents, the studies are supposed to test a certain number of potential toxic effects, such as carcinogenicity or neurotoxicity. The problem is that, assuming they are well conducted—which is far from being the rule (I will come back to this with the example of aspartame)—these studies are generally not considered as "sufficient proof" when it comes to extrapolating their results to human beings. The American epidemiologist Devra Davis points to this paradox in her masterful book *The Secret History of the War on Cancer*: "Where animal studies on the causes of cancer exist, they are often faulted as not relevant to humans. Yet when studies of almost identical design are employed to craft novel treatments and therapies, the physiological differences between animals and humans suddenly become insignificant."[22]

Impossible Proof

The fact remains that in order to be able to make a decision the experts of the regional occupational disease committees (CRRMPs) require human data: before banning a product or recognizing that a sick farmer has an occupational disease, they first want to have "counted the sick and the dead in the morgue." And that is the work of epidemiologists. According to Jean-Luc Dupupet, "Epidemiological studies are of capital importance; the MSA has relied on them to gradually recognize as occupational diseases previously neglected illnesses such as certain cancers and Parkinson's disease."

As Michel Gérin and his co-authors explain in *Environnement et santé publique*, "[E]pidemiology is traditionally defined as the study of the distribution of diseases and their determinants in human populations. . . . It does not undertake the study or definition of the mechanisms by means of which exposures act on the human organism," but it "measures their effect,"[23] in researching, for example, why some people develop cancer and others not. To do this, it has various tools that I must present briefly, because this basic knowledge is essential to understand the incredible complexity of the situation in which the unbridled industrialization of agriculture and of society as a whole has placed us. This knowledge will also help, throughout this book, to better understand the many tricks manufacturers deploy to maintain or fabricate doubt about the toxicity of their products in order to delay as long as possible their regulation or withdrawal from the market.

To determine the factors that may contribute to the emergence of a disease, epidemiologists proceed by comparison. For example, they compare a group of people suffering from a given disease, such as non-Hodgkin's lymphoma (a cancer of the lymphatic system), to a comparable group (by height or age of the participants) of healthy people. This kind of "case-control" study is retrospective, because it relies on the memory of people with whom the scientists try to reconstitute their way of life or the substances they may have been exposed to by means of questionnaires and interviews. Often disparaged by industry, which suspects patients of adapting their memories to the needs of the investigation, case-control studies are frequently used to measure the role of pesticides in the appearance of certain diseases in agricultural populations. Another type of retrospective study, a "cohort" study, consists of comparing a group of people having undergone the same exposure to a given factor (such as grain farmers practicing chemical agriculture) to a group not having undergone that exposure, to determine which diseases are more frequent among the exposed subjects.

In the two types of study, the relative risk of developing an illness (such as non-Hodgkin's lymphoma) among individuals exposed to the factor studied (such as pesticides) compared to unexposed subjects is expressed as an "odds ratio" (OR), derived from statistical calculations. If an OR exceeds the number 1, which is the normal risk of an unexposed population, it means that the study has shown an increased risk among the exposed group. For example, an OR of 4 indicates that the risk is multiplied by four among the individuals exposed to the factor studied.[24] In contrast, an OR lower than 1 indicates that the exposure protects against the disease in question.

Finally, to conclude for now this brief presentation, it should be noted that epidemiologists sometimes use a third type of study, known as "prospective." Much more costly than retrospective studies but less open to question because it does not rely on participants' memories, a prospective study begins at a time T of a population exposed to a given factor, such as a group of farm families using pesticides, and following them over several years or even decades, recording diseases when they appear. The results are compared to a control group, assumed not to be exposed to the risk factor under investigation.

This is where the principal weak point of epidemiological studies lies: whether retrospective or prospective, it is difficult to find a control group about which one is absolutely certain that it has not been exposed to the factor studied or to other factors having similar effects. "In a disease like cancer, unquestionable results are rare," according to Geneviève Barbier and Armand Farrachi; "on one hand because the process of development of cancer is long, and on the other, because, not living in a bubble, everyone is subjected to numerous carcinogenic factors that confuse the evidence. Besides, studies compare the rate of cancer in a population exposed to an 'expected' rate in the general population, a terrible term that, better than any argument, lends credence to what is known as background noise and trivializes a harm from which no one escapes. The absence of results does not prove the absence of risk, but often the impossibility of bringing those results to light."[25]

5

Pesticides and Cancer:
Consistent Studies

Wounded, she asks humanity: What use is ruin? What will desert plains produce?

—Victor Hugo, *The Earth: A Hymn*

"We have reviewed recent international publications, epidemiological studies looking for a possible connection between non-Hodgkin's lymphoma and phytosanitary products, and exhaustive research has not so far produced a positive response. . . . On the whole, we have no information that can reasonably support a certain connection between your illness and your previous occupational activity." I remember the great surprise François Veillerette expressed in Ruffec when Jean-Marie Bony read that excerpt from a letter sent to him on March 21, 2003, by Professor Jean Loriot, head of the Occupational Health Service at Lapeyronie University Hospital in Montpellier. "It's surprising he wrote that," the president of the Movement for Law and Respect for Future Generations (Mouvement pour le droit et le respect des générations futures, MDRGF) commented. "There are, even so, several farmers suffering from non-Hodgkin's lymphoma, which has been recognized as an occupational disease."

This is true. According to Dr. Jean-Luc Dupupet, in the spring of 2010, exactly three of them had been granted the priceless recognition by their regional committees of their status as victims of an occupational disease. To back their decisions the Regional Committees for the Recognition of Occupational Diseases (Comité régional de reconnaissance des maladies professionnelles, CRRMP) relied on the significant scientific literature on non-Hodgkin's lymphoma (NHL), which is "one of the most studied forms of cancer in connection with the use of pesticides," according to Dr. Michael Alavanja of the National Cancer Institute of the United States. In his often cited 2004 article "Health Effects of Chronic Pesticide Exposure: Cancer and Neurotoxicity," the epidemiologist points out that, in eighteen of the twenty studies he examined, "NHL has been associated with phenoxacetic acid-based herbicides[1] and organochlorine and organophosphate pesticides," when the risk "was doubled."[2]

Rewarded by Monsanto and Suffering from NHL

"This is exactly the kind of product I handled for more than thirty years," said Jean-Marie Bony, aged sixty-two, showing me his voluminous file in which he had recorded, year by year, the various poisons he'd been in contact with: organophosphates, organochlorines, carbamates, solvents (benzene, polyethylene esters, alkyl phenol glycol, ammonium sulfate), to cite only some generic names, because the products themselves fill a dozen pages. Until 2002, Jean-Marie Bony was the director of the Provence-Languedoc agricultural cooperative, which covers part of the departments of Vaucluse, Gard, and the Bouches-du-Rhône, a "sector rich in vines, fruit trees, market gardens, and grains" where phytosanitary products are used abundantly.

Hired by the cooperative at the age of twenty-one, the farmer's son at first handled "thousands of stitched paper bags, by hand and without gloves, because at the time we had neither forklifts nor protective equipment; the bags sometimes ripped, spilling out coated seeds[3] or powdered products," he told the Ruffec meeting. "I unloaded the trucks, stored the products in the warehouse and helped the farmers carry them to their cars." After a promotion, he supervised the collecting of treated grain, consulted on the adjustment of sprayers, and intervened on farms "when there were attacks of disease, fungi, or insects," directing the spraying of "vines, fruit trees, potatoes, grain, melons, tomatoes, asparagus, onions." "I even had the privilege of testing in farmers' fields products that were not yet licensed that the firms gave us," he noted with some bitterness. "I sprayed them on plants, then pulled the leaves apart with my bare hands to see if the insects were really dead. Later, when there were floods in Ardèche and Rhône that kept the farmers from going to their fields, I supervised spraying by helicopter. In other words, the whole thing."

After a silence, he went on: "I don't want to bite the hand that feeds me, because I did well. Since I was a very good salesman, I earned large commissions and I went on some great trips financed by Monsanto and Phyteurope: I went to Niagara Falls, I went snowmobiling in Canada, I visited Greece and Senegal. In 2001, Monsanto even organized a bus so the heads of agricultural cooperatives could go and see the first fields of transgenic corn in the Toulouse region. But in the end I paid very dearly for it, like André, the president of my cooperative, who died from leukemia."

In 1993, Jean-Marie Bony was operated on for a cancerous polyp in his colon. Nine years later, during a routine checkup, he was diagnosed with a B-cell centroblastic lymphoma, an "aggressive" form of NHL. "After chemotherapy, when Professor Jean-François Rossi, chief of hematology at Lapeyronie University Hospital in Montpellier, advised me to request recognition that I was suffering from an occupational disease, I thought the sky was falling. I had never imagined that the pesticides I had handled for years could make me sick. I trusted the manufacturers and the people who authorized their sale."

In a letter dated October 8, 2002, Professor Rossi wrote that Bony's disease has "a probable or possible link with organophosphates." He was alone in this judgment, however, because thereafter all the experts consulted said exactly the opposite. On November 5, 2004, the Agricultural Social Mutual Fund (Mutualité sociale agricole, MSA) closed Bony's file with an expected argument: "The disease from which you are suffering is not listed in the table of occupational diseases of farm workers." Bony then filed a claim with the Social Security Court (Tribunal des affaires de sécurité sociale, TASS) in Avignon, which asked Professor Bertrand Coiffier, chief of the hematology clinic at the Lyon Sud hospital center, to prepare a report. "There are no serious studies allowing us to conclude *definitively*[4] as to the involvement of pesticides in the onset of lymphoma," he wrote peremptorily on December 3, 2007.[5]

In Professor Coiffier's assertion, obviously the adverb "definitively" is the focus of attention. Yet he must know that, in the area of environmental health, "definitive" proof is impossible to obtain, except if one could require human guinea pigs to be isolated to test on them the toxicity of products. The only alternative is therefore epidemiological studies, imperfect to be sure, but they indicate a tendency and constitute the "best available evidence," to adopt the words of the American epidemiologist David Michaels.[6] But the curious thing is that in Professor Coiffier's report there is no scientific reference showing that, *at a minimum*, he was aware of the numerous epidemiological studies that had investigated the link between pesticide exposure and NHL. Hard as one looks, one can find nothing. Perhaps the professor is unaware of PubMed, the database of the U.S. National Library of Medicine that registers all the scientific studies published in the world, with references, a summary of the content, and a link to the site of the journal of publication.[7] It's in English, of course, but that shouldn't be an insurmountable obstacle.

The Difficult Work of Epidemiologists

When you enter "Non-Hodgkin's Lymphoma" and "Pesticides" in the PubMed search engine, you get 240 results. It's a large quantity, especially because you have to know how to separate the wheat from the chaff—and we will see later that that is not a simple matter—because the scientific literature is often polluted by less than rigorous and even biased studies commissioned by industry not in the search for truth but in order to muddy the waters.

To orient yourself in the fascinating labyrinth of PubMed (or MedLine, a similar database), it is advisable to rely on systematic surveys of the scientific literature carried out by researchers whose reputation is beyond question and who have rigorously examined all the studies on the subject that interests you. This was done, for example, by Michael Alavanja at the National Cancer Institute in the article I have already cited, "Health Effects of Chronic Pesticide Exposure."[8]

Similar work was done in 2004 by a group of Canadian doctors, cancer specialists, and epidemiologists, for a study titled *Systematic Review of Pesticide Human Health Effects*, often cited as a reference because of the rigor of its methodology.[9] At the request of the Ontario College of Family Physicians, the researchers located in four bibliographic databases (MedLine, PreMedLine, CancerLit, and LILACS) the studies published between 1992 and 2003 in French, English, Spanish, and Portuguese dealing with "non-Hodgkin's lymphoma, leukemia, and eight solid tumors: brain, breast, kidney, lung, ovarian, pancreatic, prostate, and stomach cancer."

After a detailed examination of 1,684 articles they had initially selected (out of a total of 12,061 dealing with pesticides), they finally retained 104 that met the quality criteria they had defined. The result was a 188-page document, presenting each study examined, with a note of evaluation (on methodology, consideration of possible bias, and so on), the populations studied (number of individuals), and the type of study (cohort or case-control). Thus, out of the twenty-seven epidemiological studies of NHL, twenty-three showed "associations between pesticide exposure and NHL, many with statistical significance."

To illustrate the working methods of epidemiologists, whose contribution is essential for the evaluation of environmental risks, I have chosen four studies. The first is a case-control study published in 1999 by Swedish scientists Lennart Hardell and Mikael Eriksson, conducted in seven counties in northern and central Sweden.[10] In their introduction, the authors note that in Sweden from 1958 to 1992, the mean age-adjusted incidence of NHL increased every year by 3.6 percent for men and 2.9 percent for women.

I take this opportunity to recall the meaning of "incidence rate," which is often confused with "prevalence rate," two fundamental tools in epidemiology that we will have occasion to refer to frequently in the course of this book. Incidence designates the number of new cases of a disease that appear in a given period (usually a year) for a defined population (generally one hundred thousand people). Prevalence measures the number of sick people at a given moment, including old and new cases. If one is interested in the progression of a disease that may become an epidemic, such as flu, for example, it is more useful to follow the development of incidence, because it provides information on peaks in which the number of individuals suffering the illness increases considerably. With respect to cancer, the fact that the incidence rate consistently grows from year to year means that carcinogenic factors are at work, which has led a growing number of people to suffer from the disease.

It was precisely some of those factors that Hardell and Eriksson tried to determine by comparing a group of 404 men who had been diagnosed with NHL between 1987 and 1990 with a control group of 741 healthy men of the same age (older than twenty-five). The participants answered a long questionnaire, supplemented by a telephone interview, about their way of life (eating habits, risk conduct—smoking, alcohol use—sporting activities), their previous illnesses, and their occupational history. Pesticide users were asked to specify where they used them (forests, crops, gardens), the type of product used (herbicides, insecticides, fungicides), the family of compounds (carbamates, organophosphates, chlorophenols), the active ingredients or the manufacturers' formulas, and the frequency and duration of use. The results showed that those who had been exposed to herbicides of the phenoxy family (chlorophenols) had a higher risk of developing NHL (odds ratio, OR: 1.6) and that the risk mounted (OR: 2.7) if the herbicide was 4-chloro-2-methylphenoxyacetic acid (MCPA). Association with fungicides practically quadrupled the risk (OR: 3.7).

Similar results were obtained by American researchers at the National Cancer Institute in Rockville in a case-control study they conducted in the farm state of Nebraska, published in 1990. It showed that the risk of developing NHL was tripled if people used 2,4-D (one of the components of Agent Orange that is also in the chlorophenol family) at least twenty days a year.[11]

Among the studies selected by the Ontario physicians for systematic review, there are some surprises, such as the retrospective cohort study conducted by University of Iowa researchers at the request of the Golf Course Superintendents Association of America. Worried by a growing number of premature deaths among its members, the association, whose mission is to maintain the legendary greens with liberal use of pesticides, made its death records available to the epidemiologists, who were able to analyze 686 deaths occurring between 1970 and 1982 in the fifty states of the union. Twenty-nine percent were due to cancer. Causes of mortality were compared to those of the general population (white men only). The results show high death rates from four types of cancer: NHL (OR: 2.37), and brain, prostate, and intestinal cancer.

To conclude, I would like to cite a prospective study conducted on a population of Danish professional gardeners (859 women and 3,156 men), who were followed from 1975 to 1984.[12] Researchers from the University of Copenhagen concluded that the use of pesticides leads to a doubling of the risk of NHL and a very significant increase of the incidence of soft-tissue sarcoma (OR: 5.26) and leukemia (OR: 2.75)

Contrary to what Professors Jean Loriot of Montpellier and Bertrand Coiffier of Lyon asserted a little hastily, a large number of epidemiological studies converge on the same assessment: there is indeed a link between pesticide exposure and NHL, and more generally, all diseases of the lymphatic system (leukemia, myeloma).

These statistical results were validated in 2009 by an extremely important study that provides a biological explanation for the observations of epidemiologists. Researchers at the National Institute of Health and Medical Research (Institut national de la santé et de la recherche médicale, INSERM) working at the Marseille-Luminy immunology center found that farmers exposed to pesticides presented "molecular markers of tumoral precursors," that is, "develop genetic anomalies that may be a source of cancer of the lymphatic system," to adopt the language of the Ligue contre le Cancer, which presented this research at the February 4, 2009, World Cancer Day meeting.

To reach these results, the scientists conducted a prospective study on a cohort of 128 farmers using pesticides, whom they followed for nine years, along with a control group of 25 unexposed farmers. Through periodic blood samples, they analyzed the development of blood lymphocytes and found that exposed farmers had "one hundred to one thousand times more translocated cells" than the control group. Translocated cells are the product of a genetic anomaly caused by an exchange of DNA fragments between chromosomes 14 and 18 (t [14;18]). Also present in healthy individuals, they can be considered a biological marker of a carcinogenic process, particularly if they begin to proliferate.

"Strikingly, although t(14;18) frequency slowly increased in the control population (+87%; P = 0.03), mostly as the result of aging, a dramatic increase was observed for the exposed cohort (+253%," the researchers note in their study, "Agricultural Pesticide Exposure and the Molecular Connection to Lymphomagenesis." They conclude: "Our results clearly demonstrate that the expanded t(14;18)+ clones, which are particularly prominent in farmers exposed to pesticides, constitute bona fide FL [follicular lymphoma] precursors standing at various stages of tumor progression."[13]

Consistent Studies on the Role of Pesticides in Certain Cancers

The results presented in the systematic review by the Ontario physicians confirm those obtained in meta-analyses like the one performed in 1992 by Aaron Blair, a colleague of Michael Alavanja at the National Cancer Institute in Bethesda and one of the most prominent epidemiologists in the area of the links between cancer and pesticides.[14] In passing, I should explain the difference between a systematic review, such as those conducted by Dr. Margaret Sanborn's group in Ontario or by Michael Alavanja, and a meta-analysis, another tool for epidemiology. The former consists of identifying and analyzing all the studies related to a subject of interest, like those dealing with "pesticides and cancer." The latter designates a statistical procedure that consists of assembling the data produced by comparable studies and putting them together to arrive at an overall conclusion. Much used in pharmaceutical research to measure the effects of new therapies, meta-analysis increases the statistical force of isolated results by augmenting the number of subjects participating in the comparison. But this is true on condition that the studies selected for this new statistical calculation are really comparable and that mediocre or frankly biased studies are eliminated in order not to distort the final result.

For his meta-analysis, Aaron Blair selected twenty-eight epidemiological studies that met the quality criteria he had established. In his introduction, he notes that farmers generally have a lower death rate from cancer and cardiovascular disease than the general population and that they have a "lower rate of lung, esophagus, and bladder cancer," because they tend to smoke less. In contrast, as shown by the results of his meta-analysis, "farmers tend to be at higher risk for cancers of the lip, melanoma, brain, prostate, stomach, connective tissue, and lymphatic and hematopoietic system than the general population." He goes on to specify that "The excesses among farmers for a few specific cancers, against a background of low risks for most cancers and nonneoplastic disease, suggest a role for work-related exposures. These patterns may have broader public health implications, since several of the high-rate tumors among farmers also appear to be increasing in the general population of many developed countries."

Was it the article's conclusion that Monsanto did not find to its liking? In any event, it asked its house epidemiologist, John Acquavella, to conduct a counter-meta-analysis. Apparently, the researcher found what he was looking for and after combining in the same pot thirty-seven carefully selected studies, he concluded unsurprisingly: "The results do not suggest that farmers have elevated rates of several cancers."[15]

In a letter to the journal *Annals of Epidemiology* that published Monsanto's meta-analysis (the multinational's name appears beneath the authors' names in the summary published online by PubMed), Samuel Milham, an epidemiologist from Washington, expresses surprise at the method used by his colleague from Saint Louis to compile his statistics: "Why were "crop/livestock" farmers considered together? [They] certainly have different kinds of exposures, and if they have different patterns of cancer mortality, lumping them will confuse the relative risk calculation. I feel that the heterogeneity of exposures in farmers is so great that meta-analysis of this type can only cloud the issue of cancer in farmers. What is needed is a finer exposure categorization."[16]

To fully understand the relevance of this comment, it should be noted that the occupation of "farmer" includes very varied activities, which depend on the type of production carried out on the farm. There is no comparison between a "grain farmer," the essence of whose work involves growing wheat or corn, and a "cattle raiser," who, as the name indicates, raises cattle. In terms of pesticide exposure, the risks are obviously not the same, the former using many more phytosanitary products than the latter. Not taking these differences into account means demonstrating ignorance of the realities of the agricultural world which might provoke a smile if it were not the act of a scientist working for a leading multinational in the global pesticide and seed market.

In substance, Milham's question points to one of the principal dangers of meta-analyses, which may lead to erroneous results if the choice of studies combined is not sufficiently rigorous—the mixing of apples and oranges. In the section discussing the methodology used in his meta-analysis, Aaron Blair particularly emphasizes this bias which must absolutely be avoided: "Since not all farmers have the same exposures, combining those with different exposures would tend to dilute the effects of relevant exposures and bias risk estimates toward the null (46). The potential magnitude of such a dilution effect can be illustrated with data from a recent study in Iowa and Minnesota.[17] Among the 698 population-based referents who ever lived on farms, 110 never used insecticides and 344 never used herbicides. . . . Approximately 40% of the farmers used phenoxy acid herbicides and 20% used organochlorine insecticides, the two most frequently used pesticide classes. Even if these chemicals were strong risk factors for a particular cancer, analyses based simply on the occupational title of farmer could seriously underestimate the relative risks."

All this would amount to nothing but a battle of experts of little interest to a lay public were there not huge stakes lying behind it, with very concrete repercussions for the lives of citizens. For instance, in the case of Jean-Marie Bony, the issue is not to question the integrity of Professors Jean Loriot and Bertrand Coiffier, especially because there is no indication that they have any conflicts of interest with pesticide manufacturers, as is sometimes the case for certain experts (see Chapters 10 and 11). But one can easily imagine that, overwhelmed with work, they were unable to spend two weeks, as I did, navigating on the sites of PubMed and MedLine. It is also possible that they came upon by chance the meta-analysis by John Acquavella, unaware that it had to be taken with some reservations because, although the name of the sponsor appears in the online summary published by PubMed, the same information is difficult to find in the article published by the *Annals of Epidemiology* (it is in small print at the bottom of the first page). Thus if the experts asked to evaluate Jean-Marie Bony's medical file were satisfied with consulting the meta-analysis by Monsanto's official epidemiologist, it is easy to understand why they found no link between pesticide exposure and NHL and, beyond that, with any type of cancer, contrary to the opinion of dozens of independent scientists who have concluded the opposite.

Bone and Brain Cancers: Farmers on the Front Lines

In general, all researchers have reached the same conclusion: although farming populations overall have lower cancer mortality than the general population, some types of cancer are more frequent among farmers. This is the case for malignant hemopathies, such as leukemia and NHL, as well as for multiple bone myeloma. Also known as Kahler's disease or simply myeloma, this cancer, which develops in bone marrow, "has been gradually increasing in most parts of the world," as Michael Alavanja points out in his systematic review, where he cites a meta-analysis that evaluated thirty-two studies published between 1981 and 1996; the meta-analysis estimated the excess risk among farmers at +23 percent.[18]

The first time I heard about this disease, which accounts for 1 percent of cancers and for which the survival rates are very low, was in Ruffec, from Jean-Marie Desdion, a corn producer who had come especially from Cher. Accompanied by his wife, he described his ordeal, which began in 2001 with the spontaneous and abrupt breaking of both humeri followed by the disappearance of half his ribs. The diagnosis was irrevocable: light-chain multiple myeloma. Hospitalized at the Hôtel-Dieu in Paris, the grain farmer underwent two bone marrow autografts, followed by burdensome treatments—chemotherapy, radiation, and corticotherapy—at Georges-Pompidou Hospital. "To conclude," he explained, "I received a gift of stem cells that were injected in a sterile room after the complete destruction of my bone marrow. It was a long and exhausting process. I'm now feeling better, but from an occupational point of view I'm in an inextricable situation: I applied for recognition of my occupational disease and, while I'm waiting, it's very hard. I received daily indemnities for three years, as provided in my insurance contract. And after that, nothing. The paradox is that I don't fit into any box: normally, after three years of sick leave, you're either dead or cured. Since I'm neither one nor the other, I have to work and keep my farm going, which is really very hard."

Encouraged by his lawyer, François Lafforgue—also Paul François's lawyer—Jean-Marie Desdion decided to file suit against Monsanto. "Paul and I have a lot of things in common," Desdion explained with a smile. "Since we're both corn producers, we used a good deal of Lasso. The difference is that he was a victim of acute poisoning and I of chronic poisoning. Yet I followed all the recommendations of the MSA, which advised spreading out treatments over the longest possible time. In general I sprayed Lasso for two to three weeks for two to three hours a day. That was a fundamental mistake."

I remember the smoldering anger that filled me when I heard Desdion tell his story. Rereading the notes I took at the time, I found a question underlined twice, angrily: How many people are dying today of cancer on the farms of France and Navarre? Will we ever know? According to Isabelle Baldi and Pierre Lebailly, two French specialists in agricultural medicine, in their 2007 article "Cancers et pesticides," "Up to now, thirty epidemiological studies have explored the risk of cerebral tumors among farmers and the majority of them show an increase in risk on the order of 30 percent."[19] They thus confirm the conclusion of the systematic review by the Ontario physicians, which noted that among solid tumors, the one that affected farmers the most was brain cancer.

Baldi, who works at the Occupational and Environmental Medicine Laboratory at the University of Bordeaux, and Lebailly, of the regional cancer study group at the University of Caen (Groupe regional d'études sur le cancer de l'université de Caen, GRECAN), know the subject well, since they participated in the CEREPHY study (on cerebral tumors and phytosanitary products), published in *Occupational and Environmental Medicine* in 2007.[20] This case-control study conducted in Gironde examined the link between pesticide exposure and diseases of the central nervous system: 221 patients with benign or malignant tumors, diagnosed between May 1, 1999, and April 1, 2001, were compared to a control group of 422 individuals without the diseases under study, randomly selected from the department's voting rolls (age and sex were obviously controlled for). Among the patients, whose mean age was fifty-seven, 57 percent were women; 47.5 percent had glioma, 30.3 percent meningioma, 14.2 percent acoustic neuroma, and 3.2 percent cerebral lymphoma.

In interviews conducted in participants' homes or in the hospital, investigating psychologists carefully evaluated modes of pesticide exposure, classifying them by categories: gardening, treatment of house plants, spraying of vines, or merely residence near treated crops. They also noted other factors that could have contributed to development of the disease, such as family background, the use of mobile phones or solvents, and so on. The results were unambiguous: winegrowers, who make massive use of phytopharmaceutical products[21]—as I confirmed when I visited the agricultural lycée in Pézenas (see Chapter 3)—have twice the risk of developing a cerebral tumor (OR: 2.16) and three times that of developing glioma (OR: 3.21). Similarly, people who regularly treat their house plants with pesticides have twice the likelihood of developing a cerebral tumor (OR: 2.21).

The incidence of brain tumors among winegrowers had already been the subject of a study published in 1998 by Jean-François Viel, an epidemiologist who had written his doctoral dissertation on the geographical associations between cancer mortality among farmers and pesticide exposure.[22] For this work he used the "geographical indices of pesticide exposure" to "test their potential link to cancer mortality among French farmers." At the time he conducted his study—the late 1980s—93,000 tons of pesticides were released annually on French territory. Relying on data supplied by the Ministry of Agriculture as well as a study conducted by the agronomist André Fougeroux,[23] he was able to develop a map of exposure according to department and crop. He found that 96 percent of straw cereal crops (which covered 17 million acres) were treated with herbicides, 31 percent with insecticides, and 70 percent with fungicides; for corn (8 million acres), 100 percent of the surface was treated with herbicides; for vines (2.5 million acres), 80 percent of the parcels were treated with herbicides, 82 percent with insecticides, and 100 percent with fungicides; for apple trees (150,000 acres) the figures were 80, 100, and 98 percent respectively. And for all cultivated land in France, the proportions were 95, 39, and 56 percent respectively.

Given the geographical distribution of the eleven principal French crops[24] and the agronomical practices involved in each type of crop (categories of pesticide used, quantities per acre, and number of treatments annually), Jean-François Viel reconstituted the distribution of chemical exposures in all French departments (except for the five most urbanized, in Île-de-France and the Territoire de Belfort). He then consulted the statistics of INSERM and the National Institute for Statistics and Economic Studies (Institut national de la statistique et des études économiques, INSEE), specifically the register of deaths occurring between 1984 and 1986 for employment categories "10" (farmers) and "69" (farm workers). The results of this vast study, called "ecological" because it focused on groups of people rather than individuals, showed an excess of mortality for pancreatic and kidney cancer in areas where crop land predominated (such as Beauce or Auvergne) and excess mortality for bladder and brain cancer in wine-growing areas (such as the Bordeaux region).

With regard to brain tumors, we should also mention a vast Norwegian cohort study published in 1996. Its authors examined the incidence of certain cancers in the offspring of farmers and other occupational pesticide users. Exceptional because of its size, the study dissected the medical history of 323,292 children born between 1952 and 1991, whose parents were at the time registered as active farmers.[25] The results showed an excess of brain tumors and NHL in children below the age of four in families of horticulturists and farmers, as well as an excess of osteosarcomas (bone tumors) and Hodgkin's disease in adolescents from families of poultry farmers—intensive poultry battery farms are large consumers of chemical disinfectants and insecticides. This corroborates the results of numerous epidemiological studies that attest to a link between parental pesticide exposure and the two forms of cancer most frequent among children: brain tumors and leukemia (see Chapter 19).

The Troubling Results of the Large "Agricultural Health Study"

This was the largest prospective study of the health effects of pesticides ever conducted in the farm environment. Called the "Agricultural Health Study," it was launched in 1993 by three prestigious American public institutions: the National Cancer Institute, the National Institute of Environmental Health Sciences, and the Environmental Protection Agency. From December 13, 1993, to December 31, 1997, 89,658 residents of the rural states of Iowa and North Carolina were enrolled in this vast cohort, which included 52,395 farmers using pesticides and 32,347 spouses, as well as 4,916 professional pesticide applicators.[26]

To be included in the study, participants had to respond to a twenty-one page questionnaire, which carefully recorded all the information concerning their medical history (earlier diseases), family background, eating habits, lifestyle (tobacco use, alcohol consumption, sports activities), and a precise description of pesticides used (families of products, exact names of formulations, quantities applied, frequency of treatments, use or not of protective equipment). In addition, when included in the cohort, participants were asked in regular follow-up interviews to communicate any change in their farming practices as well as the development of new diseases as soon as they were diagnosed by a doctor.

This exceptional study filled in a number of gaps often at issue when interpreting the results of case-control studies. First, the "data collection [concerning pesticide exposure] prior to the diagnosis of cancer precludes" biases created by uncontrollable memory lapses, according to Michael Alavanja and Aaron Blair, two of the principal authors of the study. It also avoided

the difficulty faced by most case-control studies: the lack of precise information on exposure levels and identification of the most dangerous products.
One of its strengths was precisely that it provided for each user his "exposure
for each pesticide [including] days of use per year, years of use, application
methods, and protective equipment use." Further, "the large size of the study
gives sufficient statistical power to examine the risk of exposure to a number
of specific chemical exposures."[27] In 2005, twelve years after the study began, many results had already been obtained and synthesized in some eight
scientific publications—which anyone can consult on the Agricultural Health
Study's website, an unusual example of transparency in the field.[28] One discovers, for example, that in 2005 four thousand cases of cancer developed in
the cohort: 500 cases of breast cancer, affecting essentially farmers (and not
their wives), 360 cases of lung cancer, 400 of the lymphatic system, and 1,100
of prostate cancer. Comparison with data from the general population confirmed what retrospective studies had already shown, namely, a significant
overall deficit in cancer among farmers (−12 percent) and their spouses
(−16 percent), especially for lung cancer (−50 percent) and cancer of the
digestive tract (−16 percent). In contrast, the data show an excess (+9 percent) of breast cancer among farmers (and not their wives), but a much larger
excess for ovarian cancer among women industrial applicators (the risk is
tripled), and melanoma among farmers' spouses (+64 percent). For men, the
results indicate an excess of lymphatic system cancer, as for multiple myeloma (+25 percent), as well as prostate cancer (+24 percent for farmers and
+37 percent for industrial applicators).[29]

As Alavanja and colleagues point out: "Prostate cancer is the most common
malignancy among men in the United States and in most Western countries,"
but "its etiology remains largely unknown." This is why the researchers sought
to determine whether there were specific exposures that could explain this
excess. The article they published in 2003 shows that, among the forty-five
pesticides considered, the use of methyl bromide[30] and organochlorine products considerably increased the risk (OR: 3.75).

It is interesting to note that the rate of incidence for prostate cancer found
in the vast prospective American study is very similar to the one found, for
example, by Belgian researchers Geneviève Van Maele-Fabry and Jean-Louis
Willems in a meta-analysis published in 2004. On the basis of twenty-two
retrospective studies they also observed a mean risk increase of 24 percent
(OR: 1.24), although they did not specify which pesticides were implicated
in this excess.[31]

Waiting for AGRICAN

To conclude this chapter on the links between pesticides and cancer, I would have liked to report on the first results of the AGRIculture and CANcer (AG-RICAN) cohort study, begun in France in 2005 by the MSA in collaboration with the Caen Regional Cancer Research Group (Groupe régional d'étude sur le cancer, GRECAN) and the laboratory of occupational and environmental medicine in Bordeaux, which employ respectively Pierre Lebailly and Isabelle Baldi. Unfortunately, although announced for "late 2009," the data concerning "the most common cancers," to quote the MSA, namely, prostate and breast cancer, had still not been made public one year later. Adopting the methodology of the Agricultural Health Study, AGRICAN has assembled the "largest agricultural cohort on the international level," according to the French National Cancer Institute (Institut national du cancer, INCa), which helped finance the study. From 2005 to 2007, six hundred thousand questionnaires were sent out to salaried and non-salaried farmers who had paid dues to the MSA for at least three years and lived in twelve French departments that had cancer registries.[32]

I was able to consult the model questionnaire on the MSA website. Comprising eight pages, it begins with a sentence introducing the study, whose stated purpose is to "become better aware of *occupational risks* and improve the health and safety of the agricultural world by improving *prevention*."[33] It is interesting to note that the authors carefully avoid naming the phytosanitary products whose potential effects on farmers' health are nonetheless at the source of this vast investigation. The taboo certainly has staying power. Otherwise, the document asks a series of very detailed questions on the type of farming activity (wine growing, grains, grassland, beets, cattle raising), the "fungicides or insecticides or herbicides used in the course of your working life," the farmers' "lifestyle" and "health."

With respect to this last category, one can point to a second taboo. The question: "Has a doctor already told you that you have the following diseases?" is followed by a list of fifteen illnesses, including "hay fever, eczema, asthma, arterial hypertension, diabetes, coronary infarction, Parkinson's disease, or Alzheimer's disease," but not cancer. I suppose that participants are supposed to be able to communicate that information, apparently considered too sensitive, on line H2 of the document, which leaves a blank space for specifying one's "current state of health." But for a study aimed at evaluating cancer among farmers—hence the name AGRICAN—this "omission" is nonetheless surprising.

However, review of the questionnaires allowed for the inclusion of 180,000 individuals in the AGRICAN cohort, for which "the results are expected by 2009 for the most frequent cancers (breast, prostate), and toward 2015 for the least frequent cancers," as Baldi and Lebailly wrote in 2007.[34] Although focused on cancer, it is not impossible that the study will also provide priceless information on the link between pesticide exposure and Parkinson's disease, the object of many epidemiologic studies around the world, as we shall see.

6

The Unstoppable Rise of Pesticides and Neurodegenerative Diseases

Sooner or later the risks also catch up with those who produce or profit from them.

—Ulrich Beck, *Risk Society: Towards a New Modernity*

"Don't tell me that Parkinson's disease is a disease for old people. I've had it since I was forty-six!" Now fifty-five years old, Gilbert Vendé is a former farm worker who participated in the Ruffec Appeal in January of 2010. With considerable difficulty speaking—a characteristic of Parkinson's sufferers—he told his story, triggering an emotional response from the audience. In 1998, he was working as a crop manager on a large (2,500-acre) cultivation in the Champagne Berichonne region in France, when he fell victim to acute Gaucho poisoning.

Parkinson's Disease and Gaucho: The Exemplary Case of Gilbert Vendé

Honey aficionados have undoubtedly heard about this imidacloprid-based product, manufactured by Bayer, which created "billions of victims," to quote Fabrice Nicolino and François Veillerette—referring, of course, to indispensable pollen gatherers.[1] Launched on the French market in 1991, this so-called "systemic" insecticide is, in fact, a fearsome killer. Applied to seeds, it penetrates the plant through the sap in order to poison plant bugs that destroy beets, sunflowers, and corn, but also anything that either remotely or strongly resembles a stinging or sucking insect, including bees. It is estimated that between 1996 and 2000, some 450,000 hives quite simply disappeared in France due to the use of Gaucho and other insecticide products.[2]

It took the tenacity of beekeepers' unions, who sought the court's help, and the courageous work of two scientists—Jean-Marc Bonmatin, from the National Center for Scientific Research (Centre National de la Recherche Scientifique, CNRS), and Marc-Édouard Colin, from the National Institute of Agronomic Research (Institut national de la recherche agronomique, INRA)—to secure an opinion from the Council of State to make the French Ministry of Agriculture yield.[3] The ministry would eventually ban Gaucho in 2005, despite maneuvers from some of its senior officials to fully support the product's manufacturer. These officials included Marie Guillou, director of the very powerful Directorate-General of Nutrition (Direction générale de l'alimentation, DGAL) from 1996 to 2000 (whom we previously met in the Dominique Marchal case, when she was directing the INRA in 2005— see Chapter 4), and Catherine Geslain-Lanéelle, who succeeded her at the DGAL from 2000–2003. The latter proved her quite remarkable zeal when she refused to submit Gaucho's marketing authorization dossier to Judge Louis Ripoll while he was searching the DGAL headquarters after an investigation had begun. In July 2006, the senior officer was nominated to the head of the European Food Safety Authority (EFSA) in Parma (Italy), where I would meet her in January 2010 (see Chapter 15).[4]

This brief historical reminder is necessary in order to understand to what extent the decisions—or nondecisions—of those who govern us have direct repercussions on the lives of the citizens they supposedly serve. As it happens, the dilatory maneuvers to keep Gaucho on the market—by denying its toxicity despite overwhelming proof—helped put some ten thousand beekeepers out of a job,[5] and made a number of farmers, like Gilbert Vendé, sick.

Indeed, after having "inhaled an entire day's worth of Gaucho" in October 1998, Vendé, a farm employee, suffered horrible headaches coupled with vomiting. He consulted his physician, who confirmed the poisoning; then he went back to work soon after, "as if nothing had happened." "For years, I sprayed dozens of products," he explained at Ruffec. "Of course, I was closed up in a cabin, but I refused to wear the gas mask, because it's impossible to spend hours like that, you feel like you're suffocating." A year after his poisoning, Vendé was regularly experiencing unbearable shoulder pain: "It was so bad that I would come down off the tractor to roll on the ground," he explained. In 2002, he decided to consult a neurologist in Tours who informed him that he had Parkinson's disease. "I'll never forget that appointment," the farmer said, his voice shrouded in emotion, "because the specialist bluntly said that my disease could be due to the pesticides that I'd used."

It is a safe bet that this neurologist was familiar with the "extensive litera-ture suggesting that pesticide exposure may increase risk of Parkinson's dis-ease" as Michael Alavanja has written.[6] In his 2004 systematic review, the National Cancer Institute epidemiologist cites around thirty case-control studies that show a significant statistical link between this neurodegenerative affliction and chronic exposure to "plant products" (organochlorines, carba-mates, organophosphorus compounds), namely exposure to widely used mol-ecules such as paraquat, maneb, dieldrin, and rotenone. He came to similar conclusions two years later, when he analyzed an initial set of data from the Agricultural Health Study with his colleague Aaron Blair.

Five years after their inclusion in the mega-cohort, 68 percent of partici-pants (57,251) were interviewed. In the meantime, seventy-eight new cases of Parkinson's disease (fifty-six pesticide users and twenty-three spouses) had been diagnosed, in addition to twenty-three cases recorded during "en-rollment" (sixty users and twenty-three spouses). The results of the study show that the probability of developing Parkinson's disease increased with the frequency of use (the number of days per year) of nine specific pesti-cides, the risk potentially multiplying by a factor of 2.3. In their conclusion, the authors note that "receiving pesticide-related medical care or experienc-ing an incident involving high personal pesticide exposure was associated with increased risk."[7] Reading this, I of course thought of Gilbert Vendé, since everything indicated that his acute Gaucho poisoning was an aggravat-ing circumstance that accelerated the pathological process, initiated by chronic exposure to pesticides.

The rest of his story looks strangely like those I have already told. Faced with the refusal of the Agricultural Social Mutual Fund (Mutualité sociale agricole, MSA) to grant occupational disease status, with the justification that Parkinson's disease is not found in the famous tables of occupational diseases, the farmer turned to the Regional Committee for the Recognition of Occupational Diseases (Comité regional de reconnaissance des maladies professionnelles, CRRMP) of Orléans, which issued an unfavorable opinion. He then took the matter to the Social Security Court (Tribunal des affaires de sécurité sociale, TASS) in Bourges, which eventually ruled in his favor in May 2006. The court based its decision on the favorable opinion given by the CRRMP of Clermont-Ferrand, which clearly performed a different reading of the available scientific literature than its counterpart in Orléans.

At the time, Gilbert Vendé was the second farmer for whom Parkinson's was recognized as an occupational disease. Four years later, there were "a dozen," according to the MSA's statistics, supplied by Dr. Jean-Luc Dupupet. The Berrichon farmer then left his "home country" to live in Paris, where he now works as a volunteer at the Association France Parkinson. "Why?" he asked during the Ruffec meeting. "Simply because in our capital, I live incognito, I'm free! If I were in my countryside, they'd point at me. I wouldn't be able to live . . ."

Toxins and Toxic Products at the Root of Parkinson's Disease

This neurodegenerative disease, long considered an illness related to aging, was described for the first time in 1817 by Englishman James Parkinson (1755–1824) in his short *An Essay on the Shaking Palsy,* in which he lists its symptoms: tremors, rigid and uncontrollable gestures, difficulties in speech.[8] This exceptional doctor, a geology and paleontology enthusiast, was also a political activist who used a pen name ("Old Hubert") to write pamphlets that, in light of industrial history, today appear incredibly coherent: "Workmen might no longer be punished with imprisonment for uniting to obtain an increase of wages, whilst their masters are allowed to conspire against them with impunity," he advised in *Revolutions without Bloodshed.*[9]

In his *Essay on the Shaking Palsy,* Dr. Parkinson does not give any explanations for the disease that would bear his name, but suggests that it has occupational or environmental origins. He was right; while the majority of cases today are declared "idiopathic"—from an unknown cause—a number of occupational and environmental factors have been identified. After World War II, researchers quite fortuitously discovered that toxins could trigger Parkinsonian symptoms, as Professor Paul Blanc reports in his book *How Everyday Products Make People Sick*: the researchers measured an abnormally high rate of prevalence of the disease in the aboriginal Chamorro populations on the Mariana Islands of Guam and Rota in the West Pacific.[10] They put forward the hypothesis that this excess (the rate was one hundred times higher than in the United States) was due to the seeds of a small palm tree in the *cycas* genus, which the Chamorro would eat in the form of a flour and contains a toxin called β-methylamino-L-alanine (BMAA). Some scientists contested this explanation, arguing that the quantity of BMAA present in the flour was too low to provoke such problems. Eventually, a researcher from Hawaii ended the controversy: he observed that the aborigines were fond of bats, which frequently consume *cycas* seeds. Thus, BMAA would accumulate in the flying mammals' fat, according to the process of bioconcentration (see Chapter 3). Incidentally, the extinction of bats, much appreciated for the delicacy of their flesh, would bring about a disappearance of Parkinson's disease on the Mariana Islands.

The annals of industry confirm the role of toxic substances in the etiology of the illness. Starting in the early twentieth century, occupational physicians observed that exposure to manganese dust brought about Parkinsonian symptoms in miners or laborers working in steel mills. In 1913, nine of these cases were reported in the *Journal of the American Medical Association.* As Paul Blanc ironically emphasizes, the article started on an "optimistic note," characteristic of the then budding ideology (which still lives on today) that says progress is unavoidably accompanied by "collateral damage." "A certain indication of the humanitarian trend of modern times is the ever-increasing interest in the accidents, intoxications and diseases coincident with various trades," the authors wrote, with the arrogance typical of those who would never have to suffer from the evils they bent over backward to minimize.[11]

Over the course of the twentieth century, scientific studies started to pile up throughout the world on the psychiatric effects produced by exposure to metals (notably in welding workshops), including "manganese madness," which manifests itself by hallucinations and uncoordinated gestures, considered precursor symptoms to Parkinson's disease. In 1924, a study carried out on monkeys allowed for the understanding of manganese's effect on the central nervous system: it causes the premature death of certain neurons, and the loss in turn causes a decrease in the production of dopamine, a neurotransmitter necessary for the control of motor functions.[12]

Until the 1980s, scientific literature only covered nonorganic forms of manganese—in other words, simple oxides or metal salts used in industrial applications. But in 1988, a study published in the journal *Neurology* revealed that farm workers tasked with spraying maneb, a manganese-based fungicide, developed early signs of the Parkinson's disease.[13] These results were confirmed by another study published six years later, concerning a thirty-six-year-old man who had used maneb on his barley seeds for two years before developing the disease.[14] Similar effects were observed in those using mancozeb, a similar fungicide still used today, as is maneb.

Finally, the role of toxins in the onset of the illness was confirmed by a series of observations carried out on drug addicts in California. In 1980, doctors noted that the injection of synthetic heroin, called "MPPP," triggered the disease. MPPP contains a contaminant, MPTP, a derivative of which—cyperquat—is structurally similar to the widely used herbicides paraquat and diquat. The "MPTP model," which facilitates comprehension of biological mechanisms leading to Parkinson's disease, has been the subject of multiple studies on monkeys.[15] It has been used notably to test the effects of rotenone, a natural toxin produced by certain tropical plants and present in the composition of numerous insecticides. Researchers have observed that when injected in repeated small doses, rotenone induces Parkinsonian symptoms in rats.[16]

Again, it should be noted that, like methyl bromide, rotenone was prohibited by the European Commission in 2009, but France obtained a special dispensation to use it on apples, peaches, cherries, grapevines, and potatoes until October 2011.[17] Following Rachel Carson's example in *Silent Spring*, it is now more important than ever to find an answer to the question, "Who makes this kind of decision?" Who decides that the agronomic advantages of a poison outweigh the health considerations and risks faced by those who handle them, but also, as we will see, by consumers? All the more when we can easily imagine the number of patients and deaths that had to accumulate in experimental laboratories and morgues before the European institution finally decided to act. That France systematically requests a "grace period"—to borrow the expression used by the French newspaper *Le Syndicat agricole* (The Agriculture Union) used in 2007 in relation to the prohibition of Monsanto's Lasso—is, quite simply, scandalous.[18]

A Disease of the Industrial World

"In view of the fundamental similarities between the vertebrate and invertebrate nervous systems, insecticides designed to attack the insect nervous system (organochlorines, pyrethroids, organophosphoruses, and carbamates) are clearly capable of acute and long-term neurotoxic effects in humans," the World Health Organization (WHO) writes in its prevention manual published in 2006 (see Chapter 3). The venerable institution specifies: "Symptoms may appear immediately after exposure or be delayed. They may include limb weakness or numbness; loss of memory, vision or intellect; headaches; cognitive and behavioral problems; and sexual dysfunction."[19]

Everything the WHO describes, with the clinical coldness so characteristic of "experts," has been observed in numerous epidemiological studies, which are impossible to present in their entirety. They concern Parkinson's and Alzheimer's diseases, which affect 800,000 people in France, with 165,000 new cases every year, and amyotrophic lateral sclerosis, also called "Lou Gehrig's disease." Isabelle Baldi, an epidemiologist, demonstrated in a study published in 2001 that exposure to a number of pesticides used on grapevines brought about a reduction in cognitive function (selective attention, memory, speech, ability to process abstract information) in winemakers in the Bordelais region. The investigation, named "Phytoner," dealt with 917 farmers affiliated with the MSA: 528 had been directly exposed to pesticides for at least twenty-two years; 173 had been exposed in an indirect way through contact with leaves or grapes treated with them; and 216 had never been exposed (control group). After being submitted to mental aptitude tests, the subjects directly exposed were three times more likely to respond erroneously to the questions they were asked. Another very troubling fact: the subjects exposed to pesticides in an *indirect* way answered almost as poorly as those directly exposed.[20]

This reminds me of the fate of the students at the Bonne-Terre high school in Pézanas, destined to join the family winemaking business, where they would be in contact with a multitude of poisons. In another study published in 2003, Isabelle Baldi and Pierre Lebailly showed that exposure to pesticides, used namely in the vineyards of Gironde, raised the risk of developing Parkinson's disease by a factor of 5.6 and Alzheimer's by 2.4. These results were the product of a prospective study (named "Paquid"), where 1,507 people over the age of sixty-five were followed for ten years.[21]

"What is regrettable," explained Caroline Tanner, neurologist at the Parkinson's Institute in Sunnyvale, California, where I met her on December 11, 2009, "is that all the data we've accumulated on human populations was already obtained on lab animals decades ago."

"You mean that the results of experimental studies can be extrapolated to humans and that they should be used to take action, for example by taking suspect products off the market?" I asked.

"Exactly! The ideal would even be that the products are tested *before* they go on the market to avoid painful human tragedies," the scientist answered without hesitation, employing the straightforwardness only found on that side of the Atlantic.

The author of numerous publications on Parkinson's disease, Caroline Tanner is one of the most renowned neurologists in the United States. She works in a "privileged place," since the Parkinson's Institute is "simultaneously a care and research center." In association with the interpretation of data gathered by the Agricultural Health Study, in 2009 she published a case-control study showing that exposure to pesticides significantly raised the risk of developing Parkinsonian symptoms.[22]

"We observed that the risk could be multiplied by a factor of three after exposure to three pesticides: 2,4-D and paraquat, two herbicides, and permethrin, which is an insecticide," she commented. "Our work came at just the right time for Vietnam veterans who were exposed to Agent Orange, which includes 2,4-D in its makeup. They had requested that Parkinson's disease be added to the list of diseases giving the right to compensation and medical care by the Department of Veterans Affairs, which they eventually obtained.[23] We were surprised about paraquat, because the Parkinson's Institute has worked a lot on MPTP,[24] and they are two very similar molecules. Finally, our results are worrying for permethrin, because it is an insecticide widely used in the prevention of malaria. It is found soaked into mosquito nets, military uniforms, or even basic clothing, and a lot of people can come into contact with it through the skin."

"Is exposure time an important factor?"

"According to our study, it isn't a determining factor. Incidentally, one of the surprises was that the wives of farmers also presented a higher risk than the general population. In reality, they are also exposed to the products, because they sometimes take part in the preparation of the fungicides, but also because they wash their husband's clothes, or simply because they live in a polluted environment or consume contaminated food. I took part in a study with some colleagues in Honolulu, who compared male twins where one of them had developed Parkinson's and the other hadn't. We observed that one of the risk factors was the consumption of dairy products. The hypothesis we put forward was that persistent organic pollutants, the notorious 'POPs,' some of which have neurotoxic effects, like dioxins or PCBs [polychlorinated biphenyls], have the ability to accumulate in milk fat. It would be interesting to conduct a study specifically on the subject, even more so because a recent experiment showed that the combination of paraquat and maneb, a manganese-based herbicide, considerably raised the risk of Parkinson's disease and could induce symptoms of the disease in animals that had been exposed in utero."

"They often say that Parkinson's disease is on a clear rise in industrialized countries, is that true?"

"Actually, we don't know! For a very simple reason, which is that we don't have records old enough to be able to confirm it with any certainty. I asked that question myself, and to answer it, I went to China about twenty years ago, at a time when the process of agricultural industrialization wasn't advanced and when Parkinson's disease was very rare. I directed a number of research projects there, and I can say that today the illness has become as common there as in the United States. The only explanation is that in twenty years, the country has been greatly industrialized, and ever since then they have been using the same pesticides as in Western countries."

Pesticides Widely Miss Their Target, but Don't Spare Mankind

A few days later, on January 6, 2010, I went to La Pitié-Salpêtrière hospital in Paris to meet Dr. Alexis Elbaz, a neuroepidemiologist who works for a unit at the National Institute of Health and Medical Research (Institut National de la Santé et de la Recherche Médicale, INSERM). This young researcher is a pioneer in France, one to whom Gilbert Vendé is deeply indebted. It was while reading an article in *Le Quotidien du médecin* (Physician's Daily) in 2004 that Maître Gilbert Couderc, the Berrichon farm worker's attorney, discovered that one of Dr. Elbaz's studies, which showed a positive correlation between exposure to pesticides and Parkinson's disease, had just won the Prix Épidaure.[25] "We felt reassured," Gilbert Couderc said. He hurried to share the invaluable publication with the CRRMP.[26]

At the time of our interview, Alexis Elbaz had just published a new study in *Annals of Neurology* that he had conducted in close collaboration with the MSA—further proof, if it was needed, that the mutual fund had indeed decided to shed light on the health consequences of pesticide use.[27] In this case-control investigation, 224 farmers with Parkinsonian symptoms were compared to a group of 557 healthy farmers, all originally from the same region and affiliated with the MSA.

"The occupational medicine specialists at the MSA played an integral role," the neuroepidemiologist explained. "They went to farmers' homes and meticulously pieced together with them their exposure to pesticides over their entire professional life. They gathered a large amount of information, such as the surface area of cultivations, the type of crops and the pesticides used, the number of years and annual frequency of exposure, and the method of spreading—with a tractor or with the aid of a backpack reservoir. They carried out true detective work, taking into account all the documents the farmers supplied: farm bureau or farming co-op recommendations, which are usually strictly followed; treatment calendars; invoices; empty containers that might have been kept on the farm. All these data were then evaluated by experts, who checked their validity."

"What was the result?"

"We observed that organochlorine insecticides raised the risk of Parkinson's disease by a factor of 2.4. Among those are DDT and lindane, which were widely used in France between 1950 and 1990. One of their characteristics is that they remain in the environment several years after use."

"Do you know if pesticides used in the fields can also affect residents living close to the treated areas?"

"We don't have any data on that subject, but it's true that, beyond exposure at elevated levels in a professional context, our results raise the question of consequences of exposure at weaker doses, such as that observed in the environment—in other words, in the water, in the air, and in food. To date, only one study has been able to provide a convincing answer."

Published in April of 2009, the study to which Dr. Elbaz referred was conducted by a team of researchers from the University of California in the Central Valley of California.[28] The researchers had a precious advantage, one that France unfortunately cannot claim. Since the 1970s, the richest state in America has required that all pesticide sales, including the indication of their planned place and time of use, be recorded in a centralized computer system, called the Pesticide Use Reporting (PUR) database. This makes it possible to know, down to the day, which geographical sectors were treated and with what chemicals; this was how Sadie Costello's team was able to "reconstruct the history of agricultural pesticide exposure in the residential environment" of the entire region under study, between 1975 and 1999. To do this, the study participants—368 with Parkinsonian symptoms and 341 without (control), all living in California's Central Valley—provided their addresses so their exposure level over the course of the twenty-four-year period could be calculated.

Before finding out the very troubling results of this remarkable work, it would be useful to understand its relevance, as it concerns all of us. Indeed, as David Pimentel, an American professor at the College of Agriculture and Life Sciences at Cornell, explained in 1995, "Less than 0.1 percent of pesticides applied for pest control reach their target pests. Thus, more than 99.9 percent of pesticides used move into the environment where they adversely affect public health and beneficial biota, and contaminate soil, water, and the atmosphere of the ecosystem."[29] Some observers are slightly less pessimistic, like Hayo van der Werf, an agronomist at the INRA: "Each year an estimated 2.5 million tons of pesticides are applied to agricultural crops worldwide," he wrote in 1996. "The amount of pesticide coming in direct contact with or consumed by target pests is an extremely small percentage of the amount applied. In most studies the proportion of pesticides applied reaching the target pest has been found to be less than 0.3%, so 99.7% went 'some-where else' in the environment."[30] And, he adds, "Since the use of pesticides in agriculture inevitably leads to exposure of non-target organisms (including humans), undesirable side-effects may occur on some species, communities, or on ecosystems as a whole."

As we will see, chemical agriculture is anything but an exact science, to the point that we end up wondering how and in the name of what we could have allowed the establishment of such a system of generalized poisoning on our land: "The pesticides which reach the soil or plant material in the target area begin to disappear by degradation or dispersion," van der Werf continues. "Pesticides may volatilize into the air runoff or leach into surface water and groundwater, be taken up by plants or soil organisms or stay in the soil. The total seasonal losses in runoff for soil-surface applied pesticides average about 2% of the application and rarely exceed 5–10% of the total applied; the fraction removed by leaching is generally less. In contrast, volatilization losses of 80–90% have sometimes been measured within a few days after application. [. . .] Worries about the movement of pesticides in the atmosphere have arisen during the 1970s and 1980s. Transport and redeposition of pesticides may occur over very long distances, as evidenced by the presence of pesticides in ocean fog and arctic snow."[31]

After reading about such a catastrophic scenario, it's hard not to wonder: Does this at least do something? Have the "pests" all been exterminated? No! That's what Professor David Pimentel explained as early as 1995: "World-wide, an estimated 67,000 different pest species attack agricultural crops. Included are approximately 9,000 species of insects and mites, 50,000 species of plant pathogens, and 8,000 weeds. In general, less than 5% are considered serious pests. [. . .] Despite the yearly application of an estimated 2.5 million tons of pesticides worldwide, plus the use of biological controls and other non-chemical controls, about 35% of all agricultural crop production is lost to pests. Insect pests cause an estimated 13% crop loss, plant pathogens 12%, and weeds 10%."[32]

To sum up: the poisons poured onto fields generally miss their targets—either because pests resist or escape them, or because they "go somewhere else," to use Hayo van der Werf's expression—and contaminate the environment. Hence the extremely relevant question posed by Sadie Costello's team: Can pesticides induce Parkinson's disease in people living in proximity to treated crops? The answer is clearly affirmative. Pesticide use records have indicated that maneb, the manganese-based fungicide I have already mentioned, and the inescapable paraquat are both included among the most widely used products in California's Central Valley. The study's results showed that living less than five hundred yards from a treated area increased the risk of developing the disease by 75 percent. What's more, the probability of onset of the illness before the age of sixty was multiplied by two if there was exposure to one of the two pesticides (OR: 2.27) and by more than four (OR: 4.17) if there was combined exposure, especially if the exposure took place between 1974 and 1989, that is to say when the people in question were children or teenagers.

Beate Ritz, professor of epidemiology at the UCLA School of Public Health, who supervised the University of California team's work, explained that "the new study confirms previous observations in animal studies." First, "exposure to multiple chemicals may increase the effects of each chemical," which is important, because humans are generally exposed to more than one pesticide in the environment. Secondly, "the timing of the exposure is an important risk factor."[33]

Pesticides and Immunotoxicity: Affecting Whales, Dolphins, and Seals

In a 1996 report entitled *Pesticides and the Immune System: The Public Health Risks*, which was commissioned by the prestigious World Resources Institute (WRI) in Washington, DC, Robert Repetto and Sanjay Baliga write: "The scientific evidence suggesting that many pesticides damage the immune system is impressive. Animal studies have found that pesticides alter the immune system's normal structure, disturb immune responses, and reduce animals' resistance to antigens and infectious agents. There is convincing direct and indirect evidence that these findings carry over to human populations exposed to pesticides."[34]

"That document sparked the chemical industry's wrath," explained Robert Repetto, an economist who specializes in sustainable development and who was vice president of the WRI when the report was written. "It was the first time a study had gathered all the available data on the effects of pesticides on the immune system, a subject that was completely underestimated at the time and, in my opinion, continues to be now, even though it is crucial to understanding the epidemic of cancer and autoimmune diseases that are observed, notably in industrialized countries."[35]

Indeed it is—and we will revisit this—as cancer is rarely caused by one factor alone; more often it is the result of a complex and multifactorial process, generally initiated by the action of pathogens (or of antigens), such as rays, viruses, bacteria, toxins, or chemical pollutants, and possibly favored by genetic predispositions, lifestyle, or diet. In good health, the body can defend itself against the aggression of pathogens by mobilizing its immune system, whose function is precisely to track and eliminate intruders using the action of three distinct, but complementary, mechanisms.

The first, which biologists call "nonspecific immunity," involves macrophages and neutrophils that consume invaders (the process is called "phagocytosis"), and natural killer (NK) lymphocytes, whose mission is to exterminate them. The second, named "humoral immunity," activates B lymphocytes, producing antibodies. Finally, the third, "cellular immunity," sets T lymphocytes (T4 or T8) in motion, which poison the intruders that were phagocytized by the macrophages thanks to the secretion of lymphotoxins.

In their lengthy report, Robert Repetto and Sanjay Baliga devote fifteen pages or so to the numerous in vivo (that is, directly on animals) or in vitro (on cell cultures) studies that have shown that pesticides can disturb one or more of the mechanisms that make up the immune system.[36] From this long list, of which organochlorines (DDT, lindane, endosulfan, dieldrin, and chlordecone) make up the lion's share, I chose the example of atrazine, an herbicide that was banned in Europe in 2004 but continues to be used in massive quantities, notably in the United States (see Chapter 19). When administered orally to mice, atrazine disturbs the action of T lymphocytes as well as the process of phagocytosis by macrophages.[37] In another study published in 1983, researchers demonstrated an effect on the weight of the thymus in exposed rats. (The thymus is an essential organ in the immune system, as it is where T lymphocyte maturation takes place, and it also plays a role in the protection against autoimmunity[38]—that is, the fabrication of antibodies, which, instead of attacking intruders, target immune system cells. Finally, another experiment in 1975 revealed that salmon exposed to atrazine through oral or cutaneous methods showed a lower weight of the spleen, an organ involved in controlling bacterial infections, such as pneumococcal or meningococcal ones.[39]

However, as Repetto and Baliga point out, the immune system anomalies observed in lab animals following exposure to pesticides have also been observed in wild fauna. For example, in Canada, autopsies of whales found dead on the shores of the St. Lawrence Estuary showed an elevated concentration of organochlorine pesticides and PCBs, as well as an abnormal rate of bacterial infections and cancer. Sylvain de Guise, who conducted a study on the abnormally high death rate of the cetaceans, explains that only "two factors could have contributed to such a high prevalence of neoplasms in that single population: exposure to carcinogenic compounds and decreased resistance to the development of tumors."[40]

Similarly, in the early 1990s, a strange epidemic decimated the dolphins of the Mediterranean; dozens of their corpses turned up on the coast of Valencia, in Spain. Autopsies revealed that the marine mammals had succumbed to an infection brought on by viruses they could normally overcome (such as *morbillivirus*). "We have gone back over the literature for more than a hundred years and we have found nothing like it, no other cluster of virulent epidemics like we have now," a British researcher commented.[41] In the end, studies concluded that the mass deaths had to be due to lowered immune defenses in the dolphins, whose bodies had accumulated organochlorine pesticides, PCBs, and various chemical pollutants in their bodies.[42]

The studies conducted on fauna showing the immunosuppressive effects of pesticides are numerous, but one of them is particularly impressive. It all started during the 1980s, when zoologists noticed that seals living close to ports in the Baltic and North Seas were succumbing in huge numbers to *morbillivirus* infections. Dutch researchers decided to conduct a prospective experiment. They captured baby seals off the northwest coast of Scotland, considered relatively unpolluted. The friendly mammals were divided into two groups: the first was fed with herring from the Baltic Sea, where the pollution rate is significant; the second was fed with herring caught in Iceland, where contamination is very low. It is worth noting that the herring for both groups was bought at "normal" markets—that is, destined for human consumption. After two years, the fat of the seals in the first group showed a concentration rate of organochlorine pesticides ten times higher than that of the control group. The researchers also observed that the seals fed with contaminated herring had immune defenses three times weaker than those of the control group, notably with a very clear reduction in NK cells and T lymphocytes, and lower neutrophil levels and antibody responses.

At a conference held in February 1995 in Racine, Wisconsin, where he presented his team's work, Dutch virologist Albert Osterhaus noted that this was "the first demonstration of immunosuppression in mammals as a result of exposure to environmental contaminants at ambient levels found in the environment."[43] Incidentally, it's worth noting the title of the conference: "Chemically-induced Alterations in the Developing Immune System: The Wildlife/Human Connection."

Allergies and Autoimmune Diseases: Effects on Humans

As Robert Repetto and Sanjay Baliga point out, "the immune systems of all mammals (but also of birds and fish) have similar structures," and what happens to whales, dolphins, or seals concerns us directly. They cite as evidence studies carried out on cyclosporine, an immunosuppressant medication prescribed to organ transplant recipients to stop the body from rejecting the transplant. Researchers observed that the drug "has been found to have similar toxicological and immunosuppressive properties in a wide variety of mammalian species, including rats, mice, monkeys and humans," which, in the long run, lay the grounds for cancer. Indeed, as shown by Arthur Holleb, an oncologist and former chief medical officer of the American Cancer Society, patients treated with cyclosporine are one hundred times more likely to develop lymphatic cancer, in particular leukemia and lymphoma.[44] Need we recall that these are precisely the malignant tumors for which farmers show a heightened risk?

In their report, Repetto and Baliga present several studies carried out by Soviet scientists, who scrupulously took a census of the effects of pesticides on the immune system. "It was very valuable, because at the time, Western studies were only interested in cancer and neurodegenerative diseases," Repetto explained during our phone interview. "Also, the communist bureaucracy was an advantage: as there was no profit mentality—which is different from capitalist countries, where manufacturers are interested in hiding the toxicity of their products, out of fear of seeing their sales drop—the Soviet researchers carried out what was essentially true health monitoring, by conscientiously recording all the effects observed in farming populations, with the goal of lowering the health care costs they might generate."

At the risk of seeming like an inveterate crypto-communist, I must admit that listening to these words, I thought that there was some merit to the "bureaucratic" scientific research—meaning independent from private interests— and that this outdated model should inspire regulatory agencies that generally forget to include potential medium- or long-term health care costs in their evaluation of chemical products. People will retort that the studies by "bureaucratic" researchers have not prevented catastrophic pollution of vast areas of the former Soviet Union (such as the Aral Sea), which is true. Nevertheless, as we will see later on, the explosion of chronic illnesses is tugging strongly at the purse strings of social security systems, which fall prey to a regulatory system where agro-economic considerations (the famous "benefits" pesticides supposedly offer) take precedence over health considerations (the "risks" associated with said "benefits").

In the meantime, the "bureaucratic" scientific literature has nonetheless shown that exposure to pesticides causes autoimmune reactions; it also leads to a disturbance in neutrophil and T lymphocyte activity, which contributes to the development of pulmonary and respiratory infections. Several studies conducted between 1984 and 1995 in the cotton-producing regions of Uzbekistan, where large quantities of organochlorine and organophosphorus insecticides were sprayed, showed extremely elevated rates of respiratory, gastrointestinal, and kidney infections not just in farm laborers, but also in the populations living close to the treated zones. At the same time, in the West, researchers were showing that exposure to pesticides such as atrazine, parathion, maneb, and dichlorvos triggered allergies, leading to what Dr. Jean-Luc Dupupet calls "cutaneous manifestations" (see Chapter 3), or in other words, types of dermatitis that are the expression of an immune system reaction to a chemical aggression.[45]

In its manual for pesticide poisoning prevention published in 2006, the WHO devotes a significant portion to autoimmune diseases and allergies, the prevalence of which keeps rising, especially in children.[46] The manual notes: "Allergies can have many manifestations, including hay fever, asthma, rheumatoid arthritis and contact dermatitis. The cause of allergies is a hypersensitivity response which occurs after exposure to some occupational and environmental agents. Antigens that cause allergic responses are called 'allergens.' [. . .] When the immune system loses the ability to distinguish between the body's own cells and foreign cells, it attacks and kills host cells, resulting in serious tissue damage. This condition is called 'autoimmunity'. Although it is not as common as immunosuppression or allergy, occupational exposure to certain chemicals has been associated with autoimmune responses."[47]

During our phone interview, Robert Repetto said that the report he wrote for the WRI triggered a lively (allergic!) reaction from manufacturers, whose scientists, just this once, decided to collectively author a "critique" in the journal *Environmental Health Perspectives*.[48] The selection's signatories included licensed epidemiologists from Dow Chemical (Carol Burns and Michael Holsapple), Zeneca (Ian Kimber), DuPont de Nemours (Gregory Ladics and Scott Loveless), BASF (Abraham Tobia) and, of course, Monsanto, meaning Dennis Flaherty and John Acquavella, the author of the controversial meta-analysis I discussed in Chapter 5. After offering a firm criticism of the report, notably of the Soviet studies that they deemed "difficult to evaluate," the authors end their article with remarks that are contradictory, to say the least. It is unclear whether they express embarrassment or a well-calculated conciliatory strategy. They write that they "do not find consistent, credible evidence" that there is a widespread phenomenon of immunosuppression linked to pesticide exposure. Nonetheless, the WRI report is an important document because it focuses attention on a potentially important issue for future research and brings a substantial literature of foreign language studies to the attention of Western scientists.

Here we have a perfect example of "the art of blowing hot and cold." But, as we will see, when it comes to neutralizing the impact of studies not in their favor, manufacturers' attitudes can be much more drastic, even perverse. But before examining how the regulation of chemical products that come into contact with the food chain functions, it is important to go back to the industrial history of the twentieth century, in order to understand how extremely toxic compounds managed to poison the environment and human populations, not just in the short term, but for many years to come.

PART II

Science and Industry: Manufacturing Doubt

7

The Sinister Side of Progress

Science Finds, Industry Applies, Man Conforms.
                    —Motto of the 1933 World's Fair in Chicago

"When I think of all the deaths we could have avoided in the factories if we'd taken measures as soon as we found out about the toxicity of a number of chemical products, I'm truly revolted . . ." I met with Peter Infante one day in October 2009 in his home in the Washington, DC, suburbs. He is an American epidemiologist who fought his entire career to defend a cause that was "mistreated by the ideology of progress": public health and occupational safety. "Blue collars, that is to say workers, have paid a heavy price to manufacture all the magnificent objects that consumer society provides us with every day," he explained, his voice heavy with emotion. "At the very least, public authorities should do everything they can to limit workers' exposure to dangerous chemical substances as much as possible, while guaranteeing them compensation when they become ill. Unfortunately, industry has systematically crushed all efforts to go in that direction."[1]

Peter Infante and David Michaels Versus the Chemical Industry Lobbies

At sixty-nine, Peter Infante knows what he's talking about. For twenty-four years, he worked at the Occupational Safety and Health Administration (OSHA),[2] the agency in charge of health and safety in the workplace, which was created at the same time as the Environmental Protection Agency (EPA) in 1970. It was an era when, mindful of the concerns provoked by Rachel Carson's *Silent Spring*, America was paving the way. "I came to OSHA in 1978, at a time when the agency was doing its job well," he explained. "Under

107

the direction of Eula Bingham, a toxicologist who had been nominated by President Jimmy Carter, we had succeeded in considerably reducing the Occupational Exposure Limits for lead, benzene, and cotton dust. Then Ronald Reagan, who swore by deregulation, was elected to the White House. Manufacturers had taken control of OSHA, so to speak, and I nearly lost my job."

The epidemiologist showed me a letter sent by Al Gore,[3] then chairman of the Subcommittee on Investigations and Oversight in the Congressional Committee on Science and Technology, to "the Honorable Thorne Auchter," assistant secretary for Occupational Safety and Health, Department of Labor. Written on July 1, 1981, it contested a dismissal notice for Peter Infante, whom his management reproached for having informed the International Agency for Research on Cancer (IARC) of the latest scientific work targeting formaldehyde—the IARC, which is dependent on the World Health Organization, has the mission of classifying chemical products according to their degree of carcinogenicity (see Chapter 10). Also known as methanal, formaldehyde was on a list of priority substances the IARC had announced it was evaluating. This very volatile organic compound is found (in solution) in a number of commonly used products, such as glue for plywood furniture, detergents, disinfectants, and cosmetics (nail polish, for example). As such, it is involved in a number of industrial and artisanal manufacturing processes. In November 1980, a group of scientists called upon by the National Toxicology Program concluded that it was "prudent to regard formaldehyde as posing a carcinogenic risk to humans."[4] Peter Infante decided to inform John Higginson, director of the IARC, which triggered the wrath of OSHA's management.

In his letter, Al Gore did not mince his words: "I believe that a strong case can be made that your agency's action is politically motivated. In your own statement of charges, you attach letters from the Formaldehyde Institute critical of Dr. Infante. I am highly suspicious of any personnel action that would have as its base a letter from an industrial group that obviously has a stake in finding that formaldehyde is not a carcinogen. [. . .] If OSHA succeeds in firing Dr. Infante, it will be a clear message to all civil servants who are charged with protecting the public health that those who do their jobs will lose their job."

"In the end, you weren't dismissed?" I asked, after reading the surprising letter.

"No! And the IARC classified formaldehyde as 'carcinogenic for humans' in 2006," Infante replied. "But at OSHA, our dark period was just beginning. Under the Republican administrations, first Reagan, then Bush, Sr. and Jr., we were paralyzed. The number of products we regulated is ridiculous, barely two over the last fifteen years! In 2002, I left the agency to start working as an independent consultant."

If the second part of this book starts with Peter Infante's story—which we will come back to later—it is because it is indicative of the many maneuvers the chemical industry launched over the course of the twentieth century to keep highly toxic products on the market, at the risk of poisoning those that make or consume them. The American epidemiologist David Michaels brilliantly demonstrated this in his previously cited 2008 book *Doubt Is Their Product: How Industry's Assault on Science Threatens Your Health*, which Peter Infante highly recommended to me. And for good reason: not long before I interviewed Infante in October 2009, President Barack Obama nominated David Michaels to be the head of OSHA. I would very much like to have met the renowned epidemiologist, a professor of environmental and occupational health at George Washington University, but it was not possible.

When I tried to meet with him, he was very busy with his nomination, which triggered virulent opposition from industrial lobbies prepared to do anything necessary to block the Senate's indispensable green light. To no avail. On December 3, 2009, David Michaels was confirmed for the position, which was without a doubt good news for the United States. Because if there is one thing that cannot be reproached of the new OSHA head (and the assistant secretary of labor), it is supporting—either closely or at a distance—poison manufacturers. In his book (to which I will return), he shows how those manufacturers, backed up by lies, manipulations, as well as a disregard for human life, are at the origins of an unprecedented "assault" on our health, favoring the establishment of what Geneviève Barbier and Armand Farrachi call a "carcinogenic society."[5]

Cancer, a Disease of "Civilization"

Before diving into the chemical industry's nauseating and (it must be said) criminal history, which comprises one of the key elements of my investigation, I would like to briefly review the history of medicine, as it pertains to this issue. I spent a lot of time in the libraries of Paris consulting books and doctoral theses, trying to answer this fundamental question: Is cancer, as some claim, a "disease of civilization?" And more precisely, is its development linked to that of industrial activity? From my numerous readings, I concluded that cancer is, of course, a very old disease, but that it was extremely rare until the end of the nineteenth century.

As the authors of *La Société cancérigène* (Carcinogenic Society) explain, "no discovery has ever established that a man died of cancer before the appearance of agriculture. Infectious lesions, rickets, traumas have been detected, but no cancer."[6] For his part, Jean Guilaine, a specialist in prehistory and Neolithic civilizations, notes that the chapter on "neoplasia is reduced to nothing, as no case of authentic malignant neoplasia has been found."[7] Of course, he adds that "the absence of skeletal localizations proves nothing in terms of the possible existence of malignant tumors in soft tissue" and that it remains to be seen "whether prehistoric populations paid the same cancerous toll as today's."[8]

The consensus is that the "oldest description of cancer dates back to about 1600 BC," as stated on the American Cancer Society website. It was found on Egyptian papyrus, discovered by the British surgeon Edwin Smith in 1862, and described eight cases of breast tumors, for which it was specified that there was "no treatment." According to British toxicologists John Newby and Vyvyan Howard, who have consulted a large portion of the available literature, "evidence of malignant melanoma" (skin cancer) has been found in a 2,500-year-old Incan mummy in Peru, while the discovery of traces of lymphoma in Homo erectus remains has been attributed to the Kenyan paleontologist Louis Leakey.[9]

Proving that the disease was duly identified during antiquity, the word "cancer" was invented by Hippocrates (460–370 BC), who, in observing the characteristic branching of tumors, associated their form with that of a crab (*carcinos* in Greek). In his treatises, the man nicknamed the "father of medicine" describes several types of cancer that he associates with an excess of "black bile."[10] The word *carcinos* was then translated into Latin by the Roman physician, Celsius, in the beginning of our era.

But while the disease was well known by the Ancients, it was nevertheless "remarkably rare or absent"[11] in peoples isolated from industrial development, as clearly shown in the book *Cancer: Disease of Civilization* by Vilhjalmur Stefansson (1879–1962), an Icelandic ethnologist and Arctic explorer who was a noted authority in the field.[12] In the preface of his work, René Dubos, a professor of molecular biology at the Rockefeller Institute, notes that cancer is unknown in "certain primitive people . . . as long as nothing is changed in the ancestral ways of life." This statement is confirmed by the numerous accounts of traveling physicians cited by Vilhjalmur Stefansson, such as that of Dr. John Lyman Bulkley, who reported in the journal, *Cancer*, in 1927 that "during a sojourn of about twelve years among several of the different tribes of Alaskan natives . . . he never discovered among them a single true case of carcinosis."[13] Similarly, Joseph Herman Romig, who was then "Alaska's most famous doctor,"[14] wrote in 1939 that "in his thirty-six years of contact with these people he had never seen a case of malignant disease among the truly primitive Eskimos and Indians, although it frequently occurs when they become modernized."[15] Stefansson also cites the accounts of Dr. Eugene Payne, who "examined approximately 60,000 individuals during a quarter of a century in certain parts of Brazil and Ecuador, [and] found no evidence of cancer."[16] He also cites those of Dr. Frederick Hoffman who, at the 1923 cancer congress in Brussels, said in reference to Bolivian women, "I was unable to trace a single authentic case of malignant disease. All of the physicians whom I interviewed on the subject were emphatically of the opinion that cancer of the breast among Indian women was never met with."[17]

Observations made by Anglophone scientists were corroborated by their Francophone counterparts, such as Albert Schweitzer, who in his book *On the Edge of the Primeval Forest* comments on his experience "with the indigenous people of Equatorial Africa" in 1914: "In the first nine months of my work here I have had close on two thousand patients to examine, and I can affirm that most European diseases are represented here. [. . .] Cancer, however, and appendicitis I have never seen."[18] The authors of *La Société cancérigène* also cite the account of Professor de Bovis, "one of the first doctors to take an interest in the generalization of malignant tumors," who wrote that at the beginning of the twentieth century, "primitive races were once unscathed, or nearly, by cancer. Since our civilization has penetrated into theirs, they have started to develop cancer. The word 'cancerization' has even been used in this context, in reference to primitive races."[19]

To those who would object that "it is impossible to obtain convincing statistical data on the frequency of cancer among uncivilized races such as those in Africa and the Indians in the North and South of America," Dr. Guiseppe Tallarico would rightly retort that "all of the doctors who have long practiced among these primitive races are unanimous in rarely or never witnessing cases of cancer."[20] And it isn't for lack of searching! As the French historian Pierre Darmon reports, traveling physicians identified a certain number of "exotic cancers," such as the so-called Kangri cancer, which affects the "epithelium of the anterior abdominal wall." It is "very common in Kashmir, where inhabitants shield themselves from the cold by wearing a kangri—a sort of terracotta vase containing a wood coal flame, which causes burns and chronic irritation—under their tunics." Similarly, "lip, tongue, and mouth cancers are relatively frequent in India, where women and men chew betel, a kind of mixture composed of betel leaves, tobacco, and lime."[21] Cancers linked to the chewing of betel are still common in the Indian state of Orissa, which I visited at the end of 2009, whereas all other cancers are nearly nonexistent there, though perhaps not for very much longer . . .

In reading these travel accounts written by men of science in the early twentieth century, I came to understand how they became evidence that some would work very hard to deny, even ridicule, mocking the "myth of the good savage": the incidentally observed absence of cancer in "primitive peoples" stood in stark contrast to the situation that was then prevailing in "civilized" countries, where, in the wake of the industrial revolution, cancers were increasing at an astounding rate.

An Eighteenth-Century Precursor: Bernardino Ramazzini and Occupational Diseases

"The historical period of the fight against cancer starts in 1890, the year that a collective awareness of the scourge in all its scope set in," writes the French historian Pierre Darmon, who points out the "statistical spike: 1880–1890."[22] Echoing the concerns of an era still characterized by the predominance of infectious disease,[23] the historian notes that "the take-away from early investigations is overwhelming. Year after year, cancer was gaining ground. It's clear that the raw data is incontrovertible. Between 1880 and 1900, the mortality rate of cancer per 100,000 inhabitants seems to have doubled in most countries," such as the United Kingdom, Austria, Italy, Norway, and Prussia. In England, considered the cradle of the industrial revolution, the number of deaths attributed to cancer went from 2,786 in 1840 (or 177 deaths per million inhabitants) to 21,722 in 1884 (713 deaths per million inhabitants), according to a report published in 1896 in the *British Medical Journal*.[24] "In the space of forty years, the virulence of disease thus quadrupled," Darmon concludes. He also gives the example of the "little Swedish town of Follingsbro, where cancer deaths have been documented since the beginning of the 19th century—their number went from 2.1 to 108 per 100,000 inhabitants."[25] According to the (numerous) studies published at that time, cancer affected not only the industrialized countries in Europe, but also those of the New World. "If for the next ten years the relative death-rates are maintained, we shall find that ten years from now . . . there will be more deaths in New York State from cancer than from consumption, smallpox, and typhoid fever combined," Professor Roswell Park wrote in the *Medical News* in 1899.[26]

THE SINISTER SIDE OF PROGRESS

It is interesting to note that in order to explain this troubling development, some commentators were already adopting arguments harshly criticized today by those who would like to deny the environmental origins of cancer, the prevalence of which has nonetheless increased unabated for a century. I will come back to this in more detail (see Chapter 10), but for now let's turn to Pierre Darmon's observations about the dramatic upsurge in malignant tumors at the dawn of the twentieth century: "Many authors blame longer life expectancies, flaws in old statistics and improvements in clinical medicine, which allowed an increasing number of cancers to be highlighted." This is exactly what would be written a century later at the hand of preeminent oncologists— such as Professor Maurice Tubiana in France—who continually minimize the environmental factors in the etiology of cancers. Granted, it's easy enough to blame "increased life expectancy"—it went from an average of forty-five in 1900 to nearly eighty in 2007. But as we will see, the only relevant data in measuring the unstoppable rise of cancers is the increase in prevalence rates among the general population, especially by age groups—a detail certain leading experts from the French Academy of Sciences seem to want to ignore.

These pseudo-arguments, as Pierre Darmon points out, "are often lost behind what a number of scholars consider the carcinogenic factor par excellence—the progress of civilization."[27] In fact, doctors started to establish a link between disease and certain professional activities as early as the mid-sixteenth century. For example, in 1556 the German doctor and geologist Georg Bauer (also called Georgius Agricola) published *De re metallica*, a monumental work in which he describes not only mining and metallurgic techniques, but also the many tumors and pulmonary ailments he observed in miners.[28]

However, we owe the first systematic study on the relationship between cancer and exposure to pollution or toxic substances to the Italian doctor Bernardino Ramazzini (1633–1714). In 1700, the University of Padua professor of medicine, considered the father of occupational medicine, published *De morbis artificum diatriba* (Diseases of Workers), a work in which he presents thirty or so guilds vulnerable to the development of occupational diseases, notably lung tumors. They included craftsmen working closely with coal, lead, arsenic, or metals—glassblowers, painters, goldsmiths, mirror dealers, potters, carpenters, tanners, weavers, blacksmiths, apothecaries, chemists, starch workers, fullers, bricklayers, printers, launderers, those exposed to sulfur vapors, and "those who anoint with mercurial ointment," as well as those preparing and selling tobacco. In his seminal work, which would serve as a reference for over two centuries, Bernardino Ramazzini notes that nuns have a much lower incidence of uterine cancer than other women of the era, unknowingly emphasizing the role of certain sexually transmitted viruses in the malignant disease. He states that, in contrast, single women have breast cancer more often than married women, an observation that would be confirmed four centuries later by the discovery of the protective role played by breastfeeding against the hormone-dependent disease.

Ramazzini was a curious and precise man who, simultaneously playing the sociologist, journalist, and physician, did not hesitate to visit the factory floors. He was also a humanist capable of a rare compassion for those he called "patients of the working class." In the preface to *De morbis artificum diatriba*, he cautions the physician that upon arriving "to attend some patient of the working class, he ought not to feel his pulse the moment he enters, as is nearly always done without regard to the circumstances of the man who lies sick; he should not remain standing while he considers what he ought to do, as though the fate of a human being were a mere trifle, rather let him condescend to sit down for a while with the air of a judge, if not on a gilded chair as one would in a rich man's house, let him sit, be it on a three-legged stool or a side-table. He should look cheerful, question the patient carefully, and find out what the matter is. . . . There are many things that a doctor, on his first visit to a patient, ought to find out either from the patient or from those present. For so runs the oracle of our inspired teacher: 'When you come to a patient's house, you should ask him what sort of pains he has, what caused them, how many days he has been ill, whether his bowels are working and what sort of food he eats.' So says Hippocrates in his work *Affections*. I may venture to add one more question: *What occupation does he follow?*"[29]

Ramazzini's originality lies in demonstrating that a number of serious illnesses are caused by human activity, especially activity linked to burgeoning industry. Karl Marx recognized the import of the Italian doctor's revolutionary work, and cited it in *Das Kapital*. According to Paul Blanc, Marx foresaw that "the production of illness could represent a hidden cost of industrial manufacture."[30] "Some crippling of body and mind is inseparable even from division of labor in society as a whole," the theoretician of communist thought states in the first volume of *The Process of Production of Capital*. "Since, however, manufacture carries this social separation of branches of labor much further, and also, by its peculiar division, attacks the individual at the very roots of his life, it is the first to afford the materials for, and to give a start to, industrial pathology."[31] A note follows, referencing *De morbis artificum diatriba*.

The Industrial Revolution: Source of an Epidemic of Unknown Illnesses

Strangely, as Paul Blanc remarks, the concern with illnesses developed by laborers working in factories that flourished nearly everywhere in Europe and America in the nineteenth century was not shared by those considered "progressives" at the time or, to use a more Anglophone term, "liberals." On the contrary, everything indicates that the progressive ideology that developed alongside the industrial revolution, meant to ultimately bring about universal well-being, relegated the health or environmental harm of factory activities to the background. Blanc cites the example of Harriet Martineau (1802–1876), a British militant feminist and abolitionist, journalist, and sociologist who, interestingly enough, translated the works of positivist August Comte. According to Martineau, the regulation of work safety was superfluous, as she believed it came under the sole responsibility of manufacturers, in the name of the liberal doctrine of "laissez-faire." Often compared to Alexis de Tocqueville for a study she carried out in the United States, she became famous through her heated exchanges with Charles Dickens who, in contrast, advocated for state intervention to strengthen work safety.

The *David Copperfield* author, a committed writer, inveterate adversary of poverty and industrial exploitation, maintained close relationships with physicians, whose observations on the diseases commonly found among workers of Victorian and industrial England nourished his novels. In an article published in 2006 in the *Journal of Clinical Neuroscience*,[32] Kerrie Schoffer, an Australian neurologist, demonstrates how precisely Dickens described the Parkinsonian syndromes of one of his characters, who were overcome by uncontrollable limb movements, at a time when "there was no name for that and no understanding of the biological basis of it."[33]

But while the political classes remained generally impermeable to the health consequences of the industrial revolution, doctors did not stop trying to decode the new illnesses affecting the working class. They drew their inspiration from the pioneering work carried out by the English surgeon Percivall Pott (1714–1788), who in 1775 published a study on a then little-known disease—scrotal cancer. After examining a number of chimney sweeps in a London hospital, Pott observed that they frequently developed tumors of the scrotum, due to the soot deposited in this quite delicate part of the anatomy. Pott noted that German and Swedish chimney sweeps, who had the good idea to wear leather trousers, were less affected than their British colleagues.[34] A century later, in 1892, Dr. Henry Butlin caused a sensation at a conference at the Royal College of Surgeons when he revealed that "chimney sweep cancer" also affected workers in naval shipyards who coated the hulls of ships with coal tar.[35]

But the long litany of harmful effects of coal by-products was only just beginning. Soon, various clinical reports and studies would show that laborers working in charcoal briquette factories (such as in Wales), or in workshops using creosote[36] to treat wood, were also developing skin cancer, a disease so rare at the time that it prompted the powerful dockers' union to request an official investigation. Published in 1912, this "sound epidemiological investigation," the first of its kind, confirmed the excess of melanoma among naval shipyard workers;[37] what's more, its findings "matched with an elegant set of animal experiments duplicating the same cancer link, some of the earliest laboratory work ever done in the field of chemical carcinogenesis," to quote Paul Blanc.[38]

In truth, reading the medical literature from the early twentieth century is quite chilling. For example, one finds accounts of afflictions among men and women working in matchmaking factories in Germany, Austria, or the United States, where the phosphorus industry was flourishing. In 1830, ten years after the launch of this rather profitable activity, the first medical reports pointed out the appearance of a disease as terrible as it was new: osteonecrosis of the jaw, brought on by yellow phosphorus vapors, which manifests as extremely serious lesions of the mouth's mucous membrane, erosion of the mandible bone and the progressive disappearance of teeth. As Paul Blanc emphasizes, the history of "phosphorus necrosis" perfectly illustrates the harmful effects of the laissez-faire attitude in the realm of occupational safety, as it would take until 1913 for yellow phosphorus to be banned in the production of matches, after which the industry developed less dangerous alternatives (such as red phosphorus-based solutions).

Driven Crazy by Poison

At the same time, neurological diseases were also receiving a lot of attention. Within this category—and this is but one case among many—the history of carbon disulfide is particularly terrifying. Paul Blanc devotes an entire chapter of his book to it, entitled "Going Crazy at Work,"[39] in which he speaks at length about the cynical and criminal obliviousness underpinning the industrialization of so-called civilized countries. Used in chemistry to dissolve a number of organic compounds, carbon disulfide is a highly toxic solvent that acts as an intermediate for synthesis in the manufacture of vulcanized rubber products and of medications and pesticides (in the nineteenth century, it was used to combat grape phylloxera).[40]

In 1856, Auguste Delpech, a young Parisian doctor, gave a brief statement to the Imperial Academy of Medicine (Académie impériale de médecine), in which he presented a new disease he attributed to work in rubber factories. In it, he described the case of Victor Delacroix, a twenty-seven-year-old worker whose symptoms were very similar, he said, to lead poisoning: headaches, muscle stiffness and weakness, insomnia, memory problems, mental confusion, and sexual impotence.[41] At the same time that Claude Bernard was preparing his lectures on the effects of toxic and medicinal substances, Delpech was testing the toxicity of carbon disulfide on two pigeons that died immediately, and a rabbit, which ended up paralyzed.[42] As Paul Blanc underlines, "Delpech's studies of carbon disulfide poisoning, matching narrative descriptions of human illness with an experimental model of the disease reproduced in the laboratory, fit particularly well with the scientific concerns and worldview of his medical contemporaries."

This is true, but with the exception of a few "luminaries" (like the American scientists Alice Hamilton and Wilhelm Hueper) it was rare to find doctors willing to leave the confines of their scientific milieu to appear in the public arena and denounce the occupational diseases they were diagnosing in their practices or laboratories. On the contrary, everything seems to indicate that the horrors observed were generally accepted as inevitable collateral damage of a necessary process of industrialization—an opinion shared by the majority of contemporary journals. So, in 1863, August Delpech published a lengthy article in which he detailed twenty-four cases of carbon disulfide poisoning among workers who manufactured inflatable balloons and condoms in a blown rubber factory. The article revealed that most of them suffered from hysterical fits and periods of sexual excitation followed by impotence, and that one female worker ended up killing herself by inhaling the poison's vapors.[43] The London Times commented on the impressive work: "It is one of the most dangerous substances known in chemystry [sic], but unfortunately also one of the most useful."[44]

Twenty-five years later, on November 6, 1888, the renowned professor

Jean-Martin Charcot (1825–1893), at one of his equally renowned "lectures" organized every Tuesday at the La Salpêtrière hospital, presented a patient who was the victim of acute carbon disulfide poisoning to a learned assembly of physicians in white coats. After working in a rubber factory for seventeen years, the young man had fallen into a coma after cleaning the vulcanization tanks. "This poor devil is an exceptional case of masculine hysteria," the neurologist summed up, reminding his audience that hysteria was generally considered a feminine illness. Emphasizing the role of carbon disulfide in the illness's etiology, he taciturnly explained (much like an expert examining an oddity) that, "Hygienists and clinicians are concerned with these industries because of certain accidents, principally neurological, to which its workers are subject."[45] The "lecture" would turn out to be historically significant, since a British medical dictionary in 1940 would qualify neurological problems brought on by carbon disulfide "gassing" as "Charcot's carbon disulfide hysteria."[46]

But, contrary to what might be believed, the accumulation of medical data would not bring about the prohibition, or even the regulation, of carbon disulfide use. In 1902, Dr. Thomas Oliver, a British doctor and one of Charcot's disciples, tried to sound the alarm by denouncing the limits of laissez-faire, which would have it that occupational safety be the exclusive responsibility of manufacturers. In a very well-researched study, he describes the phenomenon of addiction that accompanies the hysterical and sexual problems rubber factory workers were experiencing: "In the morning they drag themselves to the factory feeling ill and headachy, and, like people who are accustomed to the intemperate use of alcohol, they only get relief and recover their nervous equilibrium by renewed inhalation of the vapors of carbon disulfide."[47]

But this new publication would change nothing for working conditions in factories where poisons were used. Because in the meantime, their use had become even more varied with the advent of a new miracle product: viscose, also called "artificial silk" or "rayon," which had a bright future in store.[48] The synthetic fiber was fabricated using cellulose extracted from tree pulp, thanks to a chemical process in which carbon disulfide was the major chemical component. "Once again," Paul Blanc notes, "medical reports very quickly identified the hazard. The blunted response to these findings, absent any effective controls for at least several decades, demonstrates the power that economic-political forces can successfully exert in retarding public health interventions in the industrial sector."[49]

Brussels, 1936: The Congress on the Causes of Cancer

"It was perhaps the most momentous Cancer Congress ever held," Isaac Berenblum (1903–2000), a biochemist and oncologist, would later say.[50] "A veritable Manhattan Project on cancer," wrote the American epidemiologist Devra Davis in 2007, in her previously cited book, *The Secret History of the War on Cancer*.[51] The event was so significant that the magazine *Nature* decided to announce it as early as March 1936, six months before the congress opened in Brussels on September 20.[52] On that day, the two hundred top cancer specialists in the world converged on the Belgian capital. Coming from North America, South America, Japan, and all of Europe, often after long weeks of boat travel, the distinguished specialists exchanged their knowledge on a disease that was growing incessantly.

"I was stunned to see how much was known about the social and environmental causes of cancer before World War II, seventy years ago," commented Davis, who created the first center for environmentally specialized cancer research at the University of Pittsburgh. "The three volumes from this congress included surprisingly comprehensive laboratory and clinical reports showing that many widely used agents at that time were known to be cancerous for humans, including ionizing and solar radiation, arsenic, benzene, asbestos, synthetic dyes and hormones."[53]

The conference participants included William Cramer (1878–1945), a Briton who, after comparing the medical history of identical twins (that is, from the same ovule and thus sharing strictly identical genetic material), concluded (already!) that "cancer as a disease is not inherited."[54] Furthermore, after studying death records in the United Kingdom, the researcher from the Imperial Cancer Research Fund noted that the incidence of the disease had risen by 30 percent since the beginning of the century. He also specified (already!) that he had arrived at this number after deducting factors of population increase and life expectancy. On those grounds, considering that the development of tumors was the result of exposure that occurred twenty years earlier, he recommended limiting carcinogenic agents in the workplace while increasing experimental research, because, as he noted (already!), "cancer often develops in both rodents and humans in the same tissues."

In Brussels, Angel Honorio Roffo (1882–1947), an Argentinian, was also present. He presented photographs of mice that had developed tumors after regular exposure to X-rays or ultraviolet rays (already!); the risk was heightened when rodents were exposed simultaneously to hydrocarbons. James

Cook and Ernest Kennaway (1881–1958) were also in attendance: the two Britons from London's Royal Cancer Hospital had carried out a meta-analysis of thirty or so studies showing (already!) that regular exposure to the hormone estrogen led to mammary cancer in male rodents.

"How did these scientists decide what was a cause of cancer in 1936?" Devra Davis asks. "They combined autopsies with medical, personal and workplace histories of people who had come down with cancer. They reasoned that if they found tar and soot in the lungs of those who had worked in mining and showed that these same things caused tumors when placed on the skin or into the lungs of animals, that was sufficient to deem these gooey residues a cause of cancer that should be controlled."[55]

On paper, all of this seems crystal clear, or as they say, "just plain common sense." But in reading these 1936 congress proceedings, a question logically arises: If all of these researchers *already* understood that the main cause of the cancer explosion was exposure to chemical agents and if, moreover, they *already* knew how to limit the damage caused by poisons, why did no one listen to them? The answer is as simple as the question: If all these researchers' studies and recommendations were ignored, it is because starting in the 1930s industry began strategizing how to control and manipulate research on the toxicity of its products, while waging a merciless war on all the scientists wishing to maintain their independence in the name of the defense of public health. The first victim of this David-and-Goliath battle was Wilhelm Hueper, a renowned American toxicologist of German descent, considered Bernardino Ramazzini's successor, who participated in the Brussels congress a few months before being fired by his employer, the American chemical company DuPont de Nemours.

Wilhelm Hueper's Solitary Battle

Wilhelm Hueper's story is exemplary, because it captivatingly summarizes everything I discovered over the course of my lengthy inquiry. Born in Germany at the end of the nineteenth century, this young man was sent to the front at Verdun during World War I, where he saw the damage done by the poison gas invented by his fellow citizen Fritz Haber (see Chapter 2). From this experience was born an unwavering pacifism, which would remain with him his whole life. After finishing medical school, he immigrated to the United States in 1923. He worked at a Chicago medical school before joining the University of Pennsylvania's laboratory for cancer research in Philadelphia, chiefly funded by DuPont, one of the biggest chemical companies of the time. In 1932, after learning that the Deepwater, New Jersey, plant was making benzidine and beta-naphtylamine (BNA), which was used in the production of synthetic dyes, he wrote a very candid letter to Irénée du Pont (1876–1963), the company's owner, to inform her of the bladder cancer risks her workers were facing. His letter was never answered.

Wilhelm Hueper was quite familiar with the subject of synthetic dyes: an occupational health specialist, he very closely followed the medical reports that peppered the development of this booming activity, which was fortuitously born in a British laboratory. In 1856, William Henry Perkin, a chemistry student, discovered that he could transform coal tar—a by-product that had little value at the time and was obtained during the distillation of charcoal to produce gas for lighting—into a mauve solution he called "mauveine." It was the first synthetic dye in history. Young Perkin's discovery was monumental: the production of synthetic dyes would constitute the basis of the development of the organic chemistry industry, which would revolutionize the manufacture of medications (aspirin, syphilis treatment), explosives, adhesives and resins, pesticides and, of course, textiles, thanks to the use of aromatic amines, like benzidine and BNA. Very quickly, Germany muscled in on the synthetic dye market, filing hundreds of patents. However, in 1895, the German surgeon Ludwig Rehn reported that in a Griesheim factory where fuchsine (a magenta dye) was made, three workers out of forty-five had developed bladder cancer. Eleven years later, thirty-five of them had. Over the following decade, dozens of cases were reported all over Germany, and also in Switzerland.[56] In 1921, using a number of clinical reports as evidence, the International Labor Office published a position paper on aromatic amines, including benzidine and BNA, recommending that "the most rigorous application of hygienic precautions should be required."[57]

Once again, however, these reports did not accomplish much. At the end of World War I, the United States confiscated the patents held by vanquished Germany, and distributed them at low prices to American companies like American Cyanamid, Allied Chemical, Dye Corporation, and DuPont. The last immediately built its first organic chemical factory in Deepwater, called "Chambers Works," where benzidine and BNA production began in 1919. According to internal documents consulted by David Michaels, the firm's doctors detected the first instances of bladder cancer in 1931, not long before Wilhelm Hueper wrote his letter to Irénée du Pont. "For the next several years, these physicians documented the rapidly growing epidemic both at national conferences and in the scientific literature; at least 83 cases had been recognized by 1936," writes Michaels in an article on bladder cancer of occupational origins.[58]

In fact, a study published in 1936—the same year as the Brussels congress—by Dr. Edgar Evans, the chief physician at DuPont, is testimony to the firm's desire to promote transparency.[59] Two years earlier, as a belated follow-up to Hueper's letter, DuPont had even asked Hueper to join the new industrial toxicology laboratory it had created in Wilmington, North Carolina, precisely to study bladder cancer. The researcher developed an experimental protocol to test the effects of BNA on dogs. The results were incontestable: regular exposure to aromatic amines resulted in bladder tumors, as it did in humans. Deeply troubled by the human implications of his study, yet convinced of his employer's good intentions, the toxicologist requested to visit Chambers Works in order to see how workers' safety could be improved.

He detailed what followed in his memoirs:

The manager and some of his associates brought us first to the building housing this operation, which was located in a part of a much larger building. It was separated from other operations in the building by a large sliding-door allowing the ready spread of vapors, fumes and dust from the betanaphthylamine operation into the adjacent workrooms. Being impressed during this visit by the surprising cleanliness of the naphthylamine operation, which at that occasion was not actively working, I dropped back in the procession of visitors, until I caught up with the foreman at its end. When I told him 'Your place is surprisingly clean,' he looked at me and commented, 'Doctor, you should have seen it last night; we worked all night to clean it up for you.' The purpose of my visit was thereby almost completely destroyed. What I had been shown was a well-staged performance. I, therefore, approached the manager with the request to see the benzidine operation. After telling him what I had just been told, his initial reluctance to grant my request vanished and we were led a short distance up the road where the benzidine operation was housed in a separate small building. With one look at the place, it became immediately obvious how the workers became exposed. There was the white powdery benzidine on the road, the loading platform, the window sills, on the floor, etc. This revelation ended the visit. After coming back to Wilmington, I wrote a brief memorandum to Mr. Irenee Du Pont describing to him my experience and my disappointment with the attempted deception. There was no answer but I was never allowed again to visit the two operations.[60]

For Wilhelm Hueper, it was the beginning of the end. Soon after, he clashed with the company, which prohibited him from publishing his study on dogs. He was eventually fired in 1937, after the Brussels congress. Braving the wrath of DuPont, which threatened him with legal proceedings, he eventually published his study in a scientific periodical in 1938[61] and, four years later, in a book as important as Bernardino Ramazzini's was in his time. Entitled *Occupational Tumors and Allied Diseases*, it focuses on the important research carried out for more than half a century on the link between cancer and exposure to chemical products. In his autobiography, Hueper says that he had first planned on dedicating his work to "the victims of cancer who made things for better living through chemistry." It was an ironic allusion to DuPont's slogan, launched in 1935: "Better living through chemistry."[62] Fearing retaliation, he ultimately opted for a less confrontational dedication: "To the memory of those of our fellow men who have died from occupational disease contracted while making better things for an improved living for others."[63]

Despite DuPont's defamation campaign, including accusations of being "a Nazi, and later a Communist sympathizer,"[64] the scientist was recruited in 1948 by the prestigious National Cancer Institute, where he founded the first department for environmental cancer research. It was there that he met Rachel Carson, to whom he would open his archives to support her research for *Silent Spring*. As for the chemical company, it would continue to produce BNA until 1955 and benzidine until 1967, without ever truly modifying its manufacturing process. In a letter dating from June 1947 and addressed to Dr. Arthur Mangelsdorff, the medical director at American Cyanamid, Edgar Evans—the head doctor at Chambers Works and author of the 1936 study— plainly admitted: "The question of health control of employees in the manufacture of Beta Naphthylamine is indeed a grave one. [. . .] Of the original group, who began the production of this product, approximately 100% have developed tumors of the bladder."[65]

It is impossible to know today how many victims the bladder cancer epidemic claimed and continues to claim, due to use of aromatic amines, including benzidine and BNA of course, as well as ortho-toluidine (*o*-toluidine), an antioxidant widely used in the manufacture of rubber products, such as tires. This was how American health authorities, alerted by unions in the early 1990s, identified a "cluster"—that is, an abnormal concentration—of bladder cancer in a Goodyear factory in Buffalo, whose ortho-toluidine stock came from DuPont.[66] It goes without saying that this American manufacturer is far from an exception. From one product to the next, but also from one country to the next, the same story keeps repeating itself, following a pattern whose rules are invariably dictated by industry, with the tacit complicity of public powers who accept the death toll, acting only when "the human cost [becomes] so obvious that it [is] no longer acceptable,"[67] to borrow a few words from David Michaels, the U.S. assistant secretary of labor since 2009.

8

Industry Lays Down the Law

No tyranny is more cruel than that which is practiced in the shadow of the law and with the trappings of justice.

—Montesquieu

When Wilhelm Hueper fell into disfavor with DuPont and became a black sheep of chemical manufacturers, another toxicologist, Robert Kehoe, was named head of what Devra Davis calls "defensive research,"[1] meaning science designed with the sole purpose of defending those manufacturers' products. It is fascinating to compare the paths taken by these two major contemporary figures of occupational medicine who—like the two faces of Janus—embody two diametrically opposed currents in toxicology: one in service of public health, the other in service of private interests.

1924: The Groundbreaking Case of Leaded Gas in the United States

The man who would become chairman of the American College of Occupational and Environmental Medicine, as well as of the American Industrial Hygiene Association, owes his illustrious career to a veritable massacre that took place in 1923 and 1924 in several leaded gasoline refineries. In 1921, a chemist at General Motors, which was then the leader in the automobile market, had discovered that tetraethyl lead could be used as an antiknock agent in fuels. Although alternatives existed, Charles Kettering, the director of research at General Motors, encouraged the use of lead because of its low cost. The news set off rapid-fire, hostile reactions almost everywhere in the world, since, as Gerald Markowitz and David Rosner write in their book, *Deceit and Denial: The Deadly Politics of Industrial Pollution*, "By this time, no one disputed that white lead was a poison."[2] The two hundred-odd pages that the two American historians dedicate to the "mother of all industrial poisons"[3]—which builds up in living organisms and affects children in particular—clearly illustrate that its neurotoxic and reprotoxic properties have been known since the Roman empire.

The rest of the story provides supplemental evidence (if it is even needed). Despite warnings from the Public Health Service, which worried that it was a "serious menace to public health,"[4] leaded gasoline was launched on the market on February 2, 1923. General Motors, Standard Oil (now Exxon), and DuPont ensured production and collectively created a joint venture for that purpose, called the Ethyl Corporation. DuPont conducted its activities at Chambers Works where, as we saw in Chapter 7, benzidine and beta-naphthylamine (BNA) were already being manufactured. Very quickly, the ill-fated factory—where Wilhelm Hueper would be declared persona non grata twelve years later—was saddled with the nickname "the house of butterflies" due to the hallucinations its workers suffered as they were being poisoned by lead fumes, which came to be called "loony gas."[5] In a caricature published in the *New York Journal* on October 31, 1924, a hospitalized worker portrayed with bulging eyes seems to be fighting against a cloud of imaginary insects.

Admittedly, the press was particularly unrelenting against leaded gasoline that week. On October 27, the *New York Times* revealed that in only a few months, three hundred Chambers Works employees had been critically poisoned, ten of them fatally. During that same period, two slaves to "progress" had died and forty had been hospitalized following an accident at the General Motors plant in Dayton, Ohio. Similarly macabre observations were made in the Standard Bayway refinery near New York, where seven workers died and thirty-three went mad.[6] Later, it was revealed that Joseph Leslie, a young worker who made liquid lead in the factory, had been discretely interned in a psychiatric hospital (where he would die in 1964), whereas his family was told he was already dead. It was only in 2005 that the poor man's descendants discovered the truth, thanks to an article published by William Kovarik in the *International Journal of Occupational and Environmental Health* in which he writes: "The confusion in the Leslie family's history reflects a larger picture of misinformation and deception in the history of environmental and public health."[7]

At the time that Leslie disappeared from the land of the living, leaded gasoline was the subject of an intense debate, and several American cities, including New York and Philadelphia, decided to prohibit its sale, as reported the *New York Times* on October 31, 1924.[8] But these bans would not hold up for long, for leaded gasoline had its best (most poisonous) days ahead of it.[9] Incidentally, William Kovarik points out that when Chicago outlawed leaded gasoline in 1984, the same newspaper (the *New York Times*) ran an article emphasizing that this decision was "believed to be the first in the nation."[10] This anecdote is more than just a detail; as the American historian quite rightly concludes, it illustrates the "historical amnesia that is typical in the field of environment and public health policy."

Yet this "amnesia" did not spontaneously emerge. It was the result of a tireless erasure campaign carried out methodically by the industry, following a script that leaded gasoline producers were the first to write. As we will soon see, that script was still just a flawed draft, which would eventually be perfected, namely by tobacco manufacturers. Nonetheless, the events that played out in October 1924 were crucial: it was the first time that manufacturers representing three key sectors of the economy—chemistry, petroleum, and mechanics—united their efforts to carry out a plan of systematic misinformation meant to "befuddle" politicians, the press, and consumers, and muzzle independent research. The model they developed would soon be used by all poison sellers, led by pesticide, additive, and food packaging manufacturers, all of whom were ultimately members of the same family.

Leaden Silence in the Name of Science

On October 30, 1924, faced with the turmoil aroused by the poisoned workers' ordeal, General Motors organized a press conference. The journalists got the full show: Thomas Midgely, the firm's director of research, displayed a tube containing liquid lead, which he splashed on his hand; he then inhaled it for a minute. With incredible cynicism, he explained that if the workers fell ill or died, it was "caused by the heedlessness of workers in failing to follow instructions."[11] He added, "This extremely dilute product has been for more than a year in public use in over 10,000 filling stations and garages and no ill effects thus far have been reported."[12] Clearly, the demonstration paid off, for just one month later, the opinion maker that is the *New York Times* doggedly supported leaded gasoline, stating that the deaths at the Standard Oil refinery were "not a sufficient reason for abandoning the use of a substance by means of which a large economic gain could be effected. . . . As there is no measurable risk to the public in its proper use as a fuel, the chemists see no reason why its manufacture should be abandoned. That is the scientific view of the matter, as opposed to the sentimental, and it seems rather cold-blooded, but it is entirely reasonable."[13]

So there we have it. We can see, written in black and white in an article published in 1924, the two main arguments that would be hammered in throughout the twentieth century any time concerns were raised as to the safety of chemical products contaminating our environment and our food. In essence, they stated: "Don't get carried away by *emotion*, because the subject is very *complicated*, but rest assured that the *scientists*, who are *reasonable* people, know what they're doing." Of course, we would better be able to "rest assured" if the "scientists" were independent people whose sole objective was searching for the truth in order to better protect us. But unfortunately this is rarely the case, as demonstrated by Robert Kehoe, who is a veritable model for the infamous "defensive research," to use Devra Davis's terms.

In 1925, the toxicologist was recruited by General Motors and DuPont to head the medical department of the Ethyl Corporation and to direct the Charles Kettering industrial toxicology laboratory (named for the General Motors research director), which the manufacturers had just opened at the University of Cincinnati, where Kehoe was already a professor of physiology. The job was clearly important: his annual salary was fixed at $100,000, a colossal amount for the time, and largely sufficient to stifle any independent leanings. As proven by the laboratory archives Devra Davis was able to consult, his mission was to conduct experimental studies on animals on behalf of large companies such as DuPont, General Motors, U.S. Steel, Mobil Oil, Ethyl Corporation, and . . . Monsanto.

From then on, Robert Kehoe was careful not to transmit the real results of his studies. And for good reason: as Devra Davis revealed, the contracts between the laboratory and its sponsors stipulated that "the investigative work shall be planned and carried out by the University, and the University shall have the right to disseminate for the public good, any information obtained. However, before issuance of public reports or scientific publications, the manuscripts thereof will be submitted to the Donor for criticism and suggestion."[14] Note the word "donor": donor of what? Of orders or of money? Or perhaps both? In any case, everything seems to indicate that Kehoe respected to the letter the rules set out in the late 1920s, because when he retired in 1965, he issued a very meaningful memorandum, meant for his collaborators: "It is undesirable, as a rule, to refer to reports of the Laboratory made to Sponsors in papers prepared for publication, since such references bring requests for these reports. As these reports often contain confidential information, they cannot be supplied, except confidentially, to other interested persons, and unless one knows that they are suitable for issuance to others . . . they should not be mentioned in public." The memo is somewhat convoluted, but in terms of content, the message is quite clear. As Davis sums up, "The same businesses that produced the materials Kettering tested also decided what findings could and could not be made public."[15]

In the meantime, Kehoe, a zealous "scientist," was meticulously carrying out his own mission. Starting in 1926, he performed "dozens of autopsies" on babies who had died from lead poisoning. The medical reports seen by Davis are chilling. They are "the work of a very meticulous man, showing precise amounts of lead measured in the brains, livers, hearts and kidneys of poor black and white infants," writes the American epidemiologist.[16] The reports also reveal that a twenty-four-year-old mother in Waynesboro, Mississippi, lost her three children, and that the autopsy of the youngest showed a very high concentration of lead in its blood, liver, and bones. Nothing further is known, since the reports do not indicate the mother or father's occupation, or the family's living conditions. Kehoe satisfied himself with collecting the macabre data to then—adding insult to injury—publish expert articles in which he proffers his theory that lead is a "natural contaminant" and fundamentally harmless because, according to Paracelsus's age-old hypothesis, "the dose makes the poison" (sola dosis facit venenum).

The Perverse Use of Paracelsus's Principle: "The Dose Makes the Poison"

Born Philippus Theophrastus Aureolus Bombastus von Hohenheim (1493–1541), the man known in history books as Paracelsus was a Swiss alchemist, astrologer, and doctor who was as rebellious as he was mystical. But today, he must be spinning in his grave after seeing how the toxicologists of the twentieth century abused his name to justify the mass marketing of poisons. Out of all of the "cursed doctor's"[17] rants, one merits contemplation by all those charged with the protection of our health: "Who then is unaware that most of the doctors of our times have failed in their mission in the most shameful manner, by making their patients run the biggest risks?"[18] he raged in 1527. The professor of medicine had just burned the classic manuals of his discipline in front of the University of Basel, which, we can imagine, had to have earned him several robust enmities.

"Allergic to all arguments of authority"[19]—a characteristic largely forgotten by those who blindly apply the principle bearing his name—Paracelsus is considered to be the father of both homeopathy and toxicology, two disciplines which don't love each other so much. The first relies on one of his most famous maxims, which also inspired Louis Pasteur when he invented the first vaccine: "That which cures man can also harm him, and that which has harmed him can also cure him." The latter discipline prefers another, which is all in all quite complementary: "Nothing is poison, everything is poison: only the dose makes the poison."[20]

The idea that "the dose makes the poison" dates, in fact, to Antiquity. In their book *Environnement et santé publique* (Environment and Public Health), Michel Gérin and his co-authors point out that "King Mithridate regularly consumed brews containing several dozen poisons in order to protect himself from enemy assassination attempts. He succeeded so well that, when he was taken prisoner, he failed in his attempt to commit suicide by poison."[21] It is to this Greek king that we owe the word "mithridatism," or the practice of becoming accustomed to, or acquiring immunity to, poisons by exposure to increasing doses. Relying on his own observations, Paracelsus believed that toxic substances could be beneficial in small doses and that, inversely, a theoretically harmless substance such as water could turn out to be fatal if it was ingested in too large a quantity. We will see later that the principle "the dose makes the poison"—an abstract dogma used during toxicological evaluation of modern-day poisons—is not applicable to a number of substances. However, we are not quite there yet.

In any case, everything would seem to indicate that Robert Kehoe read Paracelsus, because if he made such a concerted effort to dissect the bodies of newborns, it was because he was attempting to determine a lead exposure level that *appeared* safe to him, in order to counter attacks from those demanding the prohibition of leaded gasoline. In brief, the findings from autopsies performed on the small cadavers would not be used to take measures to stop contamination but, on the contrary, to justify its continuation thanks to pseudo-scientific arguments—in other words, with reports, figures, and charts, all those things risk managers love. The reassuring "turnkey" theory that Kehoe thus delivered to these managers rested on four hypotheses that allowed poisoned gasoline to be sold for more than fifty years: "1) that lead absorption is natural; 2) that the body has mechanisms to cope with lead; 3) that below a certain threshold, lead was harmless; and 4) that the public's exposure was far below the threshold and was of little concern."[22] We will soon see (in Chapter 10) that this reasoning would serve as the basis for the establishment of what toxicologists call the "acceptable daily intake" of a poison—a pesticide, food additive, etc.—or, in other words, the amount a human being can supposedly ingest daily without becoming ill. The "ADI," as it is called in the jargon, is even used as the absolute reference value by experts tasked with regulating chemical products contaminating our food chain.

In 1966, at his Senate hearing during an investigation into air pollution, Robert Kehoe obstinately defended his theoretical notion: "During the entire history of man on this earth he has had lead in his drinking water. . . . The question is not whether lead per se is dangerous, but whether a certain concentration of lead in his body is dangerous."[23] And, to determine the concentration that could be considered "harmless," the toxicologist took drastic steps: he had no reservations about confining "volunteers" in a room and having them breathe in lead vapors for a period of three to twenty-four hours. This was an experiment he tirelessly repeated for three decades, with the support of Ethyl Corporation, DuPont, and even the U.S. Health Department.

"While human experimentation has a long and inglorious history in America and other nations," historians Gerald Markowitz and David Rosner write, "these studies were particularly pernicious because their objective was not the discovery of a therapy for those with lead poisoning but was to gather evidence that could be used by industry to prove that lead in the blood was normal and not indicative of poisoning by industry."[24] It was thus that, until the early 1980s, the standard for lead exposure in foundries was 200 milligrams per meter cubed of air, while the supposedly "safe" level of lead in the blood was 80 micrograms per deciliter for adults and 60 micrograms for children. These completely arbitrary figures, produced by Kehoe in the greatest secrecy, would turn out to be wrong, but were taken at face value by every regulatory agency in the world. "From the 1920s to the 1960s, Kehoe helped the lead industries use their economic power to define the scientific basis of lead poisoning," the historian William Kovarik writes. He cites his colleague William Graber: "So complete was the industry domination of research into and knowledge of the hazards of lead that the central paradigm for understanding lead and its effects remained that pioneered by Kehoe and his associates."[25]

But history has its ironies. Luck would have it that the two major American adversaries working in occupational health would eventually meet. In the 1960s, three workers who had developed skin cancer sued their employer, which manufactured hydrocarbon-based paraffin wax. Wilhelm Hueper was cited as an expert for the plaintiffs, while Robert Kehoe was used to support the defense. On this occasion, Hueper discovered that Kehoe had secretly continued his work on aromatic amines, which had gotten him fired from DuPont. In fact, the Kettering Laboratory archives revealed a number of reports, never published, showing that animals exposed to benzidine, BNA, paraffin oils, or hydrocarbons developed cancer.

In his memoirs, Wilhelm Hueper reports the confrontation with the industrial toxicologist, who denied the carcinogenic effects of hydrocarbons during the proceedings: "The Director of this organization in Cincinnati [Kehoe], testifying as a consultant of the oil company, had to confess that none of the results of his institute's studies with these oils had been published or had been made available to the medical profession in general or to labor organizations, because the data were considered by the oil company as 'privileged' information, i.e., the property of the oil company. When after more than a year's time, the final information became available to the court and the plaintiff, there was no longer any doubt that even in the hands of the members of the Kettering Laboratory the incriminated oils had carcinogenic properties, although its director had found it proper at the first hearing to make some snide remarks about my scientific reliability."[26]

Tobacco and Lung Cancer: The Smokescreen

"The history of tobacco is not just the history of cigarettes," Devra Davis stated at a conference at the Carnegie Museum of Natural History in Pittsburgh, on October 15, 2009. "It is also that of a model of deception that benefits all chemical manufacturers."[27] It was not easy to meet up with the American epidemiologist, who once headed the first experimental cancer research center in Pittsburgh and now lives in Washington, DC. When I contacted her in the fall of 2009, she was traveling throughout the United States, appearing at lecture halls and public talks, to present her book, *The Secret History of the War on Cancer*, all while working on a new text on the dangers of cell phones.[28]

The sixty-four-year-old researcher, a naturally gifted public speaker who seamlessly melds personal anecdotes and scientific information, knows how to win over her audience. At the Pittsburgh conference, she described, with the help of slides, her childhood in Denora, Pennsylvania, a steel industry capital, where "people came to live because there was smoke, and smoke meant that there was work. The city was covered in soot, because the blast furnaces were fueled by coal."[29] She also said that in 1986, when she was working at the National Academy of Sciences, she informed her boss, Frank Press, that she was intending to write a book on the environmental causes of cancer, and that he strongly urged her not to because "it would ruin her career." "Still," she continued, "since President Nixon declared war on cancer in 1971, the disease has kept evolving. Why? Because, since the beginning, we have been fighting with the wrong weapons, privileging research on treatments instead of prevention. I'm not saying that treatments are not important, and I'm in a position to know, since my father died of multiple myeloma and my mother from stomach cancer. But I am saying that as long as we are not attacking chemical pollutants, synthetic hormones, pesticides, or waves, we won't be able to win the war on cancer. To be able to do so, we need the courage to confront powerful interests and manufacturers' lies that hide the dangerous nature of their products, just as tobacco manufacturers did for so very long."

"Why do you say that the history of tobacco is more than just the history of cigarettes?" I asked Ms. Davis after the conference.

"Because it was the tobacco manufacturers who wrote the script used by the entire chemical industry to deny the toxicity of its products. They perfected the system established by lead manufacturers to permanently foster doubts about the dangers of tobacco by relying on scientists who were handsomely paid to publish falsified studies. It was an incredible manipulation tool which ultimately delayed preventative measures for more than fifty years!"[30]

It is impossible to reproduce here all of the elements of this vast subject, which has already been the subject of several books.[31] Instead, I will outline the major threads and focus on the "script" mentioned by Davis, since it sheds light on the methods used by the chemical industry to manipulate regulatory agencies and public opinion. If I have learned one thing over the course of my lengthy investigation, it is that only a well-developed and recurring system can explain the chemical frenzy in which humanity has been submerged for half a century.

As an early victim of tobacco use, like many teenagers of my generation, I should admit that the history of tobacco is particularly edifying in this respect. Its link to respiratory tract cancers was identified as early as 1761 by the British doctor John Hill.[32] A century later, Étienne Frédéric Bouisson, a Frenchman, observed that out of sixty-eight patients suffering from mouth cancer, sixty-three were pipe smokers.[33] But it was really in the 1930s that studies began to show that tobacco was a powerful carcinogen. One such study was carried out by the Argentinian Honorio Roffo, whom I already mentioned in relation to the Brussels congress of 1936; it showed the carcinogenic effects of solar rays, but also of hydrocarbons, which include cigarette tar.[34,35] The German epidemiologist Franz Hermann Müller explained this in Brussels, as he was working on the first case-control study on the effects of tobacco use. Published in 1939, it showed that "very heavy smokers" had sixteen times the "chance" of dying from lung cancer than nonsmokers.[36] It also revealed that out of the eighty-six victims whose histories had been reconstructed, one out of three had never smoked but had been exposed to toxic substances, such as lead dust (seventeen cases), chromium, mercury, or aromatic amines.

When Müller published his study, Nazi Germany was in the midst of launching the biggest anti-tobacco campaign of all time. As Robert Proctor, an American professor of the history of science, recounts in his compelling book, *The Nazi War on Cancer*, the campaign figured in Hitler's ideology of Aryan "racial hygiene and bodily purity," for which tobacco was "a genetic poison, a cause of infertility, cancer, and heart attacks; a drain on national resources and a threat to public health."[37] To the great displeasure of Joseph Goebbels, the minister of propaganda and a huge cigar aficionado, draconian measures were taken with the notorious efficiency of the National-Socialist machine, such as the prohibition of smoking in trains and public places and the sale of cigarettes to pregnant women. In April of 1941, the first institute for research on the dangers of tobacco (Wissenschaftliches Institut zur Erforschung der Tabakgefahren) was ceremoniously opened in Jena, Germany, which during its short existence (it was closed at the end of the war) produced seven studies on the consequences of nicotine addiction. The most important of these studies was published in 1943 by Eberhard Schairer and Erich Schöniger, who were inspired by Franz-Müller's case-control study to compare the lifestyle habits of 195 lung cancer victims to those of 700 men in good health. This study produced undeniable results: out of 109 lung cancer victims for whom family members supplied sufficient information, only three were nonsmokers (some of the smokers had *also* been exposed to asbestos or toxic industrial agents).[38]

But, for reasons most likely related to the Third Reich's criminal past, the German studies have not remained in the annals of the fight against tobacco. This honor falls upon the British epidemiologist Richard Doll (1912–2005), even though he was reportedly greatly inspired by the pioneering work of his

German predecessors. Robert Proctor writes that the young medical student, at that time a committed socialist, had attended a 1936 conference in Frankfurt on radiotherapy where the radiologist SS Hans Holfelder had given a presentation using slides that showed how X-rays, depicted as "Nazi storm troopers," destroyed "cancer cells," depicted as Jews.[39] In 1950, Richard Doll published a study in which he showed that "the risk of developing [lung cancer] increases in proportion to the amount smoked. It may be 50 times as great among those who smoke 25 or more cigarettes a day as among non-smokers."[40] Carried out in 20 London hospitals with 649 men and 60 women suffering from lung cancer, the case-control investigation made Doll "one of the pre-eminent public health authorities of the day,"[41] and earned him a knighthood in 1971—we will see later that he did not hesitate to put his notoriety at the disposal of the chemical industry, to whom he would render valuable services in exchange for cold, hard cash (see Chapter 11).

In the meantime, everything was falling apart for cigarette manufacturers: between 1950 and 1953, six studies (including Doll's) made headlines in American and European newspapers. And then, in 1954, came the final blow: Cuyler Hammond and Daniel Horn, two epidemiologists from the American Cancer Society (ACS), published the first prospective study, based on an unprecedented cohort of 187,776 white males between the ages of 50 and 69: 22,000 ACS volunteers—mainly women trained in interviewing techniques— were sent all over the country to question each participant at least twice, with a five-year interval between meetings. At the end of the period studied, smokers showed an abnormally high death rate of 52 percent.[42]

"Doubt Is Our Product"

Once they saw their sales start to lag, tobacco manufacturers got organized. In 1953, they created the Tobacco Industry Research Committee (TIRC), naming as its leader Clarence Cook Little, former director of the ACS, who appeared on the cover of *Time* magazine in 1937 with a wide smile and a pipe in his mouth.[43] He quickly set out to minimize the results of his ACS colleagues' study, wielding the argument that, from then on, would be the TIRC's leitmotiv: "The origin, nature, and development of cancer and of cardio-vascular disease are complex problems," he stated in an interview with *US News and World Report*, "offering the greatest existing challenge to creative scientific thought and to further experimentation wisely conceived, patiently executed, and fearlessly and impartially interpreted in our search for the truth."[44] "The TIRC's strategy was to create doubt," Devra Davis explained. "From then on, as soon as a study confirmed the dangers of tobacco, the institute would offer millions of dollars to universities to conduct a new study, ostensibly under its control. The influx of money would artificially maintain the illusion of scientific debate, allowing industry to say that the question of the danger of tobacco was still not answered, though it had been for a long time!"

The American epidemiologist's statement was confirmed by a secret document included in an anonymous package received by Stanton Glantz, a researcher at the University of California, in 1994. The astonishing parcel contained thousands of pages from the Brown & Williamson Tobacco Corporation. Nicknamed the "Cigarette Papers," they would serve as incriminating evidence in the big American lawsuits against tobacco manufacturers. Amidst this mine of information was a gem written by one of the company's directors: "Doubt is our product since it is the best means of competing with the 'body of fact' that exists in the mind of the general public. It is also the means of establishing a controversy. [. . .] If in our pro-cigarette efforts we stick to *well documented fact*, we can dominate a controversy and operate with the confidence of justifiable self-interest."[45]

Everything is laid out in black and white and, in reality, the industry did finance a number of rigged studies on active tobacco use and secondhand smoke, all the while using considerable resources to propagate doubt among consumers. To do this, it relied on newspapers, which relayed its messages in the form of very expensive ad inserts. The first wide-ranging initiative dates from January 4, 1954, when 448 press outlets, including the *New York Times*, published a pamphlet entitled "A Frank Statement to Cigarette Smokers." It included the following declaration:

Distinguished authorities point out:
1. That medical research of recent years indicates many possible causes of lung cancer.
2. That there is no agreement among the authorities regarding what the cause is.
3. That there is no proof that cigarette smoking is one of the causes.
4. That statistics purporting to link cigarette smoking with the disease could apply with equal force to any one of many other aspects of modern life. Indeed the validity of the statistics themselves is questioned by numerous scientists.

We accept an interest in people's health as a basic responsibility, paramount to every other consideration in our business.

We believe the products we make are not injurious to health.

We always have and always will cooperate closely with those whose task it is to safeguard the public health.[46]

In the file he compiled for one of the lawsuits against the Philip Morris company, in which he was cited as an expert, Robert Proctor (author of *The Nazi War on Cancer*) explains why the "Frank Statement" was a pioneering text: "From a historian's point of view, the 'Frank Statement' represents the beginning of one of the largest campaigns of deliberate distortion, distraction, and deception the world has ever known. The tobacco industry in effect becomes two industries: a manufacturer and seller of tobacco products, and a manufacturer and distributor of doubt about tobacco's hazards."[47] For several decades, cigarette companies would indeed repeat over and over again that tobacco's carcinogenic effects were "not a statement of fact but *merely an hypothesis*," according to a Brown & Williamson representative in 1971,[48] or that "the link between tobacco abuse and a certain number of cardiovascular diseases or cancer has never been scientifically established," to borrow the words of Pierre Millet, head of the French company Seita (the National Society for Industrial Development of Cigarettes and Matches), in 1975.[49] Because, while Seita was of course dependent on the state, it actively participated in what some have called a "conspiracy"[50] by constantly demanding more "evidence," although no one ever knew what "evidence" would suffice to finally close the case.

Exasperated by the tobacco manufacturers' repeated denials and hypocrisy, Evarts Graham, the author of one of the aforementioned six studies published between 1950 and 1953, interpreted their demands quite literally. In 1954, in *The Lancet*, he suggested conducting experiments on human guinea pigs: "(1) Secure some human volunteers willing to have a bronchus painted with cigarette tar, perhaps through a bronchial fistula. (2) The experiment must be carried on for at least twenty or twenty-five years. (3) The subjects must spend the whole period in air-conditioned quarters, never leaving them even for an hour or so, in order that there may be no contamination by a polluted atmosphere. (4) At the end of the twenty-five years they must submit to an operation or an autopsy to determine the result of the experiment."[51] The American surgeon's provocative proposal can be credited for highlighting a difficulty that I have already addressed in regards to pesticides: in the field of environmental health, it is impossible to obtain *absolute proof* that a chemical product is the *only* cause of a given illness. Nevertheless, as Christie Todd Whitman, the former administrator of the Environmental Protection Agency (EPA), so rightly said, "the absence of certainty is not a reason to do nothing."[52] This is what is called the "precautionary principle," which was established at a UN conference in Rio de Janeiro in 1992—at the precise moment when the noose was tightening around tobacco manufacturers, who decided to call on the chemical industry for help.

Junk Science, or the Sacred Alliance Between Poisoners

It all started with an intolerable "threat" to Philip Morris and its associates that, for once, came from the EPA, which issued a report in 1992 proposing the classification of secondhand smoke as "carcinogenic for humans." For "Big Tobacco," the situation was serious, as Ellen Merlo, the vice president of Philip Morris, stressed in a memo to William Campbell, the president, on January 1, 1993. She proposed the following battle plan: "Our overriding objective is to discredit the EPA report and to get the EPA to adopt a standard for *risk assessment* for all products. Concurrently, it is our objective to prevent states and cities, as well as businesses from passing smoking bans."[53] To achieve these ends, Campbell suggests forming "local coalitions to help us educate the local media, legislators and the public at large about the dangers of 'junk science' and to caution them from taking regulatory steps before fully understanding the costs in both economic and human terms."

No sooner said than done! On May 20, the leading tobacco company and its public relations firm, APCO Associates, launched an organization called The Advancement for Sound Science Coalition (TASSC), in opposition to what it called "junk science." Unbelievably, in its mission statement, the TASSC, which proved to be truly unafraid of ridicule, presented itself as "a not-for-profit coalition advocating the use of sound science in public policy decision making." In order to publicize the coalition, $320,000 was immediately applied toward sending twenty thousand letters to influential politicians, journalists, and scientists. Officially headed by Garrey Carruthers, the Republican governor of New Mexico, the TASSC was careful to hide Philip Morris's involvement, which led to some absurd situations: when Gary Huber, a professor of medicine at the University of Texas who was also a consultant for the cigarette company, received the "letter," he rushed to inform his former employer, believing it might be helpful.

Also omitted from this introductory letter was the fact that, for the purposes of this new brainwashing operation, Philip Morris had allied itself with the Chemical Manufacturers Association, the American association that had already been working for two years on a project to promote "good epidemiological practice" (in the jargon, "GEP"). Quite an unbelievable detail when we know the kind of manipulation the poison manufacturers were capable of! But the matter was even more serious than it appeared, because it would have significant repercussions on scientific practices and reinforce the legendary feebleness of regulatory agencies, which were quite literally harassed by TASSC's representatives. A 1994 letter sent to Philip Morris by attorney Charles Lister (from the firm Covington & Burling, which defended the cigarette companies during the big trials), reveals that "GEP [was] being pushed in Europe by a number of companies, including particularly Monsanto and ICI."[54]

In a very well-researched article devoted to this unbelievable maneuvering, Stanton Glantz (the lucky recipient of the anonymous package from Brown & Williamson) warns "public health professionals": "The 'sound science' movement is not an indigenous effort from within the profession to improve the quality of scientific discourse, but reflects sophisticated public relations campaigns controlled by industry executives and lawyers whose aim is to manipulate the standards of scientific proof to serve the corporate interests of their clients."[55]

Relying on a "so-called scientific orthodoxy"—to use the words of French toxicologist André Cicolella, now the spokesperson for the Environment Health Network (Réseau Environnement Santé, RES) and scientific journalist Dorothée Benoît Browaeys—the TASSC members tried to have any bothersome study eliminated by imposing new toxicological evaluation criteria for chemical products.[56] Among the "fifteen points" meant to illustrate "good epidemiological practice," there is one of which they were particularly fond: they wanted any study that presented results with an odds ratio (OR) below 2 to be discounted as not "statistically significant." As we have seen, this would mean effectively discounting most of the case-control studies carried out on pesticides, but also those on secondhand smoke (where the OR was 1.2 for lung cancer and 1.3 for cardiovascular diseases). Additionally, in an internal document, the TASSC cited the studies on secondhand smoke as an example of "unsound, incomplete or unsubstantiated science."

What's more, industrial lobbyists requested that no restrictive measure targeting a product, that is, its withdrawal from the market, be taken if the

results of animal experiments did not fulfill a condition they viewed as essential: the offending substance's mechanism of action must be "clearly identified
and understood, and the extrapolation from animals to humans verified."[57] To
fully understand the serious consequences that the implementation of such
a demand would generate, let us imagine that a study shows that product X
causes liver cancer in rats. Before deciding to take action, scientists would be
required to describe in precise detail the biological mechanism that leads to
this process of cancer formation, then demonstrate that said mechanism would
function the same way in humans. In other words, the product would have a
few good years ahead of it . . .

But that was not all! While its representatives were fighting to lay down
their laws at regulatory agencies, the TASSC was organizing slander campaigns against all the scientists who, despite the pressure, continued to carry
out their work. Their names were mercilessly broadcast on the website www
.junkscience.com, managed by Steven Milloy, a (very controversial) Fox News
personality, who is one of the leading climate change skeptics today. As early
as 1997, the list of so-called junk scientists contained more than 250 names,
including many of the scientists I met during my investigation, like Devra
Davis.

The movement against so-called junk science has, of course, European
branches, like the European Science and Environmental Movement based in
London, or the "imposters" blog (http://imposteurs.overblog.com) in France,
which, since 2007, has claimed to act "in defense of science and scientific
materialism against all intellectual charlatanisms and imposters." Led by
someone calling himself "Anton Suwalki," the site seems rather aimed at "denigrating scientists and studies whose findings do not serve the corporate cause,"
to cite David Michaels. [58] The new head of the U.S. Occupational Safety and
Health Administration adds: "Big Tobacco showed the way, and today the
manufacture of uncertainty is practiced by entire sectors of industry, because industry understood that the public is in no position to distinguish
good science from bad. Creating doubt, uncertainty, and confusion is good
for business, because it lets you buy time—lots and lots of time."

9

Mercenaries of Science

Science without conscience is the soul's perdition.

—Rabelais

"Honestly, after a career of more than forty years, I can tell you that there are well-done studies and very poorly done studies . . . Generally, studies sponsored by industry have been designed in such a way that it is nearly impossible to find harmful effects. The consequence is that the scientific literature is regularly polluted by worthless studies. It's pathetic." Peter Infante, the former U.S. Occupational Safety and Health Administration (OSHA) epidemiologist, was obviously resentful at our meeting in Washington, DC, in October 2009. He has a lot to say when it comes to listing all the schemes involving scientific ethics that he observed among colleagues working for industry. In transcribing his remarks, which would be confirmed by other experts interviewed, as we will see, I told myself that junk science does indeed exist, but it is promoted and practiced by the very people who invented the less-than-brilliant term.

"Prostituted Science"

"How can studies be designed to avoid inconvenient results?" I asked Peter Infante.

"Unfortunately, there are a number of ways. For example: you want to examine the potential carcinogenic effect of a chemical product to which workers are exposed. In this kind of study, it is very important to choose a good experimental group, or the group of exposed workers, and the control group, or the group of people who haven't been exposed, which will serve as a comparison and so as to measure the potential effect. If you include in the experimental group workers who have not been exposed or, inversely, if you put in the control group people who have been exposed to the substance, you are distorting the result, because in both cases, you will find that there are no or very few differences and conclude that the product does not cause a higher risk of cancer. This is what's called the 'dilution effect,' a well-known bias amongst epidemiologists. Another method is to underestimate the level of exposure to the substance or to mix workers who have different exposure levels. If you mix workers who were highly exposed and are therefore more likely to develop cancer, with workers who were less exposed, once again you are diluting the effect, if not actually making it disappear. This technique is often used to conclude that there is no dose–response relationship and thus that if there is an excess of cancer in a factory it must be due to something other than the suspected product."[1]

Listening to Peter Infante, I remembered my investigation into Monsanto. In that book, I wrote about how Dr. Raymond Suskind—who worked in the Kettering Laboratory, founded by Robert Kehoe (see Chapter 8)—had published three studies in the early 1980s that were rigged in order to refute the carcinogenic effects of the dioxin contained in the herbicide 2,4,5-T (one of the components of Agent Orange).[2] His "trick" was in mixing exposed and unexposed workers in *both* the experimental and control groups. He concluded that there was the same amount of cancer in both groups, so dioxin could therefore be exonerated. What followed was much bleaker: for ten years, regulatory agencies in America and Europe would use the rigged studies as the basis to conclude that dioxin was not carcinogenic. And the Vietnam veterans who had been exposed to the poison would have to wait many long years before obtaining damages and appropriate care.

Everything would seem to indicate that the "dilution effect" is a very common "trick of the trade," as David Michaels established in his book *Doubt Is Their Product*.[3] In President Bill Clinton's administration, he served as the assistant secretary for environment, safety and health in the Department of Energy. There, he worked on cases from nuclear arms factories, where a number of workers were suffering from an often fatal pulmonary disease caused by beryllium exposure (berylliosis). To obtain damages for the victims, he had to fight against the industry, which produced skewed studies in which the workers' exposure levels were poorly estimated. In a chapter entitled "Tricks of the Trade: How Mercenary Scientists Mislead You," Michaels explains that one of the recurring "tricks" used to refute a toxin's hazards consists in choosing a *restricted* cohort of exposed subjects and to study them over a *short* period. The American epidemiologist gives an enlightening il-lustration: "For example, if we know that exposure to a given chemical triples the risk of leukemia, three leukemia cases in a cohort of 100 workers in which only one case would be expected would not likely be statistically significant. We could not rule out chance distribution as the cause of the two excess cases. On the other hand, if the population is 1,000 workers, not 100, and we find thirty cases instead of the expected ten, it is very unlikely that the excess would be attributable to chance. In this case, we would say that the dif-ference between the observed and the expected was 'statistically significant,' and we would consider an alternative hypothesis: The chemical under study is the cause of the leukemia."[4] Michaels concludes: "The devil here is definitely in the details. [. . .] It is easy to see how mercenary risk assessments can be concocted. Change a few parameters that are buried deep in a mathematical model, and a hazardous chemical can be miraculously transformed into one that is not very dangerous at all. [. . .] Scientific research that industry con-ducts or funds is manipulated to *mask* rather than *find* exposure-disease relationships—that is, to protect corporations, not their workers."[5]

"How does industry find scientists willing to carry out biased studies?" I had already asked Devra Davis this question—which kept nagging me throughout my investigation—the day before meeting Peter Infante. She an-swered with a wide smile: "Imagine you're the director of a laboratory, and they come and offer you several millions of dollars to conduct a study. On top of that, they tell you you're the best and the most handsome! What do you do? Many are flattered and all too happy to cash in that sort of jackpot in such hard times for research. After that, it's like dominoes."[6]

Peter Infante's answer was even more direct: "How does industry find sci-entists to do this kind of task? It buys them, that's all! Let's be clear—it's what I call 'prostituted science' . . . The problem is that biased studies are then sent to regulatory agencies, who take them at face value. That's how highly toxic substances have been contaminating our environment, our food, our fields or our factories, for decades. That's what happened with benzene, a case I personally monitored at OSHA. And at the end of the day, it caused a lot of deaths and victims that could have been avoided."

Dow Chemical Hides Its Data on Benzene

"Risk assessment data can be like a captured spy: if you torture it long enough, it will tell you anything you want to know."[7] As brutal as he may sound, William Ruckelshaus, who was the first Environmental Protection Agency (EPA) director, rather effectively sums up the history of benzene regulation, a subject with which Peter Infante is quite familiar. I previously mentioned this omnipresent molecule—which is used as a solvent in the chemical synthesis of plastics, rubbers, paints, and pesticides, or as an additive in gasoline—when detailing the case of Dominique Marchal, the farmer stricken with a myeloproliferative disorder who eventually obtained occupational illness recognition (see Chapter 4). I explained that the links between this "new domestic poison"—to use *The Lancet*'s terminology from 1862— and leukemia had already been the subject of fifty-four scientific studies by 1939, which were inventoried in an article in the *Journal of Industrial Hygiene and Toxicology*.[8] As Paul Blanc explains in *How Everyday Products Make People Sick*, "After this publication, one would have difficulty arguing that the failure to control benzene might be attributable to a lack of sufficient hard scientific data."[9] And yet, it changed nothing! Benzene continued to be used extensively in American and European factories with a few recommendations—at most—that shifted the responsibility of protection onto the workers themselves. Blanc states in his book that in 1941, the U.S. Public Health Service distributed a prevention pamphlet for laborers and artisans working in contact with benzene. It tells the story of "Clara," a young woman working in a shoe factory, where she attaches soles with a benzol-based adhesive.[10] "A little care will keep benzol in its place—and you on the job," the authors write, not saying a single word about the risks benzene exposure presents. "Clara is one of about 30,000 American workers whose job calls for the use of benzol in some form. Thousands more are employed to manufacture this valuable solvent. A lot of people would be out of work if there were no benzol."[11]

I must admit that I felt quite revolted at numerous times during this investigation, namely at the disgraceful cynicism displayed by manufacturers and politicians. That said, the benzene affair exceeds the limits of tolerability. In 1948, the American Petroleum Institute (API)—the hydrocarbon equivalent of the Tobacco Industry Research Committee—commissioned a summary of the "best available information on the development of leukemia as a result of chronic benzene exposure" from Professor Philip Drinker at Harvard's School of Public Health. After listing all of the irreversible illnesses brought on by acute or chronic benzene poisoning, the scientist concludes: "Inasmuch as the body develops no tolerance to benzene, and there is a wide variation in individual susceptibility, it is generally considered that *the only absolutely safe concentration for benzene is zero.*"[12] In other words: the only way to protect against hydrocarbons is to prohibit them completely.

But this report would change nothing in the behavior of manufacturers, who arbitrarily decided that the standard for benzene exposure in factories, over an eight-hour workday, was an air concentration level less than 10 ppm (parts per million). And it would take until the creation of OSHA in the early 1970s for American public authorities to finally decide to review the case. At that same time, in Europe, and of course in France, inaction was the rule (rather than the exception), because back then, as I have explained, America was paving the way. "When my director, Eula Bingham, assigned me to benzene regulation, I was very enthusiastic," Peter Infante told me. "We were convinced that the exposure standard set by the industry needed to be lowered considerably, but I didn't know how difficult that was going to be."

In a 2006 article that appeared in the *International Journal of Occupational and Environmental Health*, the epidemiologist reported all the obstacles erected by manufacturers to derail the regulation project. The manufacturers did not hesitate to hide data on benzene toxicity obtained from research carried out in their own labs,[13] in total violation of the law. It was thus discovered that Dow Chemical[14] had concealed a study demonstrating that exposure at a level under 10 ppm caused chromosomal damage in its workers. What's worse, the company had forbidden its researcher, Dante Picciano, from publishing his data or sending them to OSHA. "He was so disgusted that he contacted me," Peter Infante told me. "Eventually, he resigned and braved the threats by publishing his study in 1979."[15]

But the OSHA epidemiologist was not out of the woods quite yet. Even as he was battling Dow Chemical, Infante was completing a study meant to curtail all the industry's waffling. It was carried out in two Goodyear Tire & Rubber factories where synthetic rubber (Piofilm) was produced: twelve hundred workers who were exposed to benzene from 1940 to 1949 were monitored until 1975. The results were that much more impressive because they showed an unequivocal dose–response relationship. Workers who had been exposed for one to four years showed a risk of leukemia that was twice as high as that of the control group; the risk was multiplied by fourteen when exposure lasted five to nine years and by thirty-three when exposure lasted for more than ten years.

"As a result of past failure to control benzene as a carcinogen, millions of people, without knowledge of the haemopoietic dangers, are continually being exposed to benzene at work," Infante and his colleagues write. "We hope that our findings, which demonstrate overwhelmingly an increased risk of leukaemia in workers exposed to benzene, will stimulate efforts to control [. . .] an agent known for almost a century to be a powerful bone-marrow poison."[16] The tone of these "conclusions" is clearly different from the generally more subdued character of scientific publications, but it reflects the researchers' emotions when faced with what should be called a "health disaster" of epic proportions. Convinced that urgent action was needed, OSHA decided to announce a new exposure standard for benzene in 1977 and set it at 1 ppm—ten times lower than what was (theoretically) being used by manufacturers.

Alas, the API went to the Supreme Court, which overturned the decision on July 2, 1980. In its seventy-five-page judgment, the venerable institution explained that it refused to "enforce the 1 ppm exposure limit on the ground that it was not supported by appropriate findings," and that OSHA did not show that this new exposure limit was "reasonably necessary or appropriate to provide safe or healthful employment."[17] According to the court, the OSHA researchers did not show how the new recommended standard would be likely to better protect workers than the "consensus standard" of 10 ppm. Indeed, the distinguished judges dared evoke a "consensus standard"! Quite the paradox, when we know that manufacturers imposed said "standard" completely arbitrarily and did not have to present a single study to justify it!

Historically known as the "benzene decision," the Supreme Court's judgment prompted extensive press coverage. Ultimately, it rewarded a practice that would characterize chemical risk management throughout the twentieth century, that is, that in the field of environmental health the responsibility of proof belongs to the public powers rather than to industry. It then falls upon the "plaintiffs"—in other words, regulatory agencies or presumed victims—to demonstrate the toxicity of a product, and not upon its makers to prove that it is harmless.[18] In *The Secret History of the War on Cancer*, Devra Davis points out that in the case of benzene, "the court insisted that sufficient numbers of sick or dead workers had to be assembled to provide proof that harm had already happened before allowing [OSHA] to act to prevent further harms."[19]

Industry Mercenaries

For Peter Infante, the benzene judgment meant that he had to try even harder, so he returned to the two Goodyear factories. "My colleague Robert Rinsky and I had to establish what's called a 'job-exposure matrix,'" he explained. "Since the workers we studied had worked over thirty years ago, we had to extrapolate the exposure levels from what we knew about the manufacturing processes at the time, since the factories had of course not recorded that kind of information. It was a tremendous amount of work, which allowed the industry to gain seven years." It is easy to understand how infuriating that effort must have been, especially considering OSHA's limited resources; according to the organization's new administrator, David Michaels, the staff size allows for inspections of every business in the United States once every 133 years! In the end, Infante's new study confirmed that the closer the daily exposure level was to zero, the fewer cases of leukemia there were, and that the risk could potentially increase by a factor of sixty when the exposure level rose above 10 ppm.[20] Relying on these results OSHA announced a new standard in 1987, when the World Health Organization's International Agency for Research on Cancer declared benzene "carcinogenic for humans."[21]

But the story does not end there; manufacturers were already preparing for the next battle, which involved very low doses—lower than 1 ppm—found in the environment, after the spraying of pesticides, for example, or in the areas surrounding service stations, where samples have shown that the air contained between 0.17 and 6.59 ppm of benzene.[22] After all, industry is in the position to know about such data—hadn't the Harvard scientist it discreetly consulted in 1948 concluded that "the only absolutely safe concentration for benzene is zero"? For that reason, the API contacted Dennis Paustenbach, a toxicologist working for Exponent, a consulting firm specializing in what David Michaels calls "science for hire."[23] The firm's 2003 annual report plainly states: "Many of our engagements are initiated by lawyers or insurance companies, whose clients anticipate, or are engaged in, litigation over an alleged failure of their products, equipment or services." The report also lists all of the sectors with which the firm is involved: "Exponent serves clients in automotive, aviation, chemical, construction, energy, government, health, insurance, manufacturing, technology and other sectors of the economy."[24]

Before introducing the infamous Dennis Paustenbach, who is considered one of the most "talented" "mercenaries of science," it is worth noting that Exponent and its American competitors Hill and Knowlton and the Weinberg Group—who all have European branches—are products of the criminal and deceitful maneuvering of poison manufacturers. These firms owe their existence to what some American researchers describe, in an article entitled "Maximizing Profit and Endangering Health," as a process of criminalization of industrial activity, which had to develop increasingly sophisticated strategies to "avoid regulation and liability."[25] As Dr. David Egilman and his co-authors note, these "strategies" are not a paranoid rant born of a new "conspiracy theory," but a reality brought to light thanks to the "dissemination of previously secret industry documents produced in toxic tort litigation," which revealed that "the actions of industry have been both deliberate and malign." What's more, these researchers insist that this is indeed systematic, and not a series of isolated "bad apple" incidents: "Over the course of several decades, corporations and industries have developed and refined scientific, legal, and public relations tactics to maintain their ability to make profits despite the dangers posed by their products. Viewed together, these tactics take the shape of a strategy that, although enacted differently by various industries and corporations, has enough commonality to be understood as part of the *modus operandi* of at least a large proportion of corporations in the United States." And I would add in Europe as well, because, while the model was developed in the United States, the Old World is no different—in both the globalization of modern capitalistic structures and its ideology. The authors specify: "The strategy is meant to achieve two main goals: 1) secure the least restrictive possible regulatory environment; and 2) avoid legal liability for worker or consumer deaths or injuries."

To achieve their goals, the multinational corporations work in close collaboration with businesses specializing in this sort of task, like Exponent, which states its mission as developing a panoply of recurring "tactics":

1. Conducting or hiring outside scientists to conduct research designed to show the "safety" of a particular process or product; generate controversy about its effects; and mount attacks on scientists and on scientific work that shows the dangers of the product or process.
2. Organizing in groups of industry-friendly "third-party" scientists to support industry's scientific positions in regulation-setting processes, the courtroom, and public opinion; these are frequently dubbed "scientific advisory boards" (SABs).
3. Creating and/or utilizing "front" groups, industry organizations, and think tanks to provide an appearance of legitimacy and/or to further objectives.
4. Utilizing and influencing the media to sway popular opinions.[26]

Science plays an essential role in this relentless apparatus, which, as we will soon see, managed to infiltrate the agencies charged with our safety, like the U.S. Food and Drug Administration (FDA) or the European Food Safety Authority (EFSA). And unfortunately, it is not hard to find scientists willing to put their talents and knowledge at the service of this "illegal conspiracy," to quote American historians Gerald Markowitz and David Rosner.[27] This was the case with Dennis Paustenbach, whose career "is illustrative of the problems that arise when research is conducted to specification."[28]

Known for his fearlessness (to put it simply), the toxicologist won fame for his inveterate defense of dioxin during the big environmental scandals with Times Beach and Love Canal.[29] But his name would ultimately remain associated with Hinkley, California, a small town contaminated by hexavalent chromium[30] whose misfortunes inspired the film *Erin Brockovich* (2000), directed by Steven Soderbergh. In 1996, the story's heroine—played in the film by Julia Roberts—who was working in a law office, managed to secure a conviction against Pacific Gas & Electric Company, which was proven responsible for the pollution of potable water. To prepare for this mega-lawsuit, wherein approximately 660 victims obtained $330 million in damages, the energy company called upon Dennis Paustenbach for help, who was the director for the ChemRisk firm at the time. His mission was to counteract a 1987 Chinese study showing that water and soil pollution by chromium VI caused cancer.[31] The affair was all the more urgent since the EPA had utilized the same study in order to demand the decontamination of a waste site in New Jersey. Not a problem for Paustenbach, however, who decided to contact Dr. Jian Dong Zhang, the study's author, who for $2,000 agreed to reinterpret his data and publish the new "results" in the American *Journal of Occupational and Environmental Medicine*.[32] This falsified study, considered a reference for nearly ten years, was used by industry in a number of lawsuits involving hexavalent chromium—until the *Wall Street Journal* discovered the truth,[33] which led to an official retraction of the article by the journal that had published it.[34]

Even as exposure to low doses of benzene was attracting a lot of attention, the ineffable Dennis Paustenbach was contacted by the API. In 1997, a study carried out in Chinese factories by the U.S. National Cancer Institute and the Chinese Academy of Preventative Medicine showed that the risk of leukemia was two times higher there than what had been observed by Peter Infante's team.[35] It was difficult to attack this research, because China is the ideal terrain for epidemiologists: exposure levels at every work station are recorded in minute detail, and workers can be monitored for a long time since their professional mobility is nearly zero.

To perpetuate doubt, the API asked Dennis Paustenbach to reexamine the exposure values originally estimated by Peter Infante and his colleagues in the two Goodyear factories. Note that in order to carry out their evaluation, the OSHA epidemiologists had to extrapolate from a reconstruction of production processes over the course of the 1940s and 1950s. Paustenbach's trick consisted in systematically reevaluating (and overestimating) exposure levels for various work stations, so as to conclude that only levels above 10 ppm caused leukemia.[36] As David Michaels points out, "in the regulatory arena, the studies [of this type] are useful [for industry] not because they are good work that the regulatory agencies have to take seriously but because they clog the machinery and slow down the process."[37]

The momentum worked up by manufacturers to defend their products tooth and nail, without ever taking into account—not even in the slightest— the horrible repercussions their tenacity might produce is quite fascinating. I chose to pull apart the history of benzene in particular because it exemplifies the relentless machine that places short-term profits before any other consideration, including the death or illness of thousands of innocent victims. Whether it's Monsanto, Dow Chemical, DuPont de Nemours, BASF, or Saint-Gobain, the businesses never loosen their grip, even if it means spending a fortune to "perpetuate doubt." This is fascinating, but also very worrying: Who could imagine such a plethora of "deliberate and malign" measures?[38] Someone who manages to put the puzzle pieces together risks being accused of acute paranoia, or even of brandishing a new "conspiracy theory"—an argument that manufacturing representatives unfailingly voice as soon as some wise guy manages to unmask their numerous "ploys." And therein lies the companies' strength: with never-ending double-talk, they have managed to pull all the strings in the regulation game, thanks to techniques of systematic "deceit and denial,"[39] which are hard to detect because they are literally "unimaginable."

The (provisional) end of the benzene affair is proof of this, if it is even needed. After Dennis Paustenbach's pseudo-study was published in 2003, a new publication reignited the manipulation machine: in 2004, *Science* published the supplementary results of the vast investigation carried out in Chinese factories. The article, titled "A Little Is Still Too Much," reported that examinations of workers exposed to benzene at levels below 1 ppm showed changes in their white blood cells and platelets.[40] This time, the API rolled out its heavy artillery by putting $22 million—pocket change—on the table, split between the different oil companies according to the number of barrels they produced.[41] The goal was to finance a new study in China that would invalidate the disastrous results of the first one. It was laid out in black and white in a secret document drafted by Craig Parker, an executive at Marathon Oil, which David Michaels managed to obtain: "Should the toxic effects of low-level benzene exposure reported by the original China study become widely accepted by regulators, calls would soon follow for the reformulation of gasoline, for control of emissions from refineries and marketing facilities, and for the clean-up of contamination. A nightmare for the industry. And then there is litigation."[42] In the memo, Parker clearly states the goal of the "research"(!): "Provide strong scientific support for the lack of a risk of leukemia or other hematological disease at current ambient benzene concentrations to the general population. Establish that adherence to current occupational exposure limits (in the range of 1–5 ppm) do not create a significant risk for workers exposed to benzene."

Outcry About Conflicts of Interest

For now, the results of this "study" have not yet been published, but it is highly likely that they will conform to the objectives listed above. It will be interesting to see if the sponsors appear in the publication, which is rather rare. In fact, as Susanna Rankin Bohme and her co-authors point out, until the early 2000s, "mercenaries'" conflicts of interest were never mentioned, and "their work was published in the scientific literature, without disclosure that the research had been conducted with a foregone conclusion and had been subject to editing by a task force of industry representatives."[43]

The first to denounce this very common incongruence, which casts doubt on the quality of articles published by scientific journals, was Arnold Relman, the editor of the well-respected *New England Journal of Medicine*. In 1985, he published an editorial—which dropped like a bomb since the subject was still very taboo at the time—in which he denounced the entrepreneurialism that had become "rampant in medicine": "It has long been common practice for manufacturers of pharmaceuticals and medical devices to retain the services of academic scientists as consultants or to subsidize their research studies," he writes. "But in recent years, as the commercial possibilities of new biomedical discoveries have become increasingly attractive, these connections have become more pervasive, complex, and problematic."[44] To overcome this downward spiral, he suggested requiring authors who propose scientific articles to declare their potential conflicts of interest and connections to the industry concerned by their studies. His proposition, which initially targeted studies on clinical trials for new medications before being expanded to all sectors of biomedical research, was accepted by the *New England Journal of Medicine* and adapted by thirteen major scientific journals in 2001. In a joint statement, their editorial directors pointed out that "financial relationships (such as employment, consultancies, stock ownership, honoraria, paid expert testimony) are the most easily identifiable conflicts of interest and the most likely to undermine the credibility of the journal, the authors, and of science itself. Conflicts can occur for other reasons, however, such as personal and family relationships, academic competition, and intellectual passion."[45] Since this noteworthy declaration of faith appeared, authors have been obligated to fill out a form disclosing their conflicts of interest, which they must send with their article when they submit it for potential publication in one of the thirteen journals in the pact.

This is unarguably a welcome initiative, even if it does only concern a minority of scientific journals. But, as the Center for Science in the Public Interest (CSPI)[46] stresses, "a conflict of interest disclosure policy is only as good as its enforcement"—because ultimately nothing can force authors to declare their connections with industry if no mechanism of control is put in place to verify that this requirement is being respected. To this effect, the CSPI proceeded with an investigation in 2004 into four of the signatory journals of the "pact" that were known for being particularly vigilant about conflicts of interest (the *New England Journal of Medicine*, *Journal of the American Medical Association* [*JAMA*], *Environmental Health Perspectives*, and *Toxicology and Applied Pharmacology*). To do so, the center examined the 176 articles published between December 2003 and February 2004, 21.6 percent of which had to do with studies financed by industry (40.8 percent of these were in the *New England Journal of Medicine* and 5.4 percent in *Environmental Health Perspectives*). In 163 articles, the authors declared that they had no conflicts of interest; however, in looking more closely at the profiles of the first and last authors cited, in their references, the CSPI noted that, in thirteen articles (8 percent), the researchers had "omitted" declarations of their links to industry.[47] Among the examples cited in the report was that of William Owens, a scientist from Procter & Gamble who had no qualms presenting himself as a representative of the Organisation for Economic Co-operation and Development (OECD) while extolling the virtues of a toxicity test promoted by his employer. In conclusion, the CSPI recommended that "Journal editors should adopt strong sanctions for failure to disclose conflicts of interest, such as a three-year ban on publication within the pages of that journal should an undisclosed conflict of interest be brought to light. The threat of sanctions could improve compliance in this unregulated field."[48]

Still, while observers agreed to recognize that the declaration of conflicts of interest was a "first minimal first step,"[49] they also stressed that this was not a panacea, since knowledge that an author is financially connected to the industry that oversaw his study in no way resolves the problem of "bias" that such a connection might entail. "Full disclosure is considered an important method for reporting and managing conflicts of interest and serves to highlight the potential for bias, but cannot and does not eliminate the conflicts," notes Catherine DeAngelis, former editor in chief of *JAMA*, which receives some 6,000 articles every year.[50,51] She adds, "I am not the FBI. . . . I have no ability to know what is in the minds, hearts, or souls of authors."[52] Indeed, as we have seen, there are many kinds of "bias" frequently found in studies sponsored by industry: protocol designed so as to avoid inconvenient results; "rigged" selection of experimental groups and control groups; or selective interpretation of results. To detect possible bias, *JAMA* took another step in 2005 by requiring that the scientist collecting the study's data and the scientist analyzing these data be two different people; and, above all, that the second not be "employed by the company sponsoring the research."[53]

In an article published a year after the implementation of this new requirement, Catherine DeAngelis, then *JAMA*'s editor in chief, reported a certain number of "irregularities involving for-profit companies, such as the refusal to provide all study data to the study team, reporting only six months of data in a trial designed to have twelve months of data as the primary outcome; incomplete reporting of serious adverse events; and concealing clinical trial data showing harm." She later specifies: "For-profit companies also can exert inappropriate influence in research via control of study data and statistical analysis, ghostwriting, managing all or most aspects of manuscript preparation, and dictating to investigators the journals to which they should submit their manuscripts. For example, I have been told that in response to *JAMA*'s policy requiring an independent statistical analysis by an academician for industry-sponsored studies in which the only statistician who analyzed the data is employed by the study sponsor, some companies are insisting that the researchers not submit those studies to *JAMA*. That tactic risks not only the perception that the company may have something to hide, but the reputation of any researcher willing to accede to such a company demand."[54]

Industry's Hold on Academia

Furthermore, potential conflicts of interest concern not only the authors of publications, but also reviewers, or the people who read proposed manuscripts. At large scientific journals with "peer-review"—which is universally considered to be assurance of a publication's reliability—manuscripts are submitted to peer evaluation, or reviewers whose identities are kept secret to (theoretically) avoid any sort of pressure. Generally there are three peer evaluators who are chosen in keeping with their expertise, most often from the academic arena. As David Michaels very fairly notes, "With the increased involvement of universities in commercial enterprises and collaborations, conflicts-of-interest concerns at academic institutions have grown in importance."[55] Accordingly, a systematic review of studies put online by MedLine between January 1980 and October 2002 showed that "approximately one fourth of investigators have industry affiliations, and roughly two-thirds of academic institutions hold equity in start-ups that sponsor research performed at the same institutions." The authors conclude, "evidence suggests that the financial ties that intertwine industry, investigators, and academic institutions can influence the research process."[56]

Conscious of the fact that affiliation with a university or an academic institution is no longer a guarantee of independence, the British journal *The Lancet* decided in 2003 that it would no longer entrust its valuable "readings" to academics exhibiting "substantial financial interests." With a directness rare in the domain of scientific publishing, which is generally more consensus-based, the venerable journal states: "We have taken this stance because academics have a choice—to develop their entrepreneurial skills or to maintain a commitment to public-interest science—and we do not accept that the two options are mutually compatible."[57]

Finally—and this is a "detail" of particular consequence to consumers like us—conflicts of interest are not taken into consideration by regulatory agencies such as the FDA or the EFSA in evaluating the reliability of the research upon which their decisions depend. As we will soon see, although these agencies have recently begun to require their experts to disclose their conflicts of interest, nothing similar is asked of study authors, "even though such disclosures are not only within their authority, but central to their mission," as law professors Wendy Wagner and Thomas McGarity note. They state that "regulators should also require authors of research submitted for regulatory consideration to share the underlying data collected in a study."[58] Yet this is very rarely the case, as agencies are typically content with basing their decisions on an information summary provided by industry laboratories. In a 2003 *Science* article, David Michaels laments that "the quality and independence of private research used for regulation is subject to considerably less oversight than corresponding federally funded research. Most significantly, private research submitted for regulatory purposes escapes external scrutiny if the research or the chemical under study is claimed to be confidential business information."[59] We will see that this is the case for pesticides (see Chapter 10), but also for genetically modified organisms.[60]

To be perfectly clear about this aberration: companies refuse to submit their toxicology studies' raw data to any independent body, whether it be an association or a university research laboratory, even as they play with the health of millions of consumers, arguing that they are protected by "trade secrets"! If they have nothing to hide and are sure their products are innocuous, we have the right to wonder why that is so, and to suspect that the data in question is somewhat problematic.

To conclude this critical section—which elucidates the context in which the poisons contaminating our food chain are regulated (see Chapter 11)—I will cite, once again, David Michaels: "I am convinced that conflict of interest cannot be 'managed.' It must be eliminated. Too much is at stake," the new OSHA head writes. "The pressures on scientists who receive corporate money are too great. Even with contracts that forbid the sponsor's control of full disclosure, the fear of losing the next contract will limit true scientific independence. I prefer a system in which research and testing are carried out with true independence. Any study desired by (or required of) industry would be paid for by the industry but conducted by independent researchers, under federal auspices. Subsequent publication would be completely independent of the sponsoring corporations. [. . .] Those who oppose regulation would doubtless view such a system as a nightmare. But regulation that protects the public's health and the environment must be based on the best available science, and the best science is science done by independent investigators."[61]

In the meantime, one thing is certain: the multiple tactics used by manufacturers to conceal the toxicity of their products have been fruitful because, as we will see, the poison-producers' lies are regularly disseminated by powerful academic or government institutions that, to put it quite bluntly, are very easily blinded.

10

Institutional Lies

If you want the present to be different from the past, study the past.

—Spinoza

Dear Mr. President,

In 2009 alone, approximately 1.5 million American men, women, and children were diagnosed with cancer, and 562,000 died from the disease. [. . .] The Panel was particularly concerned to find that the true burden of environmentally induced cancer has been grossly underestimated. With nearly 80,000 chemicals on the market in the United States, many of which are used by millions of Americans in their daily lives and are un- or understudied and largely unregulated, exposure to potential environmental carcinogens is widespread. [. . .] The American people—even before they are born—are bombarded continually with myriad combinations of these dangerous exposures. The Panel urges you most strongly to use the power of your office to remove the carcinogens and other toxins from our food, water, and air that needlessly increase health care costs, cripple our Nation's productivity, and devastate American lives.

Addressed to the president of the United States, Barack Obama, this letter was not written by a Greenpeace militant or an obscure ecological organization, but by Drs. LaSalle Lefall and Margaret Kripke, who chaired the 2008–2009 "President's Cancer Panel" (PCP).

Since Richard Nixon launched the panel in 1971 when he declared "war on cancer," the PCP has become a veritable institution that annually evaluates the infamous (forty-year-long!) "war" in a hefty report written under the auspices of the National Institutes of Health and the National Cancer Institute. It should be noted that the 2010 breakdown, entitled *Reducing Environmental Cancer Risk: What We Can Do Now,*[1] has the merit of being unabashedly straightforward in regards to all those working toward organized disinformation. This letter represents the first time that the PCP broke with its well-developed discourse that invariably attributed primary responsibility for the upsurge in cancer to smoking, alcoholism, sedentary lifestyles, and other poor habits, and instead focused on environmental factors. To this end, the PCP brought together forty-five experts, all "from academia, government, industry, the environmental and cancer advocacy communities, and the public," to deal with four subjects: "Industrial and Occupational Exposures," "Agricultural Exposures," "Indoor/Outdoor Air Pollution and Water Contamination" and "Nuclear Fallout, Electromagnetic Fields, and Radiation Exposure." The report's conclusions are incontestable: if we want to reduce the "burden of cancer," we must, as a priority, attack these causes first, at the risk of turning the "war on cancer" into a scene from *Waiting for Godot*—as ridiculous as it is inefficient.

The Causes of Cancer in France (2007): A Report That "Should Not Be Taken Seriously"

Reading the PCP report on cancer was a great relief. To read that "scientific evidence on individual and multiple environmental exposure effects on disease initiation and outcomes, and consequent health system and societal costs, are not being adequately integrated into national policy decisions and strategies for disease prevention, health care access, and health system reform"—words penned by official reviewers—was particularly reassuring since three years earlier, another equally "official" report—French this time—said exactly the opposite! Entitled *Les Causes du cancer en France* (The Causes of Cancer in France), the text was written by the prestigious Academy of Sciences (Académie des sciences, AS) and the National Academy of Medicine (Académie nationale de medicine, ANM), in collaboration with the International Agency for Research on Cancer (IARC), which is part of the World Health Organization (WHO).[2]

I will never forget that morning in September 2007 when radio stations seemingly everywhere proclaimed the "good news," freely citing what is now acknowledged as one of the worst frauds in recent scientific history. "This report states that in France (as in all industrialized countries and the majority of developing countries) tobacco, on the eve of the 21st century, remains the principal cause of cancer (29,000 deaths, or 33.5% of cancer-related deaths in men, and 5,500 deaths, or 10% of cancer-related deaths in women). [. . .] *Contrary to certain allegations, the proportion of cancer related to water, air and food pollution is low in France, to the order of 0.5%; it could reach 0.85% if the effects of pollution in the atmosphere can be confirmed.*"[3] Is that so? According to the esteemed "experts," only 0.5 percent of cancer is due to chemical pollution, which would constitute a sort of "French exception" the entire world should envy, and which somehow went completely undetected! The entire report adopts the same tone, with some quite dubious selections: "The westernization of lifestyle comes with other changes that seem to be hormonal in nature: a considerable increase in height (in France, 4 to 6 inches since 1938) and in shoe size, a lower age of first menses (in France, this occurs around two years earlier than in 1950). It is plausible to consider stimulation of cell growth rhythm by hormones or nutrients in Western type foods, or [by] the greater abundance in caloric intake of children and pregnant women, which would explain the correlation that has been reported between the size of newborns and the risk of breast cancer later in life."[4]

Clearly, the report deals with the question of pesticides by making such peremptory decisions that the Crop Protection Industry Association (Union des industries de la protection des plantes, UIPP) rushed to put the scholars' evaluation on its website, as evidence that "no supported scientific result would allow to conclude today that there exists a proven and significant connection between cancer and plant protection products."[5] The report reads: "Many pesticides have been accused of causing cancer in humans, but *no currently used pesticide is carcinogenic in animals or humans.* A few case-control studies showing an association between exposure and cancer have been published, but these results are likely due to a number of factors: i) because of the large number of studies carried out, it is normal that some studies are positive, as a result of statistical fluctuations; ii) subjects suffering from cancer may have memory bias, with a tendency to remember exposure that healthy subjects have forgotten. [. . .] In conclusion, *the reputed connection between pesticides and cancer does not rest on any sound information.*"[6]

Whereas the "French exception" went undetected across the Atlantic, the 2007 academic report, on the other hand, did elicit some snickers and snide remarks. Richard Clapp, an American epidemiologist specializing in public health and who collaborated with the PCP, reassured me when I met him in his office at the University of Boston: "I think the report was invalidated and shouldn't be taken seriously. It seems as if the authors didn't have access to all of the scientific literature available, or that they misinterpreted it."

"But how do you explain that institutions as prestigious as the French Academies of Medicine and Sciences continue to deny the link between chemical products and cancer?" I asked.

"There should be a closer focus on the relationships certain representatives of these institutions have with industry," the scientist replied up front. "In the United States, we have a saying for it: just follow the money."[7]

Academies Under the Influence: The Case of Dioxins and Asbestos

I admit, the above accusation is serious, and it would doubtless take an entire book to verify the origins of the two famed institutions' finances. What can be said, however, is that they have always maintained very close relationships with the manufacturers in their respective sectors, to the point of regularly being blinded by the industry's interests and lies. For proof, read the somewhat humiliating chapter on dioxin in André Cicolella and Dorothée Benoît Browaeys's book *Alertes santé* (Health Alerts). In 1994, the AS and its applications committee (Comité des applications de L'académie des sciences, CADAS) "were responsible for a report, of which no copies are now circulating: it is no longer available from the publisher, nor is it archived, or even mentioned, on the AS's website, which is curiously limited to reports published before 1996."[8] I tried in vain to consult this document online, entitled *Dioxin and Its Analogues*. And I understand the retrospective embarrassment felt by its authors, who in 1994 stated with incredible confidence: "With current knowledge and considering the low quantities at play, we have the methods of identifying and controlling the risks connected with dioxins, [which] pose no major problems for public health."[9]

André Cicolella and Dorothée Benoît Browaeys also report that André Picot, the toxicologist who took part in the Ruffec meeting (see Chapter 1), was asked to participate in the CADAS work group, a fact he confirmed at one of our meetings in Paris. With his colleague Anne-Christine Macherey, he submitted a contribution in which he wrote: "There exists a group of information that unambiguously establishes the immunotoxic character of dioxins. The fact that these compounds exert their harmful effects at very low doses of course leads one to consider that taking into account this aspect of toxicity is absolutely essential in the evaluation of the risk they can cause for public health."

"CADAS refused to include my contribution in the Academy's report," Picot explained when we met in Paris in June 2009. "And this is not surprising, since most of the members in the work group came from chemical manufacturers, such as Rhône-Poulenc or Atochem."[10]

Meanwhile, as the authors of *Alertes santé* underline, the report served to justify the current inaction and to appease worried parties, as demonstrated in a reassuring memorandum from the Ministry of the Environment on "emissions of dioxins in the atmosphere and the presence of these pollutants in the environment," which was sent to prefects in May 1997.[11] Therein lies the fundamental "virtue" of this type of report that manufacturers cherish: haloed by the AS's seal, they are regularly cited in official documents, press articles, and judicial proceedings, even if their contents turn out to be completely erroneous. Three years after the publication of the AS's report, IARC declared dioxin "carcinogenic to humans";[12] and seven years later, the Stockholm Convention of May 22, 2001, included the modern-day poison on its list of persistent organic pollutants, or "POPs," that should be urgently eliminated.

As for the ANM, it stands out for a report it published on asbestos in 1996, in which its brave experts worked to minimize the dangers of "passive exposure" to the substance, though it had been classified as "carcinogenic to humans" by the IARC since 1987. I will not enter into the details of the well-known history of asbestos, or "white gold," which was used and abused and continues to do damage in developing countries, whereas its connection to mesothelioma, a very rare form of cancer mainly affecting the pleura, has been known since the beginning of the twentieth century, and duly documented since the 1930s.[13] I will simply note that in 1982, the French company Saint-Gobain and its Swiss counterpart Éternit established the Permanent Committee on Asbestos (Comité permanent amiante, CPA), the model for which was directly inspired by the Tobacco Industry Research Committee, created in 1953, as we saw in Chapter 8, by American cigarette companies. Bringing together manufacturers, senior officials from numerous ministries (Health, Environment, Industry, Labor, Housing, Transportation), union organizers, doctors, and representatives of public research, the CPA embodied an "absolute scientific fraud," according to journalist and writer Frédéric Denhez, who points out: "As the only state spokesperson on the asbestos problem, the CPA has managed for years to drown decision-makers and journalists in a flood of well-crafted documents that very skillfully present a ban on asbestos as unthinkable, in favoring rather 'controlled use.'"[14]

As it happens, I modestly contributed to this vast manipulative operation, despite my current views. As a freshly minted journalist, I occasionally worked for an agency specializing in business news, which was how, in the late 1980s, I came to report on Everit's internal communication branch (Everit is a subsidiary of Saint-Gobain, primarily a manufacturer of slate and asbestos steel sheets). I visited their factories in Dammarie-les-Lys and Descartes many times,[15] as well as those in Manizales, Columbia, where I was meant to observe the safety measures implemented by the company in order to protect its workers from the harmful (deadly) effects of asbestos. I remember an interview I did with a scientific director on the CPA, who learnedly explained to me that if the concentration of asbestos fibers per meter cubed of air did not exceed a certain threshold, then exposure carried no risk. As proof of this, he cited scientific studies that were "above suspicion," which I naturally cited in my article. Indeed, it would have been difficult to imagine the lies and manipulations of the leading experts recruited by the CPA to "supply an incontestable scientific guarantee," to quote a scathing report that a French Senate information committee would write in 2005.[16]

The fact remains that a few months before the prohibition of the "miracle mineral" in France on January 1, 1997 (twenty years after the United States!), the ANM produced a report under the aegis of Professor Étienne Fournier, then president of the High Council for the Prevention of Occupational Risk (Conseil supérieur de la prevention des risques profesionnels, CSPRT).[17] Even though mesothelioma is extremely rare outside of cases clearly caused by exposure to asbestos—to the point that it is commonly known as "asbestos cancer"—the report maintains, without citing its sources, that "25–30% of current mesothelioma cases are not attached to any identifiable cause and have no scientifically demonstrated relation to asbestos. [. . .] Tobacco use remains the main, if not exclusive, cause of lung cancer of exogenous cause, even in those who currently work with asbestos, and public health officials should not mistake their targets in their recommendations."[18] The author then embarks on a chaotic demonstration in strict conformity to the asbestos lobby's theories: "Media-centered publications indicate figures of several tens of thousands [of deaths] by adding probable cases accumulated over thirty years. In the same timespan, 18 million French citizens were killed by other causes (300,000 on highways, 1 million from tobacco-caused lung cancer), and the number of mesothelioma cases that are not explained by earlier, massive and prolonged occupational exposure, is and will remain too low to separate spontaneous mesothelioma from mesothelioma due to low asbestos levels in the air."

My only response—as is usually said in cases of proven dishonesty—is "No comment." It should simply be noted that a nota bene, found in the report's publication under a list of experts who participated in the work group, pointed out that "J. Bignon, P. Brochard, and J.-C. Laforest are participating in a committee at INSERM [French National Institute of Health and Medical Research] on the subject and do not wish to be co-signatories of the report after its adoption by the National Academy of Medicine." And for good cause: on July 2, 1996, the aforementioned committee submitted their first conclusions to Prime Minister Alain Juppé and revealed the magnitude of the health catastrophe brought on by asbestos, estimating that it could cause one hundred thousand deaths in France by 2025.[19,20]

Confusion at IARC

"The global asbestos cancer epidemic is a story of monumental failure to protect the public health," wrote the American physician Joseph LaDou, one of the founders of the *International Journal of Occupational and Environmental Health*, in 2004. He estimated then that "white gold" could claim 10 million victims worldwide before being definitively outlawed in developing countries, where the lethal fibers were still being widely used.[21] As for the ANM, it evidently revised its judgment: ten years after the publication of its controversial pamphlet on asbestos, it placed the substance at the top of the list of products responsible for "cancer attributable to occupational exposure in France" in the aforementioned report *Les Causes du cancer en France*, which it co-signed with the AS and the IARC. In a very brief table containing only fourteen chemicals, the report highlighted some of the poisons I have already mentioned, such as benzene, chromium VI and aromatic amines.[22] But only one pesticide . . .

Intrigued by this "omission," I decided to dig into the IARC. Created by President Charles de Gaulle in 1965, the organization was established in Lyon and, as mentioned above, is part of the WHO. Since its inception, it has become an international authority in the domain of cancer studies, thanks to its famous "monographs," which are official documents classifying chemical products according to their carcinogenic potential. To this end, its experts examine the scientific literature concerning these substances—in other words, all the studies published in scientific journals. Classification has three levels. Group 1 includes agents that are "carcinogenic to humans": this is an exceptional category, because in order for an agent to be listed here, there has to be epidemiological data available, which, as we know, is very difficult to obtain. In 2010, only 107 agents were classified in group 1,[23] including asbestos, benzene, benzidine, beta-natphtylamine, dioxin, formaldehyde, tobacco, cyclosporine, and mustard gas (and here I am only citing previously mentioned substances)—the birth control pill is also on this list, which I will come back to later (see Chapter 19).

Then there is group 2A, "probably carcinogenic to humans" (58 agents in 2010), and group 2B, "possibly carcinogenic to humans" (249), which describe substances for which there exists some epidemiological and experimental (animal) data that is more or less significant. Group 3 (512 agents) designates substances that are "not classifiable," for which it is impossible to come to a conclusion given the sparse and insufficient information available. Finally, group 4, "probably not carcinogenic to humans," only included one substance in 2004: caprolactam (an organic compound used in nylon blends).

Out of the some one hundred thousand chemical products that have invaded our environment since World War II, only 935 have been evaluated by IARC, which launched its Monographs Programme in 1971. This is not very many products at all. And of course, this was the first question I asked Vincent Cogliano during our meeting in Lyon in February 2010. Cogliano, an American epidemiologist, was nominated to direct the Monographs Programme in 2002.

"In thirty years of functioning, IARC has only established 935 monographs. Why so few?" I asked him.

"The answer is very simple, because it's important to know that out of the one hundred thousand products you mention, only two thousand or three thousand have been tested with their carcinogenic potential in mind. So our program has covered a third of those."

"Does the fact that a chemical has not been classified by the IARC mean that it is not dangerous?"

"No, not in any way. In general, this means that no one has studied its potential carcinogenic effects. Sometimes it has been tested, but we haven't scheduled its evaluation yet."

"What are the consequences of classification in group 1? Does this lead to a ban of the product?"

"Not at all! This simply means that IARC has gathered a group of experts who, in light of the published scientific literature, has decided that the substance in question is carcinogenic to humans. This information is put at the disposal of national regulatory agencies, which then take measures that seem the most appropriate to them. In general, they carry out an evaluation comparing the benefits the product offers to the risks it produces. This often leads to a restriction of the product's use—for example, with stricter exposure standards or a reduction of the authorized levels of residue in food. But in all cases, the IARC does not have the power to ban chemical substances. It settles for synthesizing the toxicological or epidemiological studies available, so that governmental authorities can hypothetically take action."

"Do you know of any chemicals that have been classified in group 1 that are still present in our environment?"

"To be frank, all the substances that the IARC has declared 'carcinogenic to humans' are still used, sometimes with very strict use restrictions."

"Is this classification important for industry?"

"Of course, because classifications have repercussions, either long- or short-term, on the manner in which these products are used."

"In other words: Do manufacturers do everything they can to avoid their products being classified in group 1?"

"Yes. . . . Or in group 2, because that means that the product is placed under high surveillance."

"How many pesticides have been evaluated by the IARC?"

"I haven't really counted, but I think that we must have evaluated about twenty or thirty pesticides in the entire history of our program," Cogliano admitted with a self-conscious smile.

"But that's nothing!"

"It's true that it isn't a lot, if we compare it to the number of pesticides used. In fact, it is very difficult for us to do a serious evaluation of pesticides, because the majority of experimental studies involving them are not public. Of course, the companies producing pesticides are supposed to supply toxicological information to national health agencies, and they do tests. The studies are sent to governmental agencies, but they are never published. It is very difficult for us to have access to them, because they are protected by trade secret. The only pesticides we have been able to evaluate are very old substances so controversial that they have been the subject of numerous independent studies. For example, DDT or lindane, which are now banned for agriculture use."[24]

At this point in the interview, I should emphasize the significance of the "bombshell" that the IARC director of monographs dropped on me: he affirmed, in fact, that IARC is incapable of evaluating the carcinogenic potential of pesticides because the vast majority of them have been put on the market on the basis of toxicological information that is not "public"—that is to say that no one can verify its quality. It's quite simply incredible! Hence my next question: "How do you explain that studies carried out by industry on pesticides are not published in peer-reviewed scientific journals?"

"Um . . . It perhaps isn't in the companies' best interest to publish results that might suggest that their products can be harmful," Cogliano replied, visibly searching for his words. "In any case, they are not obligated to make their studies public." Now, it was clear: pesticide manufacturers do "tests," because they are required to do so by regulatory agencies, but they are very careful not to publish them in scientific journals, where they would be submitted to critical examination. This keeps IARC from evaluating them, which in turn allows manufacturers to proclaim loud and clear that "pesticides are not carcinogenic!" Which really is impressive sleight of hand . . . but the rest of the interview proved even more surprising.

"As you know, in 2007 the French Academies of Medicine and Sciences published, with IARC, a report entitled *Les Causes du cancer en France*," I continued. "The authors write that 'no pesticide currently used is carcinogenic to animals or humans.' I consulted your monographs, and I found at least two pesticides currently used, classified in group 2B—dichlorvos, an insecticide, and chlorothalonil, a fungicide. If they were classified as 2B, does that mean that studies have shown that they are carcinogenic, at least to animals?"

"Yes, they are still used, and I am sure they are carcinogenic to animals," Cogliano murmured, as he examined a photocopy of the two monographs I held out to him.

"Does this mean that the report's statement is inaccurate?"

"Yes, I think it is," the director of monographs eventually acknowledged, with a nervous smile.

"I interviewed Professor Richard Clapp, in Boston, and he told me that this report 'was invalidated and shouldn't be taken seriously.' Do you agree with him?" I asked, determined to pick apart the infamous report.

"Um . . . Actually, to understand the report's conclusions, one should analyze the methodology its authors thought was appropriate to use; they were only interested in chemicals that IARC classed in group 1. Now, this category consists of very few substances, due to the difficulty in obtaining solid epidemiological data. That is particularly true for pesticides because, as you know, it is very difficult to demonstrate that one pesticide in particular causes cancer in humans. This is why there are no pesticides classed in group 1. However, there are many of them in group 2, such as DDT, or those you cited—dichlorvos and chlorothalonil, which is very few, as I explained, because IARC has not been able to evaluate the vast majority of them, due to the absence of published studies. So that is how the report's authors managed to state that there are no pesticides in use that are carcinogenic to humans or animals."[25] In short, the report *Les Causes du cancer en France* is biased. And remember this is not a militant ecologist saying so, but a representative of IARC, which, once again, co-signed the infamous "report"!

Conflicts of Interest at IARC

I was impressed by Vincent Cogliano's honesty, though I must point out that he was not one of the report's co-authors, even if he was working at IARC when it was published. Rather, Paolo Bofetta, an Italian who worked at IARC from 1990 to 2009,[26] and Peter Boyle, an Englishman who was the agency's director from 2003 to 2008, signed off on it on behalf of the UN-affiliated agency. The two epidemiologists—whose actions were very controversial, even within the institution—published an article in 2009 in the *Annals of Oncology* with Maurice Tubiana, a French cancer specialist known for his systematic denial of pollution's role in the cancer explosion and who co-signed the report in question, on behalf of the AS. Together, they reaffirmed in the article that pollutants were responsible for less than 1 percent of cancer deaths in France.[27]

But before delving deeper into IARC's history, and particularly into what some call "its dark period," I wanted to know what Dr. Christopher Wild, its new director as of January 2009, thought of the report co-signed by his predecessor. "To be honest, there are two things that surprised me in the document," the British epidemiologist admitted during our meeting in Lyon in February 2010, choosing his words with noticeable caution. "First, the authors write that 50 percent of cancer is due to an unknown reason, and, for me, the real challenge is to try to understand, precisely, what factors are at the root of one out of two incidences of cancer."[28] Quite right! The report contains the following enigmatic phrase: "Causes for half of the incidences of cancer have not been found in France. We hope to find others in the future, but everything must be done to speed up this process."[29] It is odd that such a surprising admission was overlooked by the many journalists who reported on the "good news." In their defense, one has to admit that the tidbit was placed on page 47, and quite often, in a fast-paced newsroom, the press settles for reading summaries or conclusions of long, fastidious reports. "The second thing that struck me," the new IARC director continued, "was that the report's authors excluded from their study all products classified in groups 2A and 2B, which considerably reduces the impact of their conclusions."

Although Christopher Wild would say no more on the issue, he'd already said a great deal, especially given the legendary impenetrability of the WHO-affiliated agency—a UN organization whose lack of transparency was itself widely talked about during the deplorable episode of the fictional H1N1 flu pandemic in 2010. At the time of his nomination in May 2008, Wild had probably read the editorial that had appeared in *The Lancet*, which, as we saw in Chapter 9, was leading the fight against conflicts of interest. "The International Agency for Research on Cancer (IARC) is soon to appoint a new Director," writes the British journal's editorialist. "Traditionally the names of the official candidates are not publicly disclosed. At the last election in 2003 we criticized the elective process for its lack of transparency and called for a change in policy to allay concerns about political or commercial influences that could bias the selection. 5 years on, there is no change. [. . .] The choice of a new Director for IARC remains shrouded in medieval mystery."[30]

Five years earlier, as the scientific journal notes, it had indeed used the nomination of Christopher Wild's predecessor to report "accusations of industry influence on IARC, especially when carcinogens are downgraded to a lower category of risk, and the difficulties faced by non-industry observers in attending IARC meetings." The journal specified that "Paul Kleihues and Gro Harlem Brundtland, the outgoing heads of IARC and WHO, respectively, denied any such influence."[31]

It should be said that the two "outgoing heads" had ended their terms under heavy critiques, which is rather rare in the often subdued arena of UN organizations. And Dr. Lorenzo Tomatis—far from an inconsequential figure, given that he oversaw the Monographs Programme from 1972 to 1982 and then directed the agency until his retirement in 1993—was the one to declare "war." In 2002, he published an article in the *International Journal of Occupational and Environmental Health* entitled "The IARC Monographs Program: Changing Attitudes Towards Public Health," in which he wrote: "From its outset, the International Agency for Research on Cancer's (IARC's) program for the evaluation of carcinogenic risks for humans had to resist strong direct and indirect pressures from various sources to protect its independence. External experts for Monographs working groups were selected on the basis of competence and the absence of conflicts of interest. The IARC did not use unpublished or confidential data, so readers could access the original information and thus follow the groups' reasoning. The strength of the original program lay in its scientific integrity and its transparency. Since 1994, however, the IARC appears to have attributed less importance to public health-oriented research and primary prevention, and the Monographs program seems to have lost some of its independence."[32]

This article—note its tone, simultaneously measured in form but very firm in content—was the follow-up to a letter written by twenty-nine international scientists, including Lorenzo Tomatis, but also James Huff, who directed the Monographs Programme from 1977 to 1979, addressed to Gro Harlem Brundtland, the director general of the WHO. On February 25, 2002, they wrote:

We are concerned about the problems of corporate influence and undisclosed conflicts of interest in the development of documents by WHO agencies, particularly regarding the cancer-causing properties of major industrial products and pollutants. [. . .] We are also concerned about the role of "observers" at meetings of WHO agency scientific expert groups. At the IARC Task Group meeting where the carcinogenic evaluation of 1,3-butadiene was made in 1998, there was a highly unusual second vote conducted the day after the group had voted 17-13 to classify butadiene as a human carcinogen. One of the scientists who voted in the majority left the meeting that day and thus did not return the next day. Observers and panel members allied with the oil and rubber industries were that evening able to persuade two others to reverse their votes, and without any discussion of why such re-voting was justified, a second vote was allowed the next day, with the result that butadiene was downgraded to probable human carcinogen by a vote of 15-14. [. . .] In order to protect the integrity of WHO institutions, it is necessary that genuine efforts be made to assure that financial conflicts of interest are fully disclosed and analyzed. If an individual has such a conflict of interest, it should be presumed that s/he cannot be totally objective and therefore should not be a member of the scientific panel.[33]

The letter caused such a stir that it was published in the *International Journal of Occupational and Environmental Health*, which dedicated an entire series to the problem of conflicts of interest at IARC. The authors contacted in this vein notably included Dr. James Huff, who after directing the Monographs Programme from 1977 to 1979, was nominated as deputy directory of the department of chemical carcinogenesis at the U.S. National Institute of Environmental Health Sciences (NIEHS).

James Huff's Battle for Independent Research

Without the shadow of a doubt, James Huff is an extraordinary scientist. On October 27, 2009, he welcomed my team and me with a wide smile, sporting jeans and a Che Guevara T-shirt. After helping us complete the formalities required by security—"the war on terrorism" obliging—he granted us a tour of the NIEHS, an enormous complex set in the middle of the forest in Research Triangle Park (RTP) in North Carolina. Created in 1959 and extending over nearly seven thousand acres, RTP is "the largest research park in the nation," as its website proclaims, with some fifty thousand employees working in 170 public and private research centers, one of the most important being the NIEHS.

Known throughout the entire world thanks to its magazine, *Environmental Health Perspectives*, the NIEHS is an incontrovertible authority in the field of environmental health. The NIEHS supervises the National Toxicology Program, whose mission is to evaluate the toxicity of chemical agents by developing tools for the use of governmental agencies like the Food and Drug Administration, which is in charge of food and medication safety, or the Environmental Protection Agency, which is tasked with pesticide regulation, among other things.

After the tour of the impressive institution where hundreds of scientists work, James Huff showed us (with some difficulty) into his office, which over the years has been transformed into an indescribable bazaar where thousands of documents, newspapers, and journals sit in haphazard piles. "I'm very proud of my time at IARC," he told me in an enigmatic tone, as I searched for somewhere to sit. "I am especially proud of having the expression changed from 'carcinogenic to man' to 'carcinogenic to humans.' From a professional point of view, my experience at IARC led me to change specialties: I went from pharmacology and toxicology to research on chemical carcinogenesis. For the past thirty years, I've done nothing but that, despite the difficulties, because I think it is an absolute health emergency."

James Huff, who was associated with the creation of the National Toxicology Program, was one of the first to develop a research protocol for what are called "bioassays"—that is, experimental studies meant to test the effects (in this case, the carcinogenic effects) of chemicals on rodents, which are followed until their natural death. Using this method, he showed in 1979, when the battle over benzene was at its height, that the molecule provoked so-called multisite cancer, or cancer in several organs in exposed mice and rats.[34]

"Why do you say 'despite the difficulties?'" I asked, after noting the emotion in his words.

"The two Bush administrations were terrible for public health defenders. Just like my friend Peter Infante, in his time, I almost lost my job," Huff replied, his voice rising abruptly to a sob he was unable to repress.[35] To be honest, it was very moving to see this seventy-one-year-old man, author of more than three hundred publications in the most prestigious scientific journals, break down in front of my camera. Before meeting him, I had discovered via the Internet that he had become quite the "cause célèbre," to quote the magazine *Science*.[36]

In 2001, Huff had publicly expressed his opposition to the modalities of a financial agreement that the NIEHS had accepted with the American Chemical Council, which allotted a $4 million budget (a quarter of which would come from the chemical industry) to test the effects of chemical products on reproduction and fetal development. And, in July 2002, as *Science* explained, Huff received a "gag order," banning him from sending "any letters, emails or other communications that are critical of NIEHS as an Institute or its scientific work to the media, scientific organizations, scientists, administrative organizations, or other groups or individuals outside NIEHS,"[37] at the risk of being fired within five days. The situation caused quite a stir within the international scientific community, especially for Lorenzo Tomatis, the former IARC director. Tomatis declared that the warning "had the tone you would expect to find under a dictatorship."[38] It went all the way to Congress, thanks to the involvement of Dennis Kucinich, the Democratic representative of Ohio, who advised that "NIEHS should be determining the incidence of human illness caused by chemical, pollutant, and other environmental cause, not putting a gag order on one [of] its best scientists."[39]

"Is this still a very painful matter for you, even several years later?" I asked Huff.

"Yes, because it was an enormous shock," he replied, after a long sigh. "I have always fought for our institute to keep its independence from industry, but at that point I understood that industry could have my head. It had me in a vise for a long time, because it's true that I never made any concessions. If

I thought a product was very dangerous, for example benzene, I would say so, thinking my mission was to protect the public health. Fighting against industry is part of our work, but when you have to fight against your own chain of command using the same arguments as industry, well that is quite discouraging. As a result, I had to retire in January 2003, but six years later I'm still here conducting studies for the public health, despite those who tried to destroy me at the end of a career I'm very proud of."[40] And he added, "Generally, the problem is that during Republican administrations, especially under Bush, government agency directors were not chosen for their qualifications, but for their political contacts and their sympathy for industry in particular. And that's terrible, because it's the public health that incurs the costs. That's what happened during IARC's dark period as well."[41]

IARC's "Dark Period": "Biased Monographs"

"The role of the IARC and the National Toxicology Program (NTP) is simple: protect human health. Nothing else is as important. Further, their role is not to 'guess' about mechanisms or to guess about whether this chemical carcinogen or that chemical carcinogen will be 'safe' or to guess about the economic, regulatory, and political consequences of a particular carcinogenic evaluation, but to always judge the information and articulate the overall assessment from the viewpoint of public health and safety. Period."[42] This was the conclusion of an article published by James Huff in September 2002, in the special series the *International Journal of Occupational and Environmental Health* dedicated to conflicts of interest at IARC that I mentioned earlier. This was precisely one month after his conflict with the NIEHS. (Now it is easier to see why the researcher takes such pleasure in wearing the portrait of the Argentinean rebel in his starred beret.)

"The influence of industry on the IARC Monographs over the last few years is unprecedented," he writes in this very detailed study in which he showed how, starting in 1995 (after Lorenzo Tomatis's departure), the WHO agency tried to downgrade twelve chemicals—that is, to lower their classification—by going back to previous decisions. One went from group 2A to group 2B, and eleven from group 2B to group 3, including atrazine, a particularly harmful herbicide I have already mentioned and to which I will return in Chapter 19. "It was unheard of!" James Huff explained. "You have to understand that, generally, chemical agents are relatively under-classed, due to experts' extreme caution. It is then logical that IARC would regularly raise their classification gradually as new studies come in that further confirm what has already been presupposed for a long time. In this way, between 1972 to 2002, forty-six agents were eventually upgraded, such as dioxin, which went from group 2A to group 1 in 1994. Then, suddenly, after Paul Kleihues came to the head of the agency, the tendency reversed. I think a number of monographs done during that period are quite simply biased."

"How do you explain it?" I asked, already knowing the NIEHS scientist's answer, as I had obviously read his article before our meeting.

"I examined the composition of the expert groups that wrote the monographs from 1995 to 2002, under the direction of my successors Douglas McGregor and Jerry Rice, and I separated out the participants into three categories according to background: 'public health,' 'industry,' and 'unknown.' I was very cautious with the 'unknown' category, because it meant that I didn't have enough biographical elements to decide in favor of the 'industry' category. The result was that industry's influence was largely predominant."

In his article, Huff noted that 29 percent of expert committee members came from the "public health" sector, 32 percent were representatives of industry, and 38 percent were of "unknown" origin. Then the researcher looked into the backgrounds of the infamous "observers," who are authorized to attend and take part in expert group discussions, but do not participate in the final vote: 69 percent came from the industrial sector; 12 percent came from "public health"; and 20 percent belonged to the "unknown" category. If we add the backgrounds of accredited experts to those of observers, we indeed find an overwhelming overrepresentation of industry—118 people (38 percent), versus 99 representatives of public institutions (26 percent), and 119 "unknown" (35 percent).

"How would you evaluate the work IARC does today?" I asked Huff.

"They have very clearly emerged from their dark period," he replied, without hesitation. "I know Vincent Cogliano well enough to know that he does everything he can to protect the public health."

Indeed, in Lyon, the concerned party willingly acknowledged that he knew Huff's article very well, but hurried to add that "times have changed."

"What's changed?" I insisted.

"Well, IARC has changed in its understanding of conflicts of interest," the director of monographs replied. "Now, when we plan the evaluation of a substance, we organize a 'call to experts' a year before the meeting. Candidates are selected in accordance with their expertise on the product in question, and we ask them to declare their conflicts of interest. Having a conflict of interest does not lead to the exclusion of the candidate, but it is brought to the attention of other expert group participants."

"Are these declarations public?"

"No . . . But we make a summary of them that we publish in an appendix to the monographs. Then a synthesis of the monographs is published in *Lancet Oncology*, which is very meticulous about the question of conflicts of interest and verifies our information. I sincerely think things are moving in the right direction."

"And what is the role of observers today?"

"It has been clarified considerably. They can no longer take part in expert groups without being invited to do so, generally at the end of the session. I simply regret that union or consumer protection organizations don't come to meetings more often. Unfortunately, it's a question of means. Recently, I invited an association for women suffering from ovarian cancer to participate as an observer at one of our meetings. They replied that they couldn't afford sending someone to France for a week. Of course, businesses don't have this type of problem."

"Last question: Are you going to reevaluate atrazine, which went from group 2B to group 3, as if by magic?"

"I will confirm that atrazine is on the priority list of products to be reevaluated," Vincent Cogliano concluded, and said nothing more on the subject.[43]

The Fallacious Argument of the "Mechanism of Action" of Nontransposable Cancers (Rodents to Humans)

"How could IARC have justified downgrading chemicals?"

The question makes James Huff smile, though his tone became suddenly harsher: "Well, that was really the last straw! In 1999, I attended a meeting largely dominated by industry representatives, at which they explained that certain cancers—such as kidney, thyroid, or bladder—obtained in rodents after exposure to chemicals were strictly specific to that species of mammal, because they adhered to a biological mechanism that was inoperative in humans. I emphatically protested, along with my colleague Ronald Melnick, stressing that this assertion was speculation without any scientific foundation—but to no avail. The IARC adopted the argument and for several years ignored certain toxicological studies done on rats and mice, for the reason that it hadn't been proven that the carcinogenic mechanism was transposable to humans."

These are not merely technical details. For, as David Michaels recalled, "the devil is in the details"—a point particularly well understood by manufacturers, which have accordingly left nothing to chance. The argument established by their representatives is, indeed, very serious; if followed to the letter, it would mean that the most dangerous molecules would remain on the market, since IARC and regulatory agencies would have no more tools with which to evaluate them. On one hand, manufacturers ceaselessly repeat that epidemiological studies are not reliable, since they are generally based on witnesses' memories, and their results can be due to chance—as was the case with the theory used in 2007 by the authors of *Les Causes du cancer en France*, as we have seen. So, exit epidemiological studies, stage left. And if, on the other hand, experimental studies carried out on animals are of no use, because it is impossible to extrapolate their results to humans, well then long live poisons . . .

This argument is not simply theoretical, because it has led to decisions concerning us all. For example, this was why the carcinogenic quality of formaldehyde was long ignored, even though the product is omnipresent, found namely in plywood furniture in numerous households. What's more, several experimental studies have shown that inhaling formaldehyde causes sinus and nasopharyngeal cancer (as well as leukemia and brain tumors). But, as André Cicolella and Dorothée Benoît Browaeys state, these results have been brushed aside for the reason that they "have been obtained in rats, and the surface area of a rat's snout is proportionally much larger than that of a human."[44] Formaldehyde was eventually classified as "carcinogenic to humans" in 2004, but it was already too late for carpenters suffering from sinus cancer, which in France is actually sometimes called "carpenter's cancer."[45]

The fallacious argument was also used by IARC in 2000 to downgrade di-2-ethylhexyl phthalate (DEHP), a tremendously toxic substance in the phthalate family, from group 2B to group 3. Phthalates are used as softeners and are added to polyvinyl chloride (PVC) to give "plastic materials" flexibility. They are found in all flexible or semirigid plastics, such as balloons, tablecloths, rain boots, shower curtains, umbrellas, medical equipment (blood bags, catheters), food packaging (cling film) and, until 2005 in Europe, in cosmetics and toys. Classified in 2006 by the European Union as "toxic to reproduction" (category 2), DEHP is the most commonly used phthalate; it is found as a contaminant in the air, in household dust, in water, and even in breast milk. I will return to the reprotoxic properties of phthalates (which are considered endocrine disruptors, like bisphenol A [BPA]), but for now, it suffices to note that numerous experimental studies have shown that exposure to DEHP leads to cancer, especially of the liver and pancreas. Some of these studies were published as early as 1982 by James Huff, following a bioassay carried out by the National Toxicology Program, as he pointed out in a 2003 article called "IARC and the DEHP Quagmire."[46]

At the very moment when the IARC expert group was deciding to down-grade DEHP, a new study confirmed that the phthalate produced pancreatic cancer in rats.[47] And even though the study's author, Raymond David, was admitted to the discussions as an observer, he ultimately witnessed his work be simply cast aside from the final evaluation. The affair triggered a volley of outraged reactions in the *International Journal of Occupational and Environmental Health* which denounced the "exclusion" and "suppression of key studies."[48] And in a letter from April 8, 2003, addressed to Charlotte Brody, one of the signatories of the dissenting paper, IARC director Paul Kleihues made a shocking admission that the monographs did not necessarily cite all the existing literature on the subject of the evaluation, but merely the studies that the work group deemed pertinent. He conceded that the induction mechanisms of the cancer in rats and mice had been considered invalid for humans.[49]

Industry's Double Talk

"With David Rall, who directed the NIEHS for twenty years and created the National Toxicology Program, we observed that, out of a hundred products classified as carcinogenic to humans, more than a third had been the subject of experimental studies in which they initially turned out to be carcinogenic for animals," James Huff explained to me. "Likewise, all the products that were suspected of being carcinogenic to humans showed that they were also carcinogenic for animals. Contrary to what industry would have us believe, there are more similarities between humans and animals than there are differences."

The NIEHS scientist's opinion is wholeheartedly shared by Vincent Cogliano, the director of the IARC Monographs Programme. When I quoted his predecessor's words to him, he replied without hesitation: "I completely agree with what Jim is saying. Mammals have a number of common physiological, biochemical, and toxicological mechanisms. That's why (except when there's a proven exception) we have to consider that the signals observed in animals are transposable to humans—moreover, this is what the pharmaceutical industry does continuously."

My interviewees regularly brought up this last point as proof of industry's "double talk." "When the industry develops new medications, they test them on animals first," Devra Davis noted. "If they do it, it is precisely because they believe the results obtained on rodents or other mammals are a measure of predicting the effects that the molecules will have in humans. It is interesting to note that, when there are no observed effects, the industry hurries to request market authorization, arguing that the new product has no secondary effects. In contrast, when negative effects are observed, the same industry invokes the argument of a 'mechanism specific to rodents' that is not transposable to humans. It's surprising to see that this incoherence is rarely addressed by agencies in charge of regulating chemical pollutants."[50]

"Experimental medicine has been based on animal studies for decades," Peter Infante insisted. "Why should they move away from that principle, the validity of which has been essentially proven, when they test the toxicity of chemical products that can be found in our food or our environment? This kind of maneuvering needs to end—its only objective is to paralyze the regulation process!"[51] For his part, David Michaels points out, perhaps unnecessarily, that "scientists cannot feed toxic chemicals to humans to see what dose causes cancer. [. . .] Our regulatory programs will not be effective if absolute proof is required before we act; the best available evidence must be sufficient."[52]

For all of us, the "best available proof" remains that obtained from in vivo studies—that is on test subjects—or in vitro studies on cultured cells. "Epidemiology always arrives too late," stressed Richard Clapp, who is himself an epidemiologist. "When we are at the point of counting patients and deaths in the morgue, it's because the regulation process has failed from the start."

"I completely agree with what Professor Clapp said," confirmed Vincent Cogliano when I read his Boston colleague's remarks verbatim. "Every time we classify a carcinogenic product in group 1, it's proof of our failure to anticipate and act preemptively. Because when a product goes into that category, it means that it has already caused cancer in humans. The ideal would obviously be that we are able to identify bad products before humans are subjected to long-term exposure, at the risk of suffering irreversible damage."[53]

As we shall see, "irreversible damage" is already under way, because, contrary to what chemical industry leaders and their institutional go-betweens maintain, chronic illnesses have continually progressed over the course of the last fifty years, to the point that one could call them a veritable epidemic.

11

An Epidemic of Chronic Diseases

We are in danger, and the enemy is none other than us.

—Edgar Morin

On Wednesday, January 13, 2010, Sir Richard Peto seemed particularly anxious in his Oxford University office, where I was interviewing him. Through the course of my long investigation, I had never met such a seemingly nervous scientist. Still, the British epidemiologist is not just anybody—he is chair of Medical Statistics and Epidemiology at the prestigious Oxford University, a fellow of the Royal Society, and was knighted by the Queen in 1999 for his "contribution to cancer prevention." This distinction, highly valued in Great Britain, was largely thanks to a study Peto published in 1981 with his mentor, Sir Richard Doll, which became "a bible of cancer epidemiology," to use Devra Davis's words.[1] It should not be forgotten that Richard Doll was also knighted for his work establishing the connection between smoking and lung cancer, which made him "one of the preeminent public health authorities"[2] (see Chapter 8).

Doll and Peto's 1981 Study on the Causes of Cancer: A "Fundamental Reference"

In 1978, Jimmy Carter was leading a heavy-handed campaign against smoking, which he had declared "public enemy number one." That same year, Joseph Califano, Carter's secretary of health, gave a speech before Congress in which he announced that, in the near future, 20 percent of cancer cases would be due to occupational exposure to toxic agents. Devra Davis, who at the time was delighted to see a high-ranking government official use such uncommon bluntness, noted that "this shocking number sent the public relations industry into full battle mode."[3] As a result, the U.S. Congress Office of Technology Assessment commissioned Richard Doll, known for his opposition to concessions to the tobacco manufacturing lobby, to carry out a study on the origins of occupational cancer.

In 1981, assisted by a "brilliant young epidemiologist" by the name of Richard Peto, Doll submitted a one-hundred-page document entitled "The Causes of Cancer: Quantitative Estimates of Avoidable Risks of Cancer in the United States Today,"[4] which in reality had little to do with the original request. To write their report, the two epidemiologists combed though the registries of cancer deaths of white men under the age of sixty-five occurring between 1950 and 1977. They concluded from their research that 70 percent of cancer cases were due to individual behaviors, with eating habits at the top of the list, accounting for 35 percent of deaths, followed by smoking (22 percent) and alcohol use (12 percent). In their table of disease causes, occupational exposure to chemical agents only represented 4 percent of deaths, and pollution 2 percent, much less than infections (viruses or parasites), which were estimated at 10 percent.

As Drs. Geneviève Barbier and Armand Farrachi point out in their book *La Société cancérigène* (Carcinogenic Society), "the fat lady sang over thirty years ago. Doll and Peto's writings show up in every work on the subject as *the* reference, and their table set a precedent; it continues to orient judgments."[5] They are not wrong—no official text misses the opportunity to invoke "Doll and Peto's study" as proof that the main cause of cancer is tobacco, and that the role of chemical pollution is extremely negligible. Subsequently, in France, the report of the 2003 steering committee on cancer, which oversaw the "national mobilization project against cancer"—widely promoted by President Jacques Chirac—cited the two Englishmen's study no fewer than seven times.[6] And this was more than twenty years after its original publication, as if cancer research had stopped since then. The report *Les Causes du cancer en France* (The Causes of Cancer in France) of course relies on this "fundamental reference,"[7] and the infamous Crop Protection Industry Association (Union des industries de la protection des plantes, UIPP), which as we have seen unites nineteen pesticide manufacturers—posted the irrefutable results on its website. France does not appear to be an exception, however: the same patterns occur in most Western countries, such as the United Kingdom, for example, where the Health and Safety Executive, a governmental organization, made sure to cite their two fellow citizens' study in 2007 as "the best overall estimate available" concerning cancer of chemical origins.[8]

A Surprising Meeting with Richard Peto

Before looking at why the infamous 1981 study was so heavily criticized, because of both its methodological biases and also the conflicts of interest involving Richard Doll, it would be appropriate to give the floor, so to speak, to his colleague Richard Peto. As noted above, I met him in January 2010 at his Oxford University office, located in a building named "Richard Doll" in honor of the great man, who passed away in 2005. At the age of seventy-seven, the British epidemiologist had an unquestionable air of distinction, thanks largely to his silver mane that he kept tossing back throughout his long monologues, in which he repeated the same arguments on a loop. On several occasions, visibly bothered by my questions, he got up from his desk to pace around the room, under the stunned gaze of my cameraman, who no longer knew know how to film him. While reviewing the interview, I wondered if this physical and mental agitation was routine, or if it was his way of expressing discomfort with the detailed critiques that had knocked Richard Doll off his pedestal—not to mention the infamous study, though it had long been considered as "gospel," as André Cicolella writes in *Le Défi des épidémies modernes* (The Challenge of Modern Epidemics).[9]

"There is a very common belief that there is more cancer today than before, and that this is due to the numerous chemical products present in the world," said Richard Peto. "According to some, we are lucky to even come out alive from this chemical universe, but all of that is false. It is true that we are exposed to a number of chemical molecules daily. Plants, for example, produce very harmful toxins, like potatoes do in their skins, or celery, because it is the only way that they can protect themselves against insects. As plants cannot flee, they produce defensive toxins, constantly. Kiwis, a fruit we did not know about a few decades ago, also do this. Today, we eat a lot of kiwis, though they contain many chemical substances that have turned out to be toxic during laboratory tests. Plants do this constantly, and yet, it has been observed that people who consume a lot of vegetables have fewer instances of cancer than others. So, you see that it is very difficult to predict the effects of chemical products. But, in any case, the main sources of chemicals we are exposed to are natural substances contained in the plants we eat."

After this first tirade, during which he stared at his desk, Richard Peto paused and lifted his head, as if to make sure I understood what he had just said. I was so flabbergasted by his comments that I said nothing, opting instead to let him continue his incredible discourse. "Obviously," he went on, after lowering his head once more toward his desk, "there are a few big exceptions, the first of which is tobacco, of course, which carries enormous risks. As soon as there is a sharp increase in smoking somewhere, there is immediately a sharp increase in the mortality rate. On the other hand, as soon as there is a sharp decrease in smoking, there is immediately a sharp decrease in the mortality rate. Thus, apart from the considerable effects of tobacco, which really feed into the entire issue, can we say that there is an increase in the causes of cancer? If we examine the information, the answer is no."

"I imagine you are familiar with the documents from IARC [the International Agency for Research on Cancer] in Lyon, which you have often visited," I said carefully. "According to a study published by the agency, the child cancer rate in Europe has gone up 1 to 3 percent per year over the last three decades, mainly concerning leukemia and brain tumors.[10] Is smoking also the origin of this remarkable uptrend?"

"I don't necessarily agree with everything IARC says," Peto answered, squirming in his seat. "It depends on the quality of the information presented. But tobacco has very little connection, or even no connection at all, with cancer in children or with cancer that appears in early adulthood. These cancers are due rather to impaired fetal growth."

"And how do you explain this impairment?" I asked, convinced that the epidemiologist was finally going to stop stonewalling.

Alas, no. He dodged the question to better deliver his prepared speech, recycling old arguments that, as we will see, do not hold up for one second to serious examination. "I think that the visible changes are due to better cancer detection and recording capabilities," he replied, all the while scribbling words on a sheet of paper and "forgetting" that my question had to do with the causes of "impaired fetal development," which he had just mentioned. "For example, in the 1950s and 1960s, we didn't know how to properly diagnose leukemia, so when people died, they would say it was from an infection, but not leukemia. Today, we know how to diagnose cancer better, so we get the impression that there is more of it. And then there are artifacts that make it so things are detected in early childhood that look like cancer, but then disappear."

At this point in the interview, I wondered whether Richard Peto truly knew what he was talking about—his remarks were as inconsistent as they were incoherent. I nearly threw in the towel because I felt I was wasting my time. But, lifting his head back up, the epidemiologist continued his monologue: "Generally speaking, the rate of cancer-related deaths is decreasing," he said, "even though the rate of deaths related to certain kinds of cancer is increasing. Some rates decrease, others increase, so it is hard to make a definitive conclusion."

"It is true that in developed countries, the overall mortality due to cancer is decreasing in general," I retorted. "This is due to greater treatment efficiency. The incidence rate, however, keeps rising. How do you explain that?"

"Incidence is very difficult to measure," Peto responded, abruptly getting up from his chair to hold out the paper on which he had scribbled the word "diagnosis." "We live in a time when interest in cancer keeps growing, and as a result, newspapers and television talk about it more. Additionally, people live to be older and older, so it is normal that there would be more cancer and that the disease would attract more attention. When we put all of these elements together, we realize that the image of a sea of carcinogenic products leading to an increase in cancer rates is completely false, and that it only serves to detract attention from the main subject, which is mortality due to tobacco."

"So you think that your 1981 study is still valid, thirty years later?"

"Absolutely! What we said when our study came out is still true today."[11]

Sir Richard Doll's "Cookie Cutter Argument"

"How can someone claim that a study done three decades ago can help us make good decisions today?" The American epidemiologist Devra Davis had expressed her surprise during our long discussion on Doll and Peto's work when we met three months earlier, in October 2009.[12] "Especially," she added, "when the methodology they used is biased, because it is so restrictive that it considerably reduces the impact of their results. They went through records between 1950 and 1977 concerning only white men under sixty-five when they died. They thus excluded African American men, who in general are the most exposed to chemical agents, through work or at home. They excluded men with cancer but who were still living. The ignored the incidence rate and only focused on mortality. Now, given the latency period of the illness, men who died from cancer between 1950 and 1970 are people who were exposed to carcinogenic products in the 1930s and 1940s—in other words, a time when the massive invasion of chemical products into our daily environment had not yet started. That's why it would have been better to examine the evolution of the incidence rate, if we really wanted to measure the illness's progression and determine its possible causes."

When she was working at Johns Hopkins University, Devra Davis studied the evolution of cancer incidence, namely of multiple myelomas and brain tumors in men between the ages of forty-five and eighty-four. She and her colleague Joel Schwartz, a statistician who became a renowned epidemiologist at Harvard University, observed that the incidence rate of these two fatal cancers had risen by 30 percent between the 1960s and the 1980s. Published in 1988 in *The Lancet*,[13] and then two years later in an entire volume of the *Annals of the New York Academy of Sciences*,[14] the studies attracted Sir Richard Doll's attention. In *The Secret History of the War on Cancer*, Davis recounts her excitement when, in the 1980s, she had the honored privilege of "having a drink" with the illustrious scientist at the end of a conference organized by IARC. She writes: "His entry in *Who's Who* listed conversation as one of his hobbies, and sure enough, he was a captivating, engaging, and scintillating man to talk with."[15]

That night, Richard Doll gallantly explained to his "captivated" admirer that she had been led astray by a "fundamental mistake, a colossal error" in her study: the rise in the cancer incidence rate she believed she had observed was due to a simple optical illusion in connection with doctors' improved capabilities to diagnose cancer. Before, he explained, when an elderly person died, practitioners would sign a death certificate listing "senility" when they didn't know the exact cause of death; and, sometimes, they would indicate the cause of death as "cancer of unspecified site." The epidemiologist then suggested that his young colleague verify the evolution of deaths classified by "senility" or "cancer of unspecified site," assuring her that these categories had steeply decreased. Davis did this, but she found that his assertion was false. For four years, she reviewed records, from the National Cancer Institute in particular, which had begun to systematically catalog cases of cancer since January 1, 1973. With the help of her mentor, Abe Lilienfeld, a professor at Johns Hopkins University and the godfather of American epidemiology, and Allen Gittelsohn, a biostatistician, she demonstrated that there had not been a decrease in death certificates listing "senility" or "cancer of unspecified site" in older white men. It was actually the opposite. At the same time, however, she noted a high increase in the incidence rate of cancer, as well as in mortality due to specific kinds of cancers.[16]

"What do you think of the argument that the increase in cancer is in fact an artifact due to the improvement of diagnosis methods?" I asked Davis.

"The argument doesn't hold up to analysis," she replied. "I even showed in my book that it has been used systematically for more than a century! If we take the example of leukemia or childhood brain tumors, their constant increase can in no way be explained by the improvement in detection methods, because there is no program of systematic screening like there is for colon, breast, or prostate cancer. When cancer is detected in a child, it is because he is sick and we're trying to understand why, and that practice hasn't changed over the last thirty years!"

The American authors of the President's Cancer Panel (PCP) report (see Chapter 10), who carefully examined the validity of what some call a "cookie cutter argument," shared her opinion. The PCP report distinguishes between mortality and incidence rates, which, as we have seen, are two very different ideas, though certain experts (such as Richard Peto) often have the tendency to forget it. The panel writes: "Mortality from childhood cancers has dropped dramatically since 1975 due to vastly improved treatments that have resulted from high levels of participation by children in cancer treatment clinical trials. Yet over the same period (1975–2006), cancer incidence in U.S. children under 20 years of age has increased. The causes of this increase are not known, but [. . .] the changes have been too rapid to be of genetic origin. Nor can these increases be explained by the advent of better diagnostic techniques such as computed tomography (CT) and magnetic resonance imaging (MRI). Increased incidence due to better diagnosis might be expected to cause a one-time spike in rates, but not the steady increases that have occurred in these cancers over a 30-year span."[17]

The "better diagnosis" argument was obliterated in 2007 in an article in *Biomedicine & Pharmacotherapy* published in the context of a one-hundred-page dossier entitled "Cancer: Influence of Environment."[18] The authors, including Richard Clapp and French scientists Dominique Belpomme and Luc Montagnier, take as an example breast cancer, a disease for which screening programs have been implemented in sixteen European countries.[19] However, they note that early detection of breast cancer may have an influence on mortality, but not on incidence, because the same cancer would have been detected thirty years ago, though at a more advanced stage. They cite the Norwegian system, which uses one of the oldest cancer registries in Europe (1955)[20] and introduced screening measures for breast cancer (mammography) and prostate cancer (prostate-specific antigen [PSA] assays) as early as 1992. An examination of the evolution of incidence rates of breast and prostate cancers shows that they progressed between 1955 and 2006, with a slight peak in 1993, when screening techniques were introduced. The same observation can be made for thyroid cancer, the incidence of which multiplied by a factor of six over the same period, a phenomenon that started well before the introduction of ultrasound imaging.

The Aging Population Is Not an Explanation

"Another argument regularly put forward to explain the increase in chronic illnesses is the aging population. What do you think about that?" I asked Devra Davis, who gave the hint of a wide grin as soon as I finished my question.

"Unfortunately, that argument also turns out to be false," the American epidemiologist replied. "Extended life expectancy of course means that there are more elderly people likely to have cancer. But what needs to be examined is the evolution of incidence rates of cancer or neurodegenerative illnesses in the different age groups. And we've observed that the incidence rate of certain cancers has doubled in people over sixty-five years old. This is the case for non-Hodgkin's lymphoma, for example, which has doubled in older women. The aging population does not explain why, in the United States, there are more than five times as many men and women suffering from brain tumors than in Japan, or why more and more young people in Western countries get testicular or thyroid cancer. To say nothing of the increase in childhood cancer which cannot be due to extended life expectancy!"

In fact, as the French cancer specialist Dominique Belpomme and his co-authors pointed out in 2007 in the *International Journal of Oncology*, "age is not the unique factor to be considered since the rising incidence of cancers is seen across all age categories, including children."[21] Similarly, a study carried out in England and Wales showed that the average age of appearance of prostate and breast cancers, and also leukemia, continued to decrease between 1971 and 1999, which means patients are becoming younger and younger. The authors noted that over the same period the incidence rate of prostate cancer doubled, stressing that this was *before* the introduction of PSA assays.[22]

"If aging was the sole cause, evolutions would be more or less comparable for all types of cancer and for both sexes, which is very far from being the case," André Cicolella notes in his book *Le Défi des épidémies modernes* (The Challenge of Modern Epidemics). The French chemist and toxicologist points out that "between a woman born in 1913 and a woman born in 1953, the risk for breast cancer was multiplied by nearly three, while the risk of lung cancer was multiplied by five. [. . .] Between a man born in 1913 and a man born in 1953, the risk of prostate cancer multiplied by twelve, while the risk of lung cancer remained the same."[23]

The Tobacco Industry's Alibi for "Camouflaging the Carnage"

"And what about smoking, which continues to be used as the number one cause of the increase in cancer?" I asked Devra Davis next. Given the revelations made above, such a question is clearly unavoidable, as is curiosity about the source of the sudden collective obsession that reduced cancer prevention to the fight against tobacco.

"It is clear that smoking causes cancer of the mouth, larynx, lungs, or bladder," answered Davis, a committed anti-tobacco militant. "But let's be serious: it has nothing to with the number of cancers—including prostate and breast or testicular—that are currently on the rise."

Actually, there are a number of observers who stress that "the incidence of and mortality from cancers strongly related to tobacco and/or alcohol consumption have been decreasing over the last two decades, while the incidence of cancers not related to tobacco and/or alcohol consumption or to obesity, have been increasing. This figure reversal characterizes many industrialized Western countries in Europe and in the United States."[24] In France, according to a study carried out by Catherine Hill and Agnès Laplanche,[25] the number of regular male smokers decreased from 1953 to 2001, from 72 percent to 32 percent, which should have led to a decrease in bronchopulmonary cancer starting in the 1980s. However, as Geneviève Barbier and Armand Farrachi note, "lung cancer did not stop increasing between 1980 and 2000. How can this be understood? And how can it be explained that the cancers that increased the most (melanoma, thyroid, lymphoma, brain) have little to do with tobacco?"[26]

Declared the "evil of last century and the century to come," tobacco gets the lion's share of blame in all cancer prevention campaigns. Accordingly, the 2003 report of the French steering committee on cancer,[27] which inspired Jacques Chirac's "national mobilization against cancer," spent "thirty-five pages on tobacco, eleven on alcohol, six on nutrition, seven on occupational cancers, three on the environment, and two on medication." The authors of *La Société cancérigène* wonder:

> Could tobacco be responsible for more than half of our national cancer cases? The press release declares that out of the 150,000 annual cancer

deaths, 40,000 are "attributable to tobacco-related cancer," an expression that, for those who wish to read into it, allows for a few remarks. First of all, *related* to tobacco does not mean *caused* by tobacco. But not every reader knows about this nuance. That number is repeated everywhere, as if it was inscribed in the Ten Commandments. Why 40,000 deaths? If we add *all* deaths from lip, mouth, pharyngeal, laryngeal, lung, and bladder cancer in 2000, the total does not even reach 39,000. Are they all smokers? None had contact with solvents, benzene, or asbestos? Note that nasopharyngeal or salivary gland cancers, lumped in with upper aerodigestive tract cancers, have practically nothing to do with alcohol or tobacco, but a lot to do with sawdust and ionizing radiation. [. . .] There also exist numerous occupational causes of upper aerodigestive tract cancers; exposure to sulfuric acid, formaldehyde, nickel or dyes, to name only a few, affect more than 700,000 people. Moreover, if 40 percent of bladder cancer is caused by tobacco, then the coloring, rubber, metal and solvent industries make up the rest. Finally, and above all, bronchopulmonary cancer is the most common of occupational cancers. However, as there is most often no distinction between it and smoking-related cancer, and recognition of occupational cancer is particularly underdeveloped in France, tobacco was used at just the right moment to monopolize attention, camouflage the carnage and . . . finance the cancer plan.[28]

I would also add that tobacco provides a very practical alibi in masking the role of chemical pollutants and getting manufacturers off the hook for the troubling progression of chronic illnesses, as Richard Doll and Richard Peto did with their skewed study.

Richard Doll's Work for Monsanto

"When you were preparing your study on the causes of cancer, did you know that Richard Doll was secretly working as a consultant for Monsanto?"

The question made Richard Peto jump out of his chair and pace across his office before sitting back down and declaring, nearly inaudibly, "There was no secret; this is not a secret. He was consulting for Monsanto on how to organize their records so they would actually get evidence if there were any hazards sooner than they would otherwise do . . . We were offered money by the American government for doing this and again, we didn't want to take any money because we didn't want anybody to say we were doing it because we were paid. I said I wanted to give my money to Amnesty International—he said he wanted to give his money to Green College at the University of Oxford; both giving it away so that nobody could say we did this because we were paid. The American government wouldn't let me give my money to Amnesty because they had them listed as a communist organization. I think he was completely open about the payments that he received and about the fact that he gave them away and didn't take them."

"Inquiries show that the remuneration Sir Richard Doll received from Monsanto, as well as from Dow Chemical and vinyl chloride and asbestos manufacturers, was never made public," I responded. "What proof do you have of these donations?"

"In the early days," Peto replied, "it wasn't normal to disclose fees but he believed that fees should not be taken and that they should be given away and he consistently did so. He's said it under oath when he was being cross-examined by the tobacco industry, for example. It's there on record, under oath."

And for good reason: it is indeed hard to imagine that cigarette makers would pay Richard Doll to confirm the link between smoking and lung cancer. In a 2007 article entitled "Hero or Villain?" American historian Geoffrey Tweedale rightly notes "It is, of course, inconceivable that Doll would have taken money from the tobacco industry; but why did he adopt double standards by accepting undeclared money from other producers of carcinogens?"

"It's been used to taint his legacy, of course," Peto lamented, seemingly unaware of the gravity of his statement.

"That's understandable," I retorted, "even more so because he was paid by Monsanto to state that dioxin was not carcinogenic, which turned out to be a major mistake."

"I don't think there is any good evidence of a relationship between dioxin (including Agent Orange) and human cancer," Peto replied, with such aplomb that I wondered if he truly believed what he was saying or if he simply preferred to lie in order to defend his mentor's tarnished reputation.

It should be remembered that dioxin was indeed classified as "carcinogenic to humans" in 1994 by IARC—a long overdue decision, which is indeed explained by Richard Doll's involvement in the matter. This incredible story, which I briefly mentioned in *The World According to Monsanto*, says a lot about the influence certain leading scientific experts can wield if they decide to serve the interests of big companies, even at the expense of the greater good. It all started in 1973, when a young Swedish researcher by the name of Lennart Hardell discovered that exposure to herbicides 2,4-D and 2,4,5-T—

the two components of Agent Orange, made by Monsanto, among others—caused cancer. He saw a sixty-three-year-old male patient suffering from liver and pancreatic cancer, who told him that, for twenty years, his work consisted in spraying a mixture of the two weed killers on forests in the North of Sweden. Lennart Hardell then conducted a long research project, in collaboration with three other scientists, which was published in 1979 in the *British Journal of Cancer*, showing the connection between several cancers, including soft-tissue sarcoma and Hodgkin's and non-Hodgkin's lymphoma, and dioxin exposure, a contaminant in 2,4,5-T.[29]

In 1984, Lennart Hardell was asked to testify before an investigative commission implemented by the Australian government to rule on requests for reparations claimed by Vietnam War veterans. A year later, the Royal Commission on the Use and Effects of Chemical Agents on Australian Personnel in Vietnam submitted their report, which caused a huge controversy.[30] In a 1986 article published in the journal *Australian Society*, Professor Brian Martin, who was teaching in the Department of Science and Technology at Wollongong University, denounced the manipulation that led to what he calls "the acquittal of Agent Orange."[31]

The report concluded, with surprising optimism, that no veteran suffered from exposure to chemical agents used in Vietnam, and that "this is good news and it is the commission's fervent hope that it will be shouted from the rooftops." In his article, Brian Martin tells how the experts cited by the Vietnam veterans association were "attacked strongly" by the attorney for Monsanto's Australian affiliate. Yet more seriously, the report's authors copied, nearly in full, two hundred pages supplied by Monsanto to invalidate the studies published by Lennart Hardell and his colleague Olav Axelson.[32] "The effect of the copying is to present the views of the Monsanto submission as the commission's own," Martin commented. For example, in the crucial volume concerning the carcinogenic effects of 2,4-D and 2,4,5-T, "where, for example, the Monsanto submission's phrase 'it is submitted that' has been replaced in the commission's report by the phrase, 'the commission concludes,' in the midst of pages and pages of almost verbatim copying."

Lennart Hardell, who was very harshly accused in the report, which insinuated that he had manipulated his study data, went through the infamous opus himself. He was surprised to discover that "the views taken by the commission . . . were supported by Professor Richard Doll in a 1985 letter to Honorable Mr. Justice Phillip Evatt, the commissioner," as Hardell revealed in an article that appeared in 1994. The British epidemiologist judged that

"Dr. Hardell's conclusions cannot be sustained and in my opinion his work should not be cited as scientific evidence. [. . .] It is clear [. . .] that there is no reason to think that 2,4-D and 2,4,5-T are carcinogenic to laboratory animals and that even TCDD [dioxin], which has been postulated to be a dangerous contaminant contained of the herbicides is, at the most, only weakly carcinogenic in animal experiments."[33]

And then one day in 2006, Lennart Hardell made an incredible discovery. After being informed that his famous detractor (who passed away in 2005) had left his personal archives to the library of the Wellcome Trust foundation in London, which presents itself as a charity foundation dedicated to the achievement of "extraordinary improvements in human and animal health," Hardell decided to consult these records. As was announced in a 2002 article by Chris Beckett, the library's director, "the personal papers of Professor Sir Richard Doll, CH, OBE, distinguished epidemiologist, are now catalogued and available for consultation. . . . Illustrating a life-long commitment to epidemiological research, they evince a strong sense of historical continuity and public responsibility, and demonstrate very well the social and ethical nexus in which epidemiology is rooted."[34] In his panegyric, the Wellcome Trust librarian whispers not a word of the existence in these archives of several compromising documents testifying to the financial links that united the "distinguished epidemiologist" and poison makers, which Hardell discovered. Included among these was a letter on Monsanto letterhead, dated April 29, 1986. Written by a certain William Gaffey, one of the company's scientists who had co-signed several biased studies on dioxin along with Dr. Raymond Suskind (see Chapters 8 and 9), the letter confirmed the renewal of a financial agreement providing for payment of $1,600 per day. Doll, who kept a copy of the letter in his archives, responded that he appreciated the offer to extend the consulting contract and to increase the payment amount.

So, at the same time that Doll was publishing his famous study on the "causes of cancer," which minimized the role of chemical contaminants in the etiology of the disease, he was being handsomely paid by "one of the great polluters of industrial history."[35]

Doll's Industry Involvement Embarrasses the Scientific Community

Richard Doll's industry involvement was uncovered in December 2006 by the British newspaper *The Guardian*, which showed that the collaboration between the scientist and the Saint Louis firm had lasted twenty years (from 1970 to 1990).[36] The affair caused quite a commotion in the United Kingdom, where it pitted the knighted scientist's defenders against those who believed his conflicts of interest had seriously damaged his work's credibility. The American historian Geoffrey Tweedale analyzed all the newspapers, which had a field day with the embarrassing revelation. *The Observer* wrote "Doll was a hero, not a villain," who lived in a "modest house in north Oxford," stressing that "each age has its mores: we cannot expect the giants of the past to live by ours."[37] "Actually," writes Geoffrey Tweedale, "[Doll's house] is one of the better addresses in the city."[38]

The American historian reports that the epidemiologist received support from the entire scientific establishment, which invoked five arguments: "1) Sir Richard Doll had saved millions of lives by his smoking/lung cancer research; 2) conflicts of interest were undeclared in his day; 3) he donated his consultancy money to worthy causes; 4) it was somehow unseemly to attack someone unable to defend himself; and 5) the attack on his reputation was launched by 'environmentalists' or those with a personal axe to grind."

In a letter written to *The Times*, Richard Peto emphatically points out that, "To this day and in the years to come, many tens of millions of people, in the developing as well as the developed world, will owe their lives and health to his studies."[39] "Since no one had denied this," Tweedale retorts, "it is difficult to see what relevance this had for a debate about Doll's conflict of interests." The *Sunday Mirror* shared this opinion and considered that "Doll's strictly neutral and objective image is now discredited for good."[40] Even more so because the British epidemiologist never hesitated to give lessons on professional ethics. "Any scientist who may be tempted to accept support in any form from the tobacco industry should therefore recognize that the results may be used for the purposes of the industry," he declared in 1986, a year after secretly denigrating Lennart Hardell's works.[41]

Many years later, Richard Doll's involvement with the chemical industry continued to disconcert all those who claimed to follow his legacy, namely by invoking his famous 1981 study on the causes of cancer. This is the case for the directors of the American Cancer Society (ACS), for example, an institution regarded as an authority in the field of cancer studies and whose connections with the pharmaceutical industry have often been criticized. In October 2009, I had the opportunity to meet Dr. Michael Thun, vice president of the ACS from 1998 to 2008, who was in charge of epidemiological research on cancer, and who now holds an honorary position there. Shortly before my visit to the venerable association's luxurious Atlanta building, the epidemiologist had co-signed an article in the *Cancer Journal for Clinicians*, in which the authors dissertated in a somewhat contradictory fashion on "the environmental factors of cancer."[42] On the one hand, they lament that "Carcinogen testing data are not available for many industrial and commercial chemicals," and that, "Ideally, such testing should be performed before products are introduced, rather than after there is widespread human exposure." And on the other hand, they reinvoke the timeless study: "Although the contribution of environmental and occupational pollutants to the human cancer burden is significant, it is much smaller than the impact of tobacco use. [. . .] In 1981, it was estimated that approximately 4% of all cancer deaths in the United States were due to occupational exposures."

"How can you continue to cite Doll and Peto's study, when we know today that Richard Doll was a paid consultant for Monsanto," I asked Michael Thun, who clearly had not anticipated the question.

"I don't think that Doll needed this money to live," he replied, visibly ill at ease, "because he was a very wealthy man, thanks to his wife, who was a business owner. Plus, he always said that the money chemical firms gave him was used to finance Green College at Oxford."

"How do you know that?"

"It's what I always heard," the ACS epidemiologist conceded.

"Is it common for preeminent scientists involved in public health to also work for industry?"

"Unfortunately, it is very common in medicine, and that shouldn't happen," Michael Thun replied. "It would be a good idea for researchers studying medications to not receive money from pharmaceutical companies or for those giving their opinions on the effects of chemical pollutants to not be paid by the industry making them."

"Yet this is what Richard Doll did?"

"Certainly, and it is very regrettable."[43]

This is a "regret" shared by Devra Davis, but in a more accusatory tone: "I was truly very disappointed to learn that the great Richard Doll, who had been a role model for an entire generation of epidemiologists, had secretly worked for the chemical industry," she told me. "Certainly, he was not the only one; there was also Hans-Olav Adami, from the Karolinska Institute in Stockholm, or Dimitry Trichopoulos, from Harvard,[44] but Doll's case is particularly serious, because his reputation was such that the whole world took what he said as gospel. His expertise contributed to holding back politicians' interest in environmental causes of chronic illnesses, as well as the regulation of highly dangerous toxins, such as dioxin and, more importantly, vinyl chloride."

The Harmful Effects of Vinyl Chloride

The vinyl chloride affair is indeed exemplary. Historians Gerald Markowitz and David Rosner write that it constituted "evidence of an illegal conspiracy by industry"[45] to keep a product on the market that was highly toxic, with the active complicity of a great scientific name, which happened to be that of Richard Doll. This affair represents a rarely rivaled peak in the art of premeditated manipulation and misinformation, and in fact, it shattered the last of my illusions regarding manufacturers' behavior, since they will truly stop at nothing when it comes to defending their poisons, whatever the cost or dangers may be. It offers yet another illustration of the atrocious ideology summed up in 1970 by a high-ranking Monsanto executive, as regards deadly polychlorinated biphenyls (PCBs) (sales of which had to be sustained at any price), which I will never be able to cite enough: "We can't afford to lose one dollar of business."[46]

Synthesized for the first time in 1835 by the Frenchman Henri Victor Regnault (1810–78), the director of the Royal Manufacture of Porcelain in Sèvres, vinyl chloride is a toxic gas that, when compressed, is used as a propellant in various aerosols (lacquers, cosmetics, insecticides, or air fresheners). The chemical compound is as efficient as it is dangerous. In *The Secret History of the War on Cancer*, Devra Davis recounts the story of Judy Braiman, who was hospitalized in 1965 after being diagnosed with lung cancer. The young woman's lungs were covered in layers of vinyl chloride, due to the lacquer she used daily to look like the stars of the time, decked out in impeccable perms. Judy Braiman survived, and then became a figurehead of the American consumer protection movement, but it would take until the middle of the 1970s for use of vinyl chloride to be banned in cosmetic products—but not in the fabrication of plastics or, notably, of the indispensable polyvinyl chloride (PVC).[47]

When it is assembled in chains (or polymers), vinyl chloride becomes PVC, a pioneering product in modern industry that is found in a number of everyday objects, including packaging, containers, and plastic wrap. Developed in the mid-1920s by Waldo Lonsbury Semon (1898–1999), a Goodyear chemist, the polymerization of vinyl chloride is an exceedingly dangerous process that involves very toxic fumes and high-risk operations, such as "scraping" the autoclaves used in the resin's production. In 1954, the Manufacturing Chemists' Association (MCA) arbitrarily decided to set the exposure standard in factories to 500 ppm. As Henry Smyth, a Union Carbide executive, noted in a memorandum that Gerald Markowitz and David Rosner found in the MCA archives, the standard was "based largely on single guinea pig inhalation studies by the Bureau of Mines."[48]

In the early 1960s, a strange disease began to appear in PVC factories in Italy, France, and then the United States. Acroosteolysis, which manifests by the progressive destruction of the distal bony phalanges, leads to horrible and very painful stunting of the fingers. In 1964, Dr. John Creech, the doctor for a Goodrich factory (which produced tires) located near Louisville, Kentucky, identified the first case, followed quickly by three others. They all concerned workers tasked with manually cleaning the polymerization tanks. "If four people doing the same type of work, in the same room, in the same department . . . come down with a bizarre situation like this, it doesn't take a rocket scientist to link it to industry—to their workplace," Creech would later report.[49]

The physician immediately informed Goodrich's management, which rushed to cover up the affair, as all the PVC producers did, including Monsanto, Dow Chemical, and their European counterparts. Manufacturers discreetly consulted with Robert Kehoe, director of Kettering Laboratories (see Chapter 8), who, after studying several cases, learnedly concluded that it was an "entirely new" occupational disease, in a letter addressed to R. Emmet Kelly, the medical director at Monsanto.[50] The Saint Louis company gathered data just as discreetly in one of its factories, repeating the methods it had previously used for PCBs; it asked one of its physicians, Dr. William E. Nessel, to organize X-rays of the hands of all its workers, without informing the workers about the motives behind the rare medical exam. "I am sure Dr. Nessel can prepare these people with an adequate story so that no problem will exist," R. Emmet Kelly wrote to one of the factory directors.[51]

Goodrich reacted similarly: on November 12, 1964, Rex Wilson, the head of the company's medical department, asked Dr. J. Newman, the physician at the Avon Lake, Ohio factory, to examine the hands of their employees, specifying, "I would appreciate your proceeding with this problem as rapidly as possible, but doing it incidentally to other examinations of our personnel. We do not wish to have this discussed at all and I request that you maintain this information in confidence."[52] Eventually, Newman would catalogue thirty-one cases of the bizarre disease out of a total of three thousand workers.[53]

Little by little, first with acroosteolysis, then with cancer, a true conspiracy between American and European manufacturers was put in place to hide the extreme toxicity of the PVC production process, and of the product itself, in order to impede any attempt at regulation.

The PVC Conspiracy

"We feel confident . . . that 500 ppm is going to produce rather appreciable injury when inhaled 7 hours a day, five days a week for an extended period. As you can appreciate, this opinion is not ready for dissemination yet and I would appreciate it if you would hold it in confidence, but use it as you see fit in your own operations."[54] Verald Rowe, Dow Chemical's toxicologist, wrote this to William McCormick, his counterpart at Goodrich, on May 12, 1959. The letter followed a secret study carried out under Rowe's direction, showing that rabbits exposed to 200 ppm of vinyl chloride developed microlesions in the liver. At that time, the standard that had been set by manufacturers was 500 ppm, and it would remain so for fifteen long years after that.

In May 1970, the Italian scientist Pierluigi Viola made a few waves at the Tenth International Cancer Congress, which took place in Houston. He presented a study there that showed that rats exposed to vinyl chloride vapors (four hours per day, five days per week, for twelve months, at a concentration of 30,000 ppm) developed skin cancer (65 percent), lung cancer (26 percent), and bone cancer. "The results reported . . . indicated that vinyl chloride is an effective carcinogenic agent for the rat," he concluded. However, he then added that "no implications to human pathology can be extrapolated from the experimental model reported in this paper."[55] European manufacturers, led by Montedison, an Italian company, immediately asked Cesare Maltoni—a leading cancer specialist in Bologna who would later found the Ramazzini Institute in 1987, named in tribute to Bernardino Ramazzini (see Chapter 7)—to conduct a study on the effects of vinyl chloride emissions. Using a protocol that would serve to bolster the Ramazzini Institute's reputation, the Italian scientist exposed a group of 500 rats to different concentrations much lower than those used by his colleague Pierluigi Viola, ranging from 10,000 to 250 ppm. This mega-bioassay was continued until the natural death of the guinea pigs, and its results were incontrovertible: 10 percent of the rats exposed at the weakest dose developed angiosarcoma, a very rare form of liver cancer, but also kidney tumors, after only eighty-one weeks of exposure. For industry, the matter was serious, because 250 ppm was half of the standard used in factories, and it was also the concentration found in hair salons, as a secret Goodrich memorandum pointed out.[56] More troubling still was Maltoni's comment that he was not ruling out the possibility that much lower doses could cause similar effects.

Faced with the urgency of the situation, European manufacturers—including Montedison, Imperial Chemical Industries in the United Kingdom, Rhône Progil in France (an affiliate of Rhône-Poulenc), and Solvay et Cie in Belgium—organized a meeting with their American counterparts, with whom they made a "secret deal,"[57] as early as October 1972. Several MCA documents, now declassified, reveal that the European companies claimed they were willing to hand over Cesare Maltoni's study data on the condition that the Americans never make them public without their prior permission.[58]

The Americans would keep their promise, even if it meant inciting a veritable plot against the U.S. Occupational Safety and Health Administration (OSHA). In January 1974, the National Institute for Occupational Safety and Health (NIOSH) contacted the MCA in order to take stock of the dangers of vinyl chloride and of the "voluntary standard" of 500 ppm, which OSHA had adopted when the standard was established in 1971. A meeting was set with Markus Key, director of NIOSH, on July 11, 1973, at the institute's Rockville headquarters. In order to respect the deal with European industry, the manufacturers devised a veritable battle plan over the course of secret meetings, the summaries of which were classified as "confidential": they decided that they would not mention Cesare Maltoni's study to the NIOSH director if he did not bring up the subject himself.[59] If, however, he mentioned the European study, they "could not deny awareness of the project and knowledge concerning certain preliminary results."[60]

The industry's worries did not only concern exposure standards in factories, which could be reduced, but also contamination of PVC food containers, such as plastic bottles. "Some of the questions that might be asked are does vinyl chloride stay in the diet, does it react with the food, and if so, to what forms does it react,"[61] noted Theodore Torkelson, Dow Chemical's toxicologist who also acknowledged that no test had been conducted to verify this hypothesis. In the end, the meeting was a success, since Markus Key, director of NIOSH, did not ask any difficult questions—"the chances of precipitous action by NIOSH on vinyl chloride were materially lessened," read the minutes, written by Union Carbide's representative.[62] But as we will see, the respite would be very short for manufacturers.

Red Alert for PVC Manufacturers

"Between September 1967 and December 1973, 4 cases of angiosarcoma of the liver were diagnosed among men employed in the polyvinyl chloride polymerization section of a B.F. Goodrich plant near Louisville, Kentucky. [. . .] Angiosarcoma of the liver is an exceedingly rare tumor. It is estimated that only about 25 such cases occur each year in the United States. Four cases, therefore, among a small number of workers at a single plant is a most unusual event, and one which raises the possibility of some work-related carcinogen, conceivably vinyl chloride itself."[63] Published in 1974 in *Morbidity and Mortality Weekly Report,* the Center for Disease Control's weekly bulletin out of Atlanta, this article was written by John Creech, the Goodrich doctor who, ten years earlier, had sounded the alarm after identifying four cases of another "extremely rare" disease, acroosteolysis. Shortly before the article's publication, Creech informed OSHA, who immediately organized an emergency series of hearings to review the regulation of vinyl chloride.[64]

Markus Key, director of NIOSH—which is OSHA's research institute—thus discovered that he had been lied to by industry at the infamous July 11, 1973, meeting. He detailed the "deception" at his deposition for a lawsuit filed against Goodrich and Dow Chemical by Holly Smith, the widow of one of the workers who had died from angiosarcoma of the liver. His testimony was filmed on September 19, 1995, and is very interesting, because it reveals the professional and personal mechanisms that allow manufacturers to mislead regulatory agency representatives. It was discovered that Dr. Markus Key had long known Verald Rowe, Dow Chemical's toxicologist who served as a spokesperson for manufacturers at the July 1973 meeting. If he had been deceived, it was simply because he could not *imagine* that Rowe could betray his trust, lying to him in an *intentional* manner.

Steven Wodka, the deceased's attorney, led the questioning, in the presence of Maureen Donelson, the bailiff from the District of Columbia court, and attorneys from both Goodrich and Dow Chemical. In the first section, Markus Key explained that representatives from the MCA had contented themselves with presenting Pierluigi Viola's study to him, which demonstrated the carcinogenic effects of vinyl chloride at an extremely high concentration (30,000 ppm), and that they informed him a second study was being done at "more reasonable" exposure levels, the results of which were not yet known.

"At any time during this meeting with the MCA group, including Dr. Rowe, were you informed that this new European study . . . had found tumors of exposures as low as 250 parts per million?" Steven Wodka asked.

"No," responded Dr. Key.

"Now, you've told us that at the time of this meeting, that you had known Dr. Rowe for a number of years in a professional sense, in a professional manner?"

"Yes."

"At the time of this meeting, did you trust Dr. Rowe as a professional colleague?"

"Yes."

"And at the time of this meeting, was it your belief that if Dr. Rowe knew that angiosarcoma of the liver had been produced in test animals as low as 250 parts per million, that he would have told you that information?"

"Objection," one of the opposing attorneys interjected.

"You can answer the question," said Steven Wodka.

"Yes," replied the NIOSH director.[65]

Markus Key's enormous "deception" can be understood by reading the end of the cross-examination: to cover up the manufacturers' lie, Verald Rowe went so far as to claim that they had informed Markus Key of the results of Cesare Maltoni's study, and ultimately modified the meeting minutes!

In any case, in February 1974, at the end of the first series of hearings, OSHA proposed to set the new exposure standard for vinyl chloride at 1 ppm, "the equivalent of one ounce of vermouth in eighty thousand gallons of gin," as David Michaels puts it.[66] To definitively rule on the matter, the agency announced a new series of hearings for June 1974. For manufacturers, this was another red alert. To prepare for the coming battle, they called upon the services of the firm Hill and Knowlton, experts in the art of "creating doubt," which had already pedaled its talents to lead, asbestos, and tobacco producers.[67] They hatched a veritable war strategy, booking a hotel suite across from OSHA's offices, where they set up their campaign headquarters. Training sessions were organized during which manufacturers honed their arguments, prepped by the agency's "public relations specialists." The four main points of their case, which were widely distributed to the press, can be read in a now declassified document entitled "Preparations for OSHA Hearings":

1. PVC products play an important role in our society. Unnecessarily strict standards would deprive the nation of many valuable and beneficial products.
2. Should PVC be eliminated, the economic and social hardship in terms of lost production and lost jobs would be severe.
3. It is technically infeasible to reduce occupational levels to those recommended by OSHA and NIOSH.
4. It has not been demonstrated that a health hazard exists at the levels recommended by SPI.[68]

The document is very interesting, because, as we have previously seen, "PVC" could easily be replaced by "bisphenol A" or "aspartame"—indeed, defense of these poisons always follows the same strategies, devised by so-called communication (or rather misinformation) specialists, which are thoroughly removed from scientific or health concerns. The document's conclusion is also quite enlightening, as it underlines the stakes of this incredible battle: "an equally serious potential problem could be the development among the consuming public of a crisis reaction regarding the possibility of danger from PVC products in the home—and even, by implication, from all plastics consumer products." Finally, it is worth noting that there are always prestigious newspapers ready to relay industry messages (as we saw with the *New York Times* and leaded gasoline in Chapter 8): "If government allows workers to be exposed to the gas, some of them may die," *Fortune* magazine coldly wrote in October 1974. "If it eliminates all exposure a valuable industry may disappear. . . . Medical and economic considerations collide head-on."[69]

Richard Doll's Credibility "Permanently Tarnished"

But all of industry's efforts would be in vain; at the end of the June 1974 hearings, OSHA established the new standard of 1 ppm, which went into effect on April 1, 1975. And, contrary to the predictions of professional doomsayers, PVC mostly survived the decision. The heralded "economic catastrophe" did not arrive. In fact, it was quite the opposite, as was triumphantly pointed out in *Chemical Week*, the magazine for chemical manufacturers, in an article published on September 5, 1977, entitled "PVC Rolls Out of Jeopardy, into Jubilation." The author describes how producers have been raising the prices of PVC to unprecedented levels and states. "Clearly, those actions signify U.S. vinyl producers' confidence that they have solved the 'OSHA problem' that threatened the viability of their industry less than two years ago. They have installed the equipment needed to meet the worker-exposure requirements set by the Occupational Safety and Health Administration, but without inflating production costs to the point where PVC's growth might be stunted."[70]

After this admission, one would have hoped that PVC producers had buried the hatchet for good, that is, by ending their systematic obstruction maneuvers any time new scientific or medical data called the safety of the modern-day poison into question. But that was not the case. In 1979, IARC carried out a preliminary evaluation of the product and provisionally concluded that it was "carcinogenic to humans": "Several independent but mutually confirmatory studies have shown that exposure to vinyl chloride results in an increased carcinogenic risk in humans, involving the liver, brain, lung and haemo-lymphopoietic system." Eight years later, a second evaluation confirmed the first, and "polyvinyl chloride (PVC)" definitively joined group 1 of IARC's classification system.[71]

And the war machine was back in motion! The MCA asked Richard Doll to conduct a meta-analysis of the studies that examined the carcinogenic effects of PVC. Published in 1988 in the *Scandinavian Journal of Work and Environmental Health*, the analysis concluded that only angiosarcoma of the liver could be potentially associated with PVC, but no other type of cancer could.[72] As many observers would comment—namely David Michaels, Paul Blanc, Devra Davis, and Jennifer Sass, who devoted an entire article to it in 2005[73]—the distinguished epidemiologist's new "study" was biased. In order to arrive at his conclusions, he had excluded several publications demonstrating that PVC caused brain tumors (among other tumors), which he arbitrarily considered "statistically insignificant."

As Jennifer Sass points out, "Doll did not acknowledge funding sources in his article." And yet, he should have; in 2000, although he was cited as an expert by industry in a suit filed by a worker suffering from a brain tumor, he eventually admitted that he had been paid "12,000 British pounds" (about $21,000) by the MCA to conduct his 1988 meta-analysis.[74] What he did not mention was that he was also being paid by Monsanto at that time.

"The vinyl chloride affair was the coup de grâce for Richard Doll's reputation," Richard Clapp, the Boston epidemiologist, told me. "It permanently tarnished his credibility as an authority in the field of environmental health. It's time to open our eyes to the fundamental role chemical pollution plays in the unprecedented increase in cancer—but also of neurodegenerative diseases and reproductive disorders—that characterizes the industrialized world."

An Epidemic in Industrialized Countries

"We, Scientists, Medical Doctors, Jurists, Ethicists and Citizens, convinced of the urgency and seriousness of the present situation, solemnly declare that: The development of numerous current diseases is a result of the deterioration of the environment; Chemical pollution represents a serious threat to children and to Man's survival; As our own health, that of our children and future generations, is under threat, the Human race itself is in serious danger."[75] Historically known as the "Paris Appeal," this "international declaration on the health dangers of chemical pollution" was sent to the United Nations Educational, Scientific and Cultural Organization (UNESCO) on May 7, 2004, during the "Cancer, Environment, and Health" symposium organized by Professor Dominique Belpomme's Association for Therapeutic Anti-Cancer Research (Association pour la recherche thérapeutique anticancéreuse, ARTAC).[76] The signatories included several leading figures we have already come across in this book: Richard Clapp, André Picot, Jean-François Narbonne, André Cicolella, Luc Montagnier, and, of course, Dominique Belpomme, who was the first French cancer specialist to publicly declare that cancer is, above all, an "environmental disease created by man."[77]

It suffices to visit the IARC website to see that the "crab with the golden claws"[78] has prospered the most in so-called developed countries—that is in Europe, North America, and Australia. According to data from Globocan 2008, which uses maps and graphics to present "incidence and mortality rates of cancer around the world," France is leading the international pack with an annual incidence of 360.6 new cases of cancer for every 100,000 people, just ahead of Australia (360.5), but far ahead of Canada (335), Argentina (232), China (211), Brazil (190.4), Bolivia (101), India (92.9), and Niger (68.6). The same French "distinction" is found in breast cancer (99.7), which is also the kind of cancer that is growing the most worldwide every year, even if industrialization levels vary enormously: incidence is 21.4 in Burkina Faso, 21.6 in China, and 27.2 in Mexico. The same disparities occur for prostate cancer, whose incidence rate is 118.3 in France, 83.8 in the United States, 82.7 in Germany, and only 3.7 in India. And for colon cancer: France (36), Germany (45.2), India (4.3), Bolivia (6.2), and Cameroon (4.7). And then there's thyroid, testicular, lung, brain, and skin cancers, not to mention leukemia, which has an incidence ten to twenty times higher in industrialized countries than in the developing world.

According to a study published by IARC in the European Union, then made up of twenty-five countries, 3,191,600 cases of cancer were diagnosed in 2006 (53 percent in men and 47 percent in women), or an increase of 300,000 new cases in relation to 2004.[79] What's more, childhood cancers are on the rise—proof that the phenomenon is not merely an effect of the aging population, as Richard Peto claims. Another IARC study further supports this in an analysis of sixty-three European cancer registries. Over the past three decades, the annual growth in incidence has been 1 percent for children from 0 to 14 years and 1.5 percent for adolescents (15–19 years). The phenomenon is relentless, and worsens from one decade to the next: for children, the rate increases by 0.9 percent between 1970 and 1980, but by 1.3 percent between 1980 and 1990. For adolescents, the increase is 1.3 percent between 1970 and 1980 and 1.8 percent between 1980 and 1990.[80] The situation is so troubling that in September 2006, the World Health Organization (WHO) sounded the alarm, demanding that a strategy be implemented to "control" what it calls "an *epidemic* of *preventable* diseases."[81] The use of the term "epidemic" to describe the overwhelming propagation of cancer, which is not, however, an "infectious communicable disease," according to the *Petit Robert* dictionary, marks a turning point in the generally very diplomatic language used by the UN organization. In choosing this word, which undoubtedly rubbed some the wrong way, the WHO stressed the exceptional and abnormal character of the disease's spread.

In France, the "epidemic" was the subject of a 2008 collective expert assessment from the French National Institute of Health and Medical Research (Institut national de la sante et de la recherche medicale, INSERM), requested by the French Agency for Environmental and Occupational Health Safety (Agence francaise de sécurité sanitaire de l'environnement et du travail, AFSSET). INSERM bravely countered the 2007 report *Les Causes du cancer en France* (The Causes of Cancer in France) (see Chapter 10). Incidentally, it is worth applauding the enormous amount of work accomplished by the thirty-three experts called upon to write this massive 889-page report entitled *Cancers et Environnement* (Cancers and the Environment), which, right from the introduction, squashes the weak arguments proffered by Richard Peto, as well as other respectable scholars: "An increase in the incidence of cancers has been observed for about twenty years. If we take into account demographic changes (increased age of and rise in the French population), the increase of the incidence rate since 1980 is estimated at +35% in men and +43% in women."[82] The authors point out that "environmental changes could possibly be responsible for the increase observed in certain cancers." The tone is admittedly cautious, but nevertheless, the report marks a divergence from its predecessors, while systematically minimized, or even completely ignored, the role of chemical pollution.

To conduct their assessment, INSERM researchers identified "nine sites of cancer for which incidence has continually increased over the last twenty-five years: lung cancer, mesothelioma, malignant hemopathies, brain tumors, breast, ovarian, testicular, prostate, and thyroid cancers."[83] Then, they analyzed the data from the international scientific literature, focusing exclusively on "environmental factors" defined as "physical, chemical, or biological agents present in the atmosphere, water, soil, or food, exposure to which is undergone and not generated by individual behaviors." The experts thus excluded "active smoking," whose role in the etiology of certain cancers is incontestable, and focused exclusively on "general environmental factors" (such as pesticides, dioxins, PCBs, certain heavy metals, particles resulting from automobile traffic, etc.) and "those present in the occupational environment." In their conclusions, they recommend, "reinforcing epidemiological, toxicological, and molecular research in the domain of environmental risks of cancer, [because] it is an important issue in terms of public health, [which] concerns a large portion of the population."

"We estimate that 80 percent to 90 percent of cancer is linked to the environment and lifestyle," IARC director Christopher Wild told me. "It has been proven by studies on people who immigrate from one region of the world to another, where exposure to chemical pollutants and lifestyle vary, so they adopt, so to speak, the pattern of cancers present in the regions they move to." Many of the studies cited by Wild involve Japanese immigrants who moved to Hawaii. They show how in one or two generations, the immigrants "adopted" the profile of American cancers, demonstrating that "the risks of cancers of the prostate, corpus uteri, colon, thyroid, breast, ovary, and testis were elevated,"[84] the incidence of which is much lower in Japan. As André Cicolella and Dorothée Benoît Browaeys underline in *Alertes santé* (Health Alerts), "their genetic heritage did not change, but their environment did."[85]

Another way to estimate the impact of environmental factors on the etiology of chronic illnesses consists in comparing the health evolution of what are called "monozygotic twins," who are from one single fertilized ovum and who thus have the exact same genetic makeup. Indeed, "if cancer was a purely genetic disease, true twins would have the same types of cancer," yet "this is far from being the case."[86] This was clearly demonstrated in a 2000 study that examined the medical situation of 44,788 pairs of twins monitored in Sweden, Denmark, and Finland, in order to evaluate the risks of twenty-eight possible cancer sites. The conclusion was irrefutable: "Inherited genetic factors make a minor contribution to susceptibility to most types of neoplasms. This finding indicates that the environment has the principal role in causing sporadic cancer."[87]

A European Parliament resolution on May 6, 2010, also came to this conclusion—evidence that things are starting to change. Entitled "Action Against Cancer," it stressed the role of environmental factors in the disease's origins, specifying that "environmental factors include not only environmental tobacco smoke, radiation and excessive UV exposure but also exposure to chemical contaminants in food, air, soil and water due to inter alia industrial processes and agricultural practices." The resolution's authors then ask the European Commission to encourage "reducing occupational and environmental exposure to carcinogens and other cancer-producing substances."[88]

To accomplish this, as we will see in the third part of this book, the regulation process of chemical substances would have to be reviewed from top to bottom: as it is now, it protects producers much more than it does consumers and citizens.

PART III

Regulation at Industry's Beck and Call

12

The Colossal Scientific Masquerade Behind Poisons' "Acceptable Daily Intakes"

Science has become the protector of a global contamination of people and nature.

—Ulrich Beck

"The regulatory system, which is supposed to protect public health against the effects of carcinogenic products, does not work. If it was efficient, the incidence rate of cancer would have decreased, but this isn't the case. I think that the principle of the acceptable daily intake, which is meant to be the main tool in the regulation of toxic products contaminating the food chain, protects industry more than consumer health." Erik Millstone, who was originally a physician before converting to philosophy and the history of science, is a British professor of "science policy," a unique position in Europe. In concrete terms, he is interested in the manner in which public authorities establish their policy in the fields of health and the environment, and, more particularly, in the role science plays in the decision-making process. One snowy day in January 2010, I traveled to Brighton in southern England, to meet Millstone at the University of Sussex. He welcomed me to his office, which was filled with books and documents carefully labeled according to the research projects to which he has devoted the last thirty years of his career: "Lead pollution," "Bovine spongiform encephalopathy," "Genetically modified organisms," "Pesticides," "Food additives," "Aspartame," "Obesity," "Acceptable daily intake."

The ADI: A "Black Box"

Known for his outspokenness and artful dissections of the most complex cases, Erik Millstone is one of the top European specialists on the regulatory system that governs food safety, but is also one of its most dreaded critics. "I defy you to find any scientific study that justifies the principle of acceptable daily intake, because there are none," he earnestly explained to me. "Consumer safety rests on the use of a concept that was thought up at the end of the 1950s and has become intangible dogma, even though it is completely outdated and no one can explain its scientific credibility."[1]

As a matter of fact, I had spent weeks trying to reconstruct the genesis of the "acceptable daily intake" (or "admissible daily intake")—"ADI" in the technical jargon. The ADI is used to set standards for exposure to chemical products that come into contact with our food—pesticides, additives, and food plastics. And while a web search does quickly provide a definition, stating in essence that "ADI is the quantity of a chemical substance that can be ingested daily and throughout an entire lifetime without posing any health risks," this is not accompanied by any scientific reference that would allow understanding of how the concept was developed. And when we question these people who make daily use of this tool to determine, for example, what quantity of pesticides can be tolerated in our food, they generally give evasive and somewhat discomfited answers. For example, when I questioned Herman Fontier, the head of the pesticides unit at the European Food Safety Authority (EFSA), during our meeting in Parma in January 2010, he replied: "I have been working on the authorization of plant protection products for twenty-three years, and I have always been familiar with the concept of the daily acceptable intake, but I must admit that I never asked how this instrument that regulates the ingestion of chemical substances was devised. What is certain is that there is a *consensus* in the scientific world that it is necessary to set an ADI to protect consumers."[2]

Listening to the European expert's brief explanation, I reflected on my investigation into Monsanto, during which I had tried in a similar way to determine the origins of the "principle of essential equivalence," which also elicited a "consensus" on the regulation of genetically modified organisms (GMOs). I discovered that this concept—sanctioned in 1992 by the U.S. Food and Drug Administration (FDA), and which stated that a transgenic plant is "substantially similar" to the conventional plant from which it came—is not based on any scientific data; rather, it stems from a political decision, largely influenced by the commercial interests of the world leader in biotechnologies. Nevertheless, the idea was so well established in international regulatory agencies that they continue to invoke it to justify the absence of serious scientific evaluation of transgenic plants put on the market.

Everything seems to indicate that this is also the case for "acceptable daily intake," which strongly resembles what sociologist and philosopher of science Bruno Latour calls a "black box," that this, the forgetting of how scientific or technical gains—which were then adopted as proof, particularly after heated controversies—were actually made.

In his fascinating work, *Science in Action*,[3] Latour explains how once an original discovery—such as DNA's double helix structure or the Eclipse M V/8000 computer, the fruit of a long process of experimental and theoretical research—becomes a "cold stable object" or an "established fact," no one, including the scientists using it as a tool, is in a position anymore to understand its "inner workings" or unravel the "endless links" that led to its creation. In a similar fashion, the principle of the ADI, referenced endlessly by toxicologists and chemical risk managers, has become heavily abridged "tacit knowledge," the history of which has been so lost in the annals of time that it "could have been known for centuries or handed down by God Himself together with the Ten Commandments."

"The problem," Erik Millstone stressed, "is that the ADI is a black box that is very different from those Bruno Latour uses as examples. If the DNA double helix is an established scientific reality other researchers use as support to advance knowledge—for the human genome, for example, it is still possible, for those who have the ability and the time, to piece together the multiple steps that led James Watson and Francis Crick to make that discovery. But for ADI, there is nothing like it, because it was the result of an arbitrary decision instituted as a pseudo-scientific concept to protect manufacturers and politicians who need to hide behind experts to justify their actions. The acceptable daily intake is an indispensable artefact for those who have decided that we have the right to use toxic chemicals, including in the process of food production."

"And we really don't know who invented the concept?" I insisted.

"According to the World Health Organization, ownership can be traced to a French toxicologist by the name of René Truhaut," Millstone answered, "although in the United States they prefer to attribute it to Arnold Lehman and Garth Fitzhugh, two Food and Drug Administration toxicologists doing similar work."

The Pioneer René Truhaut: French Toxicologist and Paracelsus Enthusiast

Stubborn as a mule, I traveled to Geneva to consult the World Health Organization (WHO) archives. And in the impressive documentation center's subject index, I indeed found several references to René Truhaut (1909–94), who chaired the toxicology department at the French Faculty of Medicine in Paris and is considered a pioneer in French cancer studies. This "indefatigable and tenacious worker"—who authored a doctoral thesis entitled "Contribution to the Study of Endogenous Carcinogens"—became a food toxicology specialist who attempted to "elucidate the future of a large number of chemical substances in the body and interpret their mechanism of action," to quote Belgian scholar Léopold Molle, from a 1984 tribute.[4] "A pioneer in toxicokinetics,"[5] Truhaut directed the toxicology laboratory of the French Faculty of Pharmacy in Paris, where he focused on the "evaluation of toxic possibilities, including carcinogenic possibilities, of chemical agents likely to be incorporated, voluntarily or involuntarily, into food, such as pesticide and anabolic residue, preservatives and emulsifiers and natural and synthetic colorings."

I watched one of the rare interviews given by René Truhaut, which was included in a documentary produced in 1964 by Jean Lallier (1928–2005). Entitled *Le Pain et le Vin de l'an 2000* (Bread and Wine in the Year 2000), the film was already asking all of the (right) questions I am trying to answer in this book, fifty years later. It speculated namely on the efficiency of the then burgeoning regulation of chemical products contaminating the food chain, and on the role toxicologists played in the process. On screen, René Truhaut appears in a white coat, seated in his Faculty of Pharmacy laboratory. "If you will allow me to make a comparison," he explains with obvious pedagogical curiosity, "last century, when Pasteur, citizen of the world, discovered the danger of bacteria—in the field of food specifically, a great deal of importance was given to the microbiological testing of food, and an entire series of laboratories was founded to carry out that testing. Well, it should be the same in the context of the inspection of chemical agents added to food, because their dangers, while less insidious—less spectacular, if you wish—are certainly not, in my opinion, any less serious."[6]

A member of the French National Academy of Medicine (Académie nationale de medicine) and the Academy of Sciences (Académie des sciences), René Truhaut was welcome by all the major international authorities, as his impressive resume shows: he was a member of the Permanent International Committee on Occupational Diseases, the International Labor Bureau (secretariat of the International Labor Organization), the International Union Against Cancer, the International Union of Pure and Applied Chemistry, as well as a number of scientific committees in the European Community, including the Committee on the Ecotoxicity and Toxicity of Chemical Products, which he oversaw. But his name is mostly associated with WHO, with which he was regularly involved for over thirty years. It was under the wing of this UN institution that he developed the idea of acceptable daily intake, as he claimed in an article published in 1991: "My position is relatively unassailable, I believe, in claiming to be the true instigator of the concept of the acceptable daily intake (ADI)—as, indeed, has been acknowledged in numerous articles [. . .] written by experts who, like myself, were actively involved in the field during the period from 1950 to 1962," he wrote with a certain reserve, which could be due to either caution or modesty. "Unfortunately and paradoxically, I had published nothing in the scientific journals at this time."[7]

This is indeed a shame, for there is no additional information about the scientific genesis of the infamous principle, which, from reading the French toxicologist, does not seem to derive from a duly approved experimental model, but rather from a theoretical idea—granted a brilliant and fertile one—that he developed over the course of his research: "Since the beginning of my career, I have been committed to the toxicological evaluation of long-term exposure of chemical agents, which man faces in different domains. I have always considered Paracelsus's principle, written five centuries ago, as a golden rule: 'Sola dosis facit venemum' (the dose makes the poison). It has led me to give principal importance to the establishment of dose–response relationships in the methodology of toxicological evaluation, so as to be able to set acceptable limits."

The "father of toxicology" Paracelsus notably played a role in the work conducted by Robert Kehoe on lead toxicity (see Chapter 8). The Kettering Laboratory director, who was paid by manufacturers, performed autopsies on the cadavers of newborns who were victims of lead poisoning, and carried out experiments on "volunteers" to determine an exposure amount that seemed safe to him, which he could then use to counter the attacks from opponents of leaded gasoline. Kehoe succeeded in imposing a theory based on four principles, which curiously enough resembles the concept of ADI: "1) lead absorption is natural; 2) the body has mechanisms to cope with lead; 3) below a certain threshold, lead [is] harmless; 4) the public's exposure [is] far below the threshold and [is] of little concern."

1961: The "Scientific" Ratification of the "Somewhat Nebulous" ADI Concept

It is a safe bet that René Truhaut was familiar with the work of the toxicologist lured in by poison manufacturers because, like Robert Kehoe, he was interested in the effects of occupational pollutants. Truhaut had promoted "acceptable limits of toxins in work atmospheres and/or in biological fields of exposed subjects" at the Permanent International Committee on Occupational Diseases, which met in Helsinki in 1957. His research in the field of occupational health won him the Yant Award in 1980 from the American Industrial Hygiene Association, of which Kehoe was president.

However, in the documents I found at the WHO, the "originator of the concept of acceptable daily intake," as he calls himself, says nothing about the work that inspired his invention or about the studies he might have conducted to support it. He simply lays out a timeline of the events that led the WHO and the Food and Agriculture Organization (FAO) to adopt his proposition. In a 1981 text, he writes, "in 1953, the Sixth World Health Assembly [the organ that determines the policies of the WHO] expressed the view that the increasing use of various chemical substances by the food industry had in the last few decades created a new public health problem which might usefully be investigated."[8] Meanwhile, FAO noted the "serious lack of data regarding many food additives in relation to both their purity and to the health hazards involved in their use."

So, in September 1955, the two UN organizations decided to create a committee of experts tasked with studying the multiple aspects of the problems linked to the use of food additives, in order to establish guidelines or recommendations for public health authorities and other governmental agencies in the different countries around the world. This seminal conference thus dealt only with "food additives," which it defined as the time as "non-nutritive substances which are added intentionally to food, generally in small quantities, to improve its appearance, flavor, texture, or storage properties."[9] The initiative led to the creation of the Joint FAO/WHO Expert Committee on Food Additives (JECFA), whose first session was held in Rome in December 1956. The experts, who were appointed by FAO and WHO and included René Truhaut, adopted the principle of "positive lists," by which "the use of any substance not authorized on an adequate toxicological basis is prohibited."[10] Concretely, this recommendation meant that no new food additive could be used by the food industry without having previously undergone toxicological tests that had to be submitted for JECFA evaluation (or evaluation by a national agency). In essence, it was a spectacular improvement, moving things distinctly in the direction of consumer protection. But we will see with the example of aspartame (see Chapters 14 and 15) how this system of evaluation would be regularly derailed by industry for its own profit.

The experts also stressed that one "must pay constant attention to the technological usefulness of the additive being toxicologically evaluated."[11] This remark is interesting, because it allows us to understand the ideological context surrounding René Truhaut and his colleagues' approach. At no moment did they question the social necessity to use chemical substances in food production, even if these substances were theoretically toxic, as Truhaut himself acknowledged in the second televised interview I watched: "A consumer who absorbs, for example, a small quantity of coloring over two weeks, over two months, over one or two years, may not have any harmful effect," he declared in his high-pitched voice. "But it should be expected that these small doses repeated for a long time, day after day, for an entire lifetime, may sometimes carry extremely insidious risks, and sometimes even irreversible risks, as there are certain colorings, for example, that have proven, at least in animals, to be capable of causing malignant proliferations—that is to say, cancers."[12]

René Truhaut, who appeared genuinely worried about the public health risks in connection with chemical adjuvants in food, expressed his concern (which was rather rare at the time) about the "risks of progress." That said, he in no way intended to cast doubt on the idea that these innovations could have a "technological utility"; for him, it was not a matter of demanding an outright ban on carcinogenic substances "intentionally added to food" in the economic interest of producers alone, but to manage as best as possible the risk they generate for consumers by trying to reduce it as much as possible. So during the second JECFA session, which took place in Geneva in June 1957, the experts discussed at length the type of toxicological studies that had to be required of manufacturers in order to determine the dose of a poison that could be tolerated in food. And I do mean "poison," for if the substance in question was not suspected of being one, the JECFA would have no reason to exist, nor would the idea of the ADI, for that matter.

To truly understand the approximative (to say the least) quality of the process, it is important to note the account René Truhaut would write in 1991: "I contributed to introducing a new chapter in the final report, 'Evaluation of concentrations that are *probably harmless* to humans,' with the following sentences: 'Based on these various studies, in each case, one could set the maximum dose that does not cause, in the animals used, *any discernible effect* (hereafter called the *maximum ineffective dose*, for brevity's sake). When this dose is extrapolated to humans, it is useful to provide for a certain safety margin.'" And he adds, with shocking frankness, "This was as yet somewhat *nebulous*."[13]

Indeed, this is the least that can be said, but that did not keep the JECFA from adopting the concept of acceptable daily intake at its sixth session in June 1961, where the experts decided that a "dose which [induces] no effect of toxicological significance [. . .] should be expressed in mg/kg body weight per day."[14] Before detailing what exactly this cabalistic unit of measurement means, I will stress, yet again, the lucidity of the "father of the ADI," who admits in the same breath the limits of his creation: "When speaking of toxicologically ineffective doses in experimentation we must never forget that only a zero dose is truly without effect; *all other doses do have an effect, be it ever so little*."[15] In other words, the ADI is not a cure-all, but it allows limiting the damage that ingested substances would inevitably cause, as would be the case with food additives, as well as with pesticide residues.

In 1959, when the first JECFA sessions took place, the FAO proposed the creation of a similar committee, charged with studying the hazards to consumers arising from pesticides residues on food and feedstuffs.[16] This new initiative is proof, if it is even needed, that before this time, no one had seriously worried about the effects that pesticides could have on human health, even though agricultural poisons had already widely conquered farmers' fields. Three years later, when Rachel Carson's *Silent Spring* was making headlines worldwide, the FAO held a congress to formulate and recommend a plan for future action concerning scientific, legislative, and regulatory aspects of the use of pesticides in agriculture, as René Truhaut, who was one of the main players in these meetings, would report in 1981.[17]

He describes in particular his participation in a work group "on control of the olive fly—the olive being, as you know, a key crop in the Mediterranean economy." He specifies: "I was confronted with the problem of fixing maximum residue limits for certain organophosphorus insecticides—notably parathion—in olive oil for human consumption.[18] The generally accepted concentration limit for olive oil worldwide was 1 mg/kg oil. Toxicologically speaking, however, the decisive factor is the daily quantity of oil consumed. The Greek shepherd, surrounded by olives, dips his bread into that oil and can ingest up to 60 g/day. He therefore absorbs far more parathion than consumers whose olive oil intake comes only from salad dressing. Reasoning from this example, my line of thought found further support in the idea that we needed to turn the problem on its head and fix an intake which could be used to calculate the tolerances to be fixed for any given food according to the average quantity consumed in a particular region."[19] What the French toxicologist was describing in 1991 corresponds exactly to the task assigned to the Joint FAO/WHO Meeting on Pesticides Residues (JMPR), the expert committee instituted by WHO and FAO in October 1963 to establish the ADI of pesticides, but also what is called "maximum residue limits" (MRL), or the quantity of pesticide residue permitted on each treated agricultural product (see Chapter 13).

The Manufacturing Lobby: Active ADI Proponent

"I hope I have shown how the use of the ADI concept has made a powerful contribution to protecting the health of people the world over, and oiling the wheels of international trade,"[20] René Truhaut concludes simply in his retrospective article—which was in fact the transcription of a speech given at a workshop called "The ADI Concept: A Tool for Ensuring Food Safety," organized in October 1990 in Belgium by the International Life Sciences Institute (ILSI).[21]

That last detail is interesting, because ILSI has long been an active proponent of the idea of the ADI, promoting it via various symposia and publications. Yet, this "institute" is far from neutral, since it was founded in Washington, DC, in 1978 by big food companies (Coca-Cola, Heinz, Kraft, General Foods, Procter & Gamble), which were then joined by many other leading firms not only in that sector (Dannon, Mars, McDonald's, Kellogg, and Ajinomoto, the main producer of aspartame), but also from the pesticide sector (such as Monsanto, Dow AgroSciences, DuPont de Nemours, BASF) and the pharmaceutical sector (Pfizer, Novartis).[22] With the exception of the pharmaceutical industry, all these businesses prospered thanks to the advent of the green and food industry revolutions—they produce or use chemical products that contaminate our food.

The ILSI Europe website,[23] which presents the institute as a "non-profit organization," states that its "mission" is to "play a catalytic role in identifying and addressing critical scientific issues related to nutrition, food safety and the environment," with the aim of providing "coherent scientific answers to scientific issues of public interest through scientific programs that are of mutual concern to industry, government and academia," with the "ultimate goal of . . . the improvement of public health." But behind these purported good intentions lurks a much more prosaic reality.

Until 2006, ILSI had exceptional status at the WHO, as its representatives could participate directly in working groups pursuing the establishment of international health standards. The UN institution revoked this privilege after it was revealed that the industrial organization, under cover of pseudo-independence, was engaged in lobbying to promote its members' interests.[24] It was thus discovered that it had funded a report on carbohydrates, published by the WHO and FAO, which concluded that there was no direct link between overconsumption of sugar and obesity or any other chronic illness.[25] Similarly, in 2001, an internal WHO report denounced the "political and financial connections" between the ILSI and the tobacco industry,[26] for which the institute had funded a certain number of studies minimizing the health impact of secondhand smoke, just when International Agency for Research on Cancer was planning on classifying it as carcinogenic to humans. This exposé was based on seven hundred declassified documents from the "Cigarette Papers" (see Chapter 8), which attested to sixteen years of intense collaboration between 1983 and 1998.[27]

Additionally, in 2006, the Environmental Working Group in Washington revealed that the Environmental Protection Agency (EPA) had based its exposure standards for perfluorocarbons (PFCs)—notably present in the composition of Teflon, which is found in nonstick pans, for example—on a report supplied by the ILSI.[28] The ILSI had concluded that cancers produced in rats by these highly toxic substances could not be extrapolated to humans and that they could thus be considered harmless. Eventually, the EPA filed a suit in July 2004 against DuPont, an ILSI member and the main producer of Teflon, which was sentenced in December 2006 to a fine of $16.6 million for having concealed—for over twenty years—experimental studies showing that exposure to PFC "causes cancer, birth defects and other serious health problems in animals."[29]

As the American biologist Michael Jacobson (co-founder in 1971 of the Center for Science in the Public Interest) stressed in 2005, the ILSI boasts its desire "'to work toward a safer, healthier world.' The question is, safer and healthier for whom?"[30] What is certain is that the institute has significant financial means at its disposal, allowing it to "sponsor conferences and send scientists to government conferences to represent industry's take on controversial issues." Included among these is the ADI, to which the ILSI dedicated an entire "monograph" in 2000, proof that it did indeed value René Truhaut's creation a great deal.

Diane Benford: "Why We Need the ADI"

A monograph entitled *The Acceptable Daily Intake, a Tool for Ensuring Food Safety*[31]—which was also the title of the workshop in which René Truhaut participated ten years earlier—is quite the rare item because, as we have seen, the ADI is a "black box" created ex nihilo and for which reference studies are scant. The ILSI asked Diane Benford, who directs the chemical risk department at the Food Standards Agency (FSA) in the United Kingdom, to write the text. Note that in order to praise the merits of a tool favored by toxicologists and manufacturers, the ILSI called upon a public authority representative responsible for monitoring consumer health. And I must admit that it was not easy to get a meeting with the British toxicologist, who I suspect must have googled me and likely wanted to avoid my troubling questions. And yet, Angelika Tritscher, secretary of the JECFA and of the JMPR at the WHO (whom we will encounter soon), had given me permission to use her name as a reference. Tritscher frequently attended ILSI events and had informed me of the ILSI monograph. Finally, after numerous e-mail exchanges, Diane Benford agreed to meet me, on the condition that I first send her the questions I planned to ask—which was not really a problem, as I intended to ask her, a licensed ADI specialist, to explain exactly how the ADI is calculated.

During my trip to London in the Eurostar, I carefully combed through her text, which starts with this introduction: "The concept of the Acceptable Daily Intake, the ADI, is internationally *accepted* today as the basis for estimation of safety of food additives and pesticides, for evaluation of contaminants and by this, for legislation in the area of food and drinking water. The public concerns for safety of foodstuffs has led to a requirement for more *transparency* in the expert evaluations of chemicals in relation to human health. [. . .] Understanding the ADI concept will improve the *transparency* and the *confidence* in the evaluations."[32]

In this sort of document, where each word has been carefully weighed, we have to read between the lines, and here, everything seems to indicate that the ILSI's commission is responding to its members' desire to defuse recurring critiques regarding the opacity of the poison regulation system, for which the ADI is the cornerstone. These critiques are not new, as proven by this surprising admission from René Truhaut, written in the first person plural: "We are fully aware that, because of the multiplicity and of the complexity of the problems, the approaches taken are far from being perfect in every case. Consequently, we understand and sometimes we agree with criticisms expressed against the doctrine applied till now by the Joint FAO/WHO Expert Committee. The corollary is the necessity to keep an open mind to new knowledge permitting to correct or to improve the methodology of toxicological evaluation. Further research in this typically multidisciplinary field must be encouraged and supported."[33]

To be honest, the French toxicologist's "confession" definitively reconciled me with him, because he suddenly appeared to be a man of good faith who wanted to avoid the heralded health disaster, incapable of imagining to what point the embryo of a system that he had helped to implement would be misappropriated by manufacturers, whose sole objective was precisely to impede its "correction" or "improvement" for the benefit of consumers (which is without a doubt what Truhaut would have wanted). Thus, if the ILSI asked Diane Benford to write a monograph on the ADI, it was because its very generous sponsors feared that the valuable "doctrine," which had served their interests so well, would succumb to critiques about the lack of transparency of the system it embodied.

After her foreword, the British toxicologist addresses the banalities of industry in a section called "Why We Need the ADI," where the "We" refers to consumers, for whom the "monograph" is clearly meant: "Throughout the twentieth century there has been an increasing trend towards the use of stored and processed foods. Initially this was a response to industrialization and the need to provide food for large numbers of people living in cities. [. . .] The processes involved in producing and storing foods frequently require the addition of chemicals (either natural or man-made) to improve the safety (microbiological safety) or to preserve nutritional quality. An additional benefit is increased palatability and attractiveness of foodstuffs to the consumer. Clearly the safety of such chemicals has to be assured and their use controlled in order to avoid harmful effects." Benford then recalls the role of René Truhaut, the "father of the ADI," before citing the incontrovertible Paracelsus: "All substances are poisons; there is none which is not a poison. The right dose differentiates a poison and a remedy."

Falsified Studies and "Good Laboratory Practice"

"The basic concept underlying the ADI or any chemical risk assessment is the Paracelsus Principle—the dose makes the poison. Is that true—can you explain to me what that means?" I asked the head of the British health standards agency.

To which Diane Benford replied: "With increasing dose it becomes more likely that you will have harmful effects occurring. In principle, with absolutely anything—even with things like water and oxygen which we can't live without, but if we have too much of them then they can be harmful to us as well. As you go down to lower levels, with most things it becomes less likely that they will have any kind of adverse effects."

"Certainly," I said, slightly surprised by the comparison. "But between water and a pesticide designed to kill, there is nevertheless a difference, isn't there?"

"Yes, but. . . . Generally speaking, with most elements, the weaker the dose, the lower the probability of having negative effects."

"What you toxicologists call the 'dose–response relationship?'"

"That's right. The number of individuals responding and also the severity of the response will increase as the dose increases."

"That means that all this assessment is based on the assumption that a chemical substance is supposed to produce a harmful effect and we are just trying to find a level where we won't have any effect?"

"Yes," said the British toxicologist after a long pause. "The toxicological studies are looking for effects. They are looking to identify the range of different effects that a chemical might have, and looking for the doses that don't cause those effects and then applying this safety factor in order to establish the ADI."

"It's a very complicated system, isn't it?"

"Yes! There's a lot of information that needs to be assessed and we do the best we can to protect consumers."

"And who carries out toxicological studies?"

"Industry. These toxicological studies are very expensive. It would be a very large burden on tax if they were publicly funded, and clearly there might be concern that if it's in the industry's interest to get a product on the market—you might question whether or not they are conducting the studies properly—and because of that there has been development of guidelines that define the protocols for the way they are done and also define the quality assurance processes that mean that everything is recorded properly, people are properly trained, studies are conducted properly, and then it's possible to reconstruct the results of the study on the basis of the report and the raw data and there will be external inspection to make sure that they have been conducted properly."

"This is what is called 'good laboratory practice'?"

"Yes."

"The guidelines were created because of some big scandals in the U.S. at the end of the seventies where big laboratories working for industry just cheated."

"Yes, they did. And the guidelines have been introduced now and there are lots of inspections to make sure that everything is conducted properly."[34]

In my book *The World According to Monsanto*, I wrote about a trial that made headlines at the end of the 1980s: it involved Industrial Bio-Test Labs (IBT), a private laboratory in Northbrook, Illinois, one of whose directors was Paul Write, a former Monsanto toxicologist recruited in the early 1970s to supervise studies on the health effects of polychlorinated biphenyl (PCB), as well as a number of pesticides. In combing through the laboratory's archives, inspectors from the EPA discovered that dozens of studies presented "serious deficiencies and improprieties" and "routine falsification of data" designed to conceal "countless deaths of [tested] rats and mice."[35] Among the implicated studies were thirty tests carried out on glyphosate (the active ingredient in Roundup).[36] It was "hard to believe the scientific integrity of the studies," noted an EPA toxicologist, particularly "when they said they took specimens of the *uterus* from male rabbits."[37]

In 1991, Craven Laboratories was accused of falsifying studies that were supposed to evaluate the effects of pesticide residues, including Roundup, present on fruits and vegetables, as well as in water and soil.[38] "The EPA said the studies were important in determining the levels of a pesticide that should be allowed in fresh and processed foods," wrote the *New York Times*. "As a result of the falsification . . . the EPA declared pesticides safe when they had never been shown to be."[39] The widespread fraud resulted in a five-year prison sentence for the owner of the laboratories, whereas Monsanto and the other chemical companies that had benefited from the falsified studies were never brought to justice.

The Key Concept of NOAEL: "No-Observed-Adverse-Effect Level"

Obviously, all of this is far from reassuring, especially when, as we have seen throughout the preceding chapters, industry is willing to go very far in order to keep its products on the market, however toxic they may be. So logically, one could fear the same when it comes to obtaining approval of said products. In concrete terms, "toxicological studies" are led on laboratory animals, since, as Diane Benford writes, "it would not be ethical to give a chemical to human volunteers unless there were a reasonable degree of confidence that they would not be harmed."[40] This remark is significant, because it underlines the first approximation—what some would call an "absurdity"—that characterizes the evaluation system of toxic products that were deliberately introduced onto our plate, in the name of a certain idea of "progress." As René Truhaut explained, "Undoubtedly, adequate evidence from human studies would be the most satisfactory for the assessment of human hazard [. . .] but, many difficulties and limitations arise in regard to the enforcement of this ideal approach and, for this reason [. . .] without evidence to the contrary, man is assumed to behave like the most sensitive species tested. It is obvious that it would be more adequate to select the animal species most comparable to man."[41] Here we have a statement that is, at the very least, "nebulous," to use the word of the "father of the ADI," even more so given that no experimental model has been developed to determine which animal species is the most likely to behave like humans in the case of poisoning by chemical products. Failing that, rodents (mice, rats, rabbits) are generally used, and in the most delicate cases, dogs and monkeys.

First, subjects are exposed to a high dose of the substance being tested, generally orally, to determine what is called the "lethal dose," or in the jargon, the "LD50" or "median lethal dose," meaning the dose that kills half of the animals. It should be remembered (see Chapter 2) that the notorious "LD50" is a derivative of Haber's rule, named for the German chemist who invented poisonous gases to be used in combat. The rule expressed a relation between the concentration of a gas and the exposure time necessary to cause the death of a living being—the smaller the product of the two factors, the bigger the lethal power of the gas. It is the same for LD50, which is an indicative value of the degree of toxicity of a pesticide, for example. And the "father of chemical warfare" had observed that exposure to a weak concentration of poisonous gas over a long period often had the same lethal effect as exposure to a high dose for a short time. Strangely, not only regulatory agencies, but also the JECFA and the JMPR, seem to have ignored these conclusions, as their experts are obstinate in believing that it is possible to find a dose that is harmless in the long term, even when the substance proves to be lethal in a strong dose.

"What you are doing is looking for a range of possible adverse effects," Diane Benford explained. "So, for example, you are looking to see if it causes damage to the tissues and the organs, looking for effects on the nervous system, the immune system—always very interested in the possibility of causing cancer because that is of course something that's a concern to people. So you are looking at a wide range of different possible adverse effects in animals."

In fact, when you read the monograph the toxicologist wrote for the ILSI, the list of toxicological studies that manufacturers are supposed to supply to regulatory agencies seems impressive. The "effects" they are required to investigate involve "functional changes (e.g. reduced weight gain, laxation), morphological changes (e.g. organ enlargement pathological abnormalities), mutagenicity (heritable changes in DNA, genes and chromosomes with the potential to cause cancer or fetal abnormalities), carcinogenicity (cancer), immunotoxicity (sensitization [leading to hypersensitivity or allergy]), depression of the immune system [leading to increased susceptibility to infection]), neurotoxicity (behavioral changes, deafness, tinnitus, etc.) and reproductive toxicity (impaired fertility, embryotoxicity [spontaneous abortion], teratogenicity [fetal deformities])."

According to the type of effect researched, the length of the studies varies between two weeks (short-term toxicity) and two years (carcinogenicity), during which subjects ingest a certain dose of the poison daily, as the objective of these tests is to measure the chronic toxicity and thus the effects triggered by prolonged, repeated exposure. Experiments are carried out until a dose is obtained that *apparently* does not cause any effect on animals—this is the "NOAEL" (no-observed-adverse-effect level).

"Could you say that the NOAEL is a safety threshold?" I asked Diane Benford.

"You can't guarantee absolute safety with anything in life and it will depend on the quality of the studies that are being conducted with the animals. So if you have done a fairly poor study you might not have picked up effects that would have been seen in a very good study. These sorts of things should also be taken account of using expert judgment when the ADI is established."

Safety Factors: "Shoddy" Work That Is "Absolutely Unacceptable"

"The NOAEL is a vague measurement, which isn't extremely precise," stated Ned Groth, a biologist who was an expert for twenty-five years at the Consumers Union, the main consumer organization in the United States. As such, he regularly participated in forums organized by the FAO and WHO on food safety. "This is why risk managers use what they call a 'safety' or an 'uncertainty' factor. The standard approach used by toxicologists for fifty years consists in dividing the NOAEL by a factor of a hundred. In fact, they first apply a factor of ten to consider the differences that might exist between animals and humans, because we are not sure that man would react the exact same way to the chemical as animals do. Then, they apply a second factor of ten to take into account the sensitivity differences between humans themselves—because this, of course, varies whether one is a pregnant woman, a child, an elderly person or seriously ill. The question is to know if it is enough. Many maintain that a factor of ten to account for human variability is much too low. The effect could be nothing for certain people at a certain dose, but it could be enormous for others."

"But do we know on what scientific basis this factor of one hundred has been set?" I asked.

"It was decided by the BOGSAT method—a Bunch Of Guys Sitting Around a Table!" the environmental expert replied. "That's what Bob Shipman, who used to work for the Food and Drug Administration, said at a conference I attended. He said, 'It was in the 1960s, we had to find a way of determining what level of a toxic product we could authorize in foods. So we met up and we did.'"[42]

The American expert's story was confirmed by none other than René Truhaut, who, in his 1973 article, acknowledged that the infamous "safety factor," which was supposed to serve as the last line of defense against the toxicity of poisons, resulted from pure empiricism: "A somewhat arbitrary safety factor of 100 has been widely accepted and this figure was recommended by the Joint FAO/WHO Expert Committee on Food Additives in its second report. But it would be unreasonable to apply this figure rigidly."[43] Diane Benford makes the exact same observation in her monograph: "By convention, a default safety (uncertainty) factor of 100 is normally used. Initially, this was an arbitrary decision."[44] In passing, she points out that the main source "of variation and uncertainty" of the evaluation process lay in the difference that exists between laboratory animals, which are raised in conditions of maximum hygiene and exposed to one single chemical molecule, and the human population, which presents huge variability (genetics, illnesses, risk factors, age, sex, etc.) and is subjected to multiple exposures.

An unfailingly outspoken Englishman, Erik Millstone states with refreshing clarity: "The safety factor, which is supposed to be one hundred, is a figure that fell from out of nowhere and was scribbled on the corner of a tablecloth! Besides, in practice, experts regularly change the factor's value as they need to. Sometimes, they use a factor of a thousand, when they believe a substance presents very troubling safety concerns; sometimes, they reduce it to ten, because if they applied a factor of a hundred, it would make it virtually impossible for industry to use the product. The reality is that they use all sorts of safety factors they pull out of a hat in an opportunistic and absolutely unscientific way. This sort of shoddy work is absolutely unacceptable, when we know that consumer health is in play."[45]

This view is shared by the American attorney James Turner, who is also the president of the Citizens for Health association and a renowned specialist on questions of environmental and food safety. "Applying the famous 'safety factor' does not follow any rule," he explained at our meeting in Washington, DC. "For example, the EPA currently uses a factor of a thousand for pesticides that cause neurological damage or behavioral problems in children. In fact, the determination of the safety factor depends entirely on the experts who carry out the evaluation—if they are sensitive to environmental and health protection, they will advocate for a factor of a thousand, why not a million! If they are more on industry's side, they will apply a factor of a hundred, even ten. The system is completely arbitrary and has nothing to do with science, because it is, in fact, exceedingly political."[46]

So to summarize, and thereby properly understand the incredible amateurism of the regulatory system that should be protecting us against the harmful effects of chemical poisons that come into contact with our food: in order to establish supposedly "safe" exposure standards, scientists conduct animal experiments, trying to find a dose "with no observed effect," which is somewhat random as it depends on the species used and on the competencies—to put things plainly—of industry's private laboratories; then, the dose obtained is divided by a safety factor that varies according to the experts' background. Finally, the ADI is a value expressed in milligrams of product per kilogram of body weight. Take, for example, a pesticide with an ADI of 0.2 mg. If the consumer weighs 60 kg, he is supposed to be able to ingest 60×0.2 mg, or 12 mg, of the pesticide per day for his entire life, without his health being affected. But this lovely construct, on the whole very bureaucratic, does not account for the fact that we are exposed, every day, to hundreds of chemical substances that might interact with each other, or have a harmful effect at extremely low doses—as do endocrine disruptors, which only very high-performance tools can detect; but we are not quite at that point yet (see Chapter 16).

The Driving Force of "Risk Society"

"Do you consider the ADI as a scientific concept?" The unavoidable question seemed to surprise Angelika Tritscher, the secretary of the JECFA and the JMPR.

"Of course it is a scientific concept," she answered without hesitation, "because it is the result of the evaluation of all the scientific data we have on a chemical product. Using this data, we determine the dose that has no effect, and we divide it by an uncertainty factor. It is a totally scientific process."[47]

I received a similar answer from Herman Fontier, the head of the pesticides unit at the EFSA: "I dare hope that the ADI is a scientific concept!" he exclaimed with a large smile.[48] And from David Hattan, the FDA toxicologist in charge of food additives: "I truly believe that it is a scientific concept that protects consumer health," he assured me with unflinching calm.

It would be tempting to think that these experts working for national or international agencies are all liars or imposters. I must admit that this thought occurred to me on many occasions as I discovered the indigence of a regulatory system meant to protect us from the harms of chemical poisons. The truth is, of course, much more complicated, much like the inextricable situation in which we find ourselves now, thanks to politicians—largely driven by manufacturers' thirst for profit but also by a certain vision of "progress"—who accepted the idea that it was justifiable to introduce an incommensurable number of poisons into our environment. However, we are all somewhat responsible for this evolution, as Rachel Carson noted as early as 1962 in *Silent Spring*: "The chemical agents of cancer have become entrenched in our world in two ways: first, and ironically, through man's search for a better and easier way of life; second, because the manufacture and sale of such chemicals has become an accepted part of our economy and our way of life."[49]

It was in reading Ulrich Beck's *Risk Society* that I truly understood the political and social repercussions of what is called "mass consumption" and, consequently, the unsustainable position constricting the "experts" responsible for limiting the damage such a model provokes sui generis. In this crucial work, the German sociologist explains how, in fifty years, we went from "class society," which was characterized by "scarcity" and the fundamental issue of division of "socially produced wealth," to "risk society," which is the mark of "advanced modernity" where "the social production of *wealth* is systematically accompanied by the social production of *risks*."[50]

"Class societies remain related to the ideal of *equality* in their developmental dynamics," he writes authoritatively. "Not so the risk society. Its normative counter-project, which is its basis and motive force, is *safety*. [. . .] Whereas the utopia of equality contains a wealth of substantial and *positive* goals of social change, the utopia of the risk society remains peculiarly *negative* and *defensive*. Basically, one is no longer concerned with attaining something 'good,' but rather with *preventing* the worst; *self-limitation* is the goal which emerges. The dream of class society is that everyone wants and ought to have a *share* of the pie. The utopia of the risk society is that everyone should be *spared* from poisoning."[51] Certainly, Beck continues, "risks" have always existed, but those that characterize the industrial machine of progress are quite different than those faced by Christopher Columbus when he embarked on an unlikely voyage or by peasants menaced by plague, because they "escape perception": "They are 'piggy-back products' which are inhaled or ingested *with* other things. They are the *stowaways of normal consumption*. They travel on the wind and in the water. They can be in anything and everything, and along with the absolute necessities of life—air to breathe, food."[52]

That is why, in this "new paradigm of risk society," the fundamental problem politicians must resolve is: "How can the risks and hazards systematically produced" that take "the shape of 'latent side effects' [. . .] be limited and distributed away so that they neither hamper the modernization process nor exceed the limits of that which is 'tolerable'—ecologically, medically, psychologically, and socially?"[53]

In reading these lines, I finally understood why regulatory texts involving food and environmental safety systematically reference notions that appeared after World War II, namely "risk evaluation" and "risk management." These new public policy concepts are themselves the sole raison d'être of the many "agencies" that have flourished in France in the last few decades (but also in other "developed" countries), such as the French Agency for the Safety of Health Products (Agence de sécurité sanitaire pour les produits de santé, AFSSAPS) or the French Food Safety Agency (Agence de sécurité sanitaire des aliments, AFSSA), or the French Agency for Environmental and Occupational Health Safety (Agence française de sécruité sanitaire de l'environnement et du travail, AFSSET). In this vein, Dr. Jean-Luc Dupupet (see Chapter 3) introduces himself as the "physician in charge of chemical risk" at the Agricultural Social Mutual Fund (Mutualité sociale agricole, MSA), a function similar to that held by Diane Benford, who directs the "chemical risk department" at the FSA in the United Kingdom.

In the text she wrote for the ILSI, Benford devoted a large section to the development of the notion of "risk," which she relates to that of "hazard." Note that this is especially fascinating since her monograph has to do with food. "The Codex definition of hazard is 'a biological, chemical or physical agent with the potential to cause an adverse health effect,'" she writes. "The likelihood or risk of that hazard actually occurring in humans is dependent upon the quantity of chemical encountered or taken into the body, i.e., the exposure. The hazard is an inherent property of a chemical substance, but *if there is no exposure, then there is no risk* [*sic!*] that anyone will suffer as a result of that hazard. Risk assessment is the process of determining whether a particular hazard will be expressed at a given exposure level, duration and timing within the life cycle, and if so the magnitude of any risk is estimated. Risk management may involve attempting to reduce the risk by reducing the exposure."[54]

Benefits Versus Health

"The ADI looks like a scientific tool, because it is expressed in milligrams of product per kilogram of body weight, a unit able to reassure politicians, since it seems very serious," Erik Millstone told me, smiling out of the corner of his mouth, "but it is not a scientific concept! First, because it is not a value that is representative of the extent of the risk, but of its *acceptability*. Yet, 'acceptability' is an essentially social, normative, political or commercial notion. 'Acceptable,' but for whom? And behind the notion of acceptability, there is always the question: Is the risk acceptable in relation to the supposed benefit? Now, those profiting from the use of chemical products are always businesses, and not consumers. So it is consumers taking the risk, and businesses reaping the benefit."

In fact, if politicians endeavor to obtain mountains of figures from their "experts"—and we shall see with the "maximum residue limit" (see Chapter 13) that the scale of the task goes beyond anything one could imagine—it is because they deem that the technological or economic "benefits" that chemical poisons are supposed to bring about are well worth some human risks. This "important concept of benefits *versus* risks" constitutes the foundation of the system developed by René Truhaut, as he very crudely acknowledged in a rather shocking sentence: "It is obvious that, with a malnourished population with an expectation of life less than 40 years, it is justified to be prepared to take greater risks than for populations with an overabundance of food."[55]

More prosaically, one need only read the preamble of the European directive of July 15, 1991, "concerning the placing of plant protection products on the market" to gauge the economic ideology inherent in health politics and at what point the "benefits" supplant the "risks" in our leaders' list of priorities: "Whereas plant production has a very important place in the Community; Whereas plant production yields are continually affected by harmful organisms including weeds; whereas it is absolutely essential to protect plants against these *risks* to prevent a decline in yields and to help to ensure security of supplies; Whereas one of the most important ways of protecting plants and plant products and of improving agricultural production is to use plant protection products; Whereas these plant protection products can have non-beneficial effects upon plant production; whereas their use may involve *risks* and hazards for humans, animals and the environment, especially if placed on the market without having been officially tested and authorized and if incorrectly used . . ."[56]

The phrasing of this text is so incredible that I had to reread it several times to understand what was so deeply shocking. The word "risk" is used twice: once to designate the risk to vegetables due to "harmful organisms"; and then to evoke the risk threatening human health. To the European lawmaker, there is clearly no fundamental difference between these two forms of "risk." What's worse, the second type of risk is justified by elimination of the first—a recycled argument originally wielded by chemical agriculture supporters and pesticide producers, who are the sole beneficiaries of the use of "plant protection products."

The "benefits" argument is also the basis of a French parliamentary report presented in April 2010 by Claude Gatignol, the Union for a Popular Movement (Union pour un Mouvement Populaire, UMP) deputy from Manche, and Jean-Claude Étienne, UMP senator from Marne, entitled *Pesticides et Santé* (Pesticides and Health), whose "two hundred pages are so tendentious they are laughable," to quote the French periodical *Libération*.[57] After hearing the authors of the infamous report *Les Causes du cancer en France* (The Causes of Cancer in France, see Chapter 10), who swore that "the health risks of insecticides currently authorized in France and more generally plant protection products are often very overestimated, while their advantages are very underestimated," the two national representatives responded with a quite pathetic (to put things politely) warning cry. "Your reviewers would like to note the benefits of pesticide use and invite the public powers to anticipate the consequences of an overly harsh decrease in pesticide use in France."[58]

More seriously, however—the two representatives' report is a travesty, and not worth further attention—we find the same "risks/benefits" rhetoric in the EPA text that was the basis in 1972 for the approval of pesticides and the placement on the market of all substances that did not pose "unreasonable risk to man or the environment, taking into account the economic, social and environmental costs and benefits of the use of any pesticide."[59]

"From a politician's point of view, the danger of an environmental pollutant should be measured by its economic value," James Turner explained to me. "Deep down, I am not opposed to doing an evaluation of the benefits and risks implied by the use of a chemical product, on the condition that health be the only yardstick by which results are measured. Yet, measurements are never done in terms of health versus health, but health versus economic benefit. What's more, there is a commonly used rule that states that a product is considered safe if it does not kill more than one in a million people every year. That tells you how warped the system is."

This information was confirmed by Michel Gérin and his co-authors in their manual *Environnement et santé publique* (Environment and Public Health): "Although the idea of risk and of acceptable levels is very controversial," they write, "it has been agreed that a risk in the amount of 10^{-6} (one instance of cancer per million people exposed) is acceptable in the case of chemical products qualified as carcinogenic to animals."[60] Applied to just the French population, this "quota" means sixty deaths per year for just *one* chemical product. Given that thousands of carcinogenic products (as well as neurotoxic or reprotoxic ones) are currently in circulation, it is easy to imagine the scope of the damage and understand the unease of "experts" whose mission is to conceal the fallout with columns of "acceptable daily intakes" and other "maximum residue limits," as we soon shall see.

13

The Unsolvable Conundrum
of "Maximum Residue Limits"

[W]hat the public is asked to accept as "safe" today may turn out tomorrow to be extremely dangerous.

—Rachel Carson

"I spoke with WHO management, and you are authorized to film the beginning of the JMPR [Joint Food and Agriculture Organization (FAO)/World Health Organization (WHO) Meeting on Pesticide Residues] experts' session, but without sound." As I was not entirely sure I understood the instructions from Angelika Tritscher, the secretary of the Joint FAO/WHO Expert Committee on Food Additives (JECFA) and of the JMPR at the WHO, I insisted, somewhat amused, "But I'm doing a documentary for television, not a written article. I need image and sound!"

"I know," the UN organization representative replied, "but you know that work meetings for WHO and FAO committees take place in seclusion and are closed to all outside observers. It's already a huge privilege to be able to film a few things, even if it's without sound. Take it or leave it, because I can't get anything more. I will also ask you not to reveal the identities of the experts before we have published the summary of their work, because as you know, their names must not be released until they have completed their assessment."

"But my film is going to air over a year from now."

"In that case, no problem. You can film the cards showing their names," concluded Angela Tritscher.

Geneva, September 2009: An Exceptional Visit to the JMPR

It's true that I was exceptionally lucky to penetrate the WHO walls with my camera on that September 2009 day when the JMPR experts were holding their annual meeting. The authorization came after three months of intense negotiations, punctuated by e-mail exchanges and a long telephone conversation with Angelika Tritscher, a German toxicologist who worked in the United States for a long time, and without whom my filming in Geneva would have been impossible. I understood intuitively that she knew of my work on Monsanto, but she never told me her thoughts on it, or if it had helped me gain exceptional authorization to attend the JMPR work meetings or, conversely, if it had been a handicap.

As we saw in the previous chapter, the JMPR was created in 1963, using the JECFA as a model. These two bodies are charged with supplying toxicological evaluations to the Codex Alimentarius Commission, an organization created in 1963 by WHO and FAO to write recommendations and guidelines as far as food safety is concerned. The opinions produced by the Codex Alimentarius Commission have no regulatory power, but they may be adopted by national governments to set their own health standards.

As its name indicates, the mission of the JMPR is to evaluate the toxicity of pesticides by establishing an acceptable daily intake (ADI) level, but also to define the authorized "maximum residue limits" (MRLs) for every agricultural product in use, whether in a pure or processed state.[1] Setting ADIs is entrusted to experts chosen by WHO, while setting MRLs falls to those appointed by the FAO. Before examining how these experts are selected to make up the panels, we need to know exactly what these MRLs are, as they are supposed to serve as the last line of defense against the noxiousness of agricultural poisons. To this end, I will take the example of chlorpyrifos-methyl, which is part of the chlorpyrifos family, an organophosphate insecticide known for its neurotoxic properties and suspected of being an endocrine disruptor. Produced by Dow AgroSciences, it has been the subject of numerous scientific studies, as the PubMed database attests: on January 29, 2011, typing the word "chlorpyrifos" into its search engine gave 2,469 results, 1,032 with "chlorpyrifos toxicity," and 139 with "chlorpyrifos neurotoxicity."[2]

At the JMPR session that took place at WHO from September 16 to 25, 2009, chlorpyrifos-methyl was included among the five pesticides submitted for reevaluation. The LD50 of the substance, which is sold under the name Reldan, is 2,814 mg per kg of body weight (for mammals, orally), and its ADI is 0.01 mg/kg. According to a fact sheet published on the European Union's website,[3] this ADI was obtained following a neurotoxicity study carried out over two years on rats, then applied with a safety factor of one hundred.[4]

A Complex—and Far from Reassuring—Process

As we have seen, a poison's ADI represents the maximum quantity consumers are supposed to be able to ingest daily, for their entire life, without becoming ill. The problem is that the poison could be used to treat a multitude of fruits, vegetables, or cereals. This is especially the case with chlorpyrifos-methyl, an insecticide used in the treatment of citrus fruit (lemons, mandarins, oranges, bergamot), all sorts of nuts (pecans, pistachios, coconuts, etc.), and other fruit (apples, pears, apricots, peaches, plums, berries, grapes, etc.).[5] The question risk managers are facing is thus: How can we keep a consumer from reaching his or her ADI level of chlorpyrifos-methyl quite simply because he or she has the habit of inconsiderately(!) eating several foods treated with the pesticide?

To avoid this catastrophic scenario, the JMPR's founders followed René Truhaut's recommendations (see Chapter 12): they decided a MRL must be calculated for each agricultural product likely to have been sprayed with pesticides. MRLs, which are expressed in milligrams of pesticide per kilogram of foodstuffs, are set at the end of a complex—and far from reassuring—process. The first step in the assessment of a given pesticide consists in measuring the quantity of its residue (and potentially of its metabolites, or what is produced when it is broken down) left on each agricultural product after harvesting. Then, experts estimate consumers' potential exposure, based on investigations aiming to determine which fruits, vegetables, and cereals—and in what quantity—are eaten every day, taking into account the fact that food habits vary from one country or continent to another. The result: millions of figures enabling the establishment of MRLs, food by food.

"In concrete terms, how do you proceed?" I asked the Dutch toxicologist Bernadette Ossendorp, who oversaw the FAO panel at the JMPR session in September 2009.

"First, we examine the data obtained from field tests where the pesticide has been applied on crops according to what are called 'good agricultural practices'—that is, according to the method of use recommended by the manufacturer. This allows us to establish what quantity of pesticide residue is found on the treated food and to set a maximum limit."

"Who conducts these tests?"

"Manufacturers," Ossendorp explained. "And in doing this, they must respect very precise specifications: tests must be conducted on different types of agricultural products and, if possible, repeated over at least two seasons, so there is no bias in connection with climate conditions, for example."

"How can you be sure of the quality of the data?" I then asked. "Because industrial history is overflowing with insufficient, even fixed, studies . . ."

Surprisingly, the question did not seem to shock her. "We require very detailed reports that present, for example, the analytical methods used," she replied. "We also verify that the dose of the pesticide used corresponds correctly to what is recommended to farmers or that the spray took place at the right time. If the manufacturer sprays two months before the harvest, whereas agricultural practice would have this take place two weeks earlier, there will be a lower rate of residue than what is found in the real world. It does happen—we have questioned the suitability of certain data, and asked the manufacturer for explanations. If they do not convince us, we reject the data and the product cannot be evaluated, which can impede its placement on the market."

"Except that, quite often, the products you evaluate are already on the market?"

"That's true, but in the long run, it isn't good for the manufacturer to have been 'failed' by the JMPR."

"But if, in the real world, the farmer does not follow the levels recommended in the instructions for use, all this work doesn't do much, does it?"

"That does not fall under our purview," Ossendorp admitted, "but rather that of the public authorities who have to verify that farmers are respecting pesticide use standards."

"After examining the data on residues, you have to evaluate potential consumer exposure. How do you go about that?"

"We estimate, for each agricultural product, the average consumption by inhabitants according to thirteen dietary models that correspond to the habits of five continents, with specificities such as vegetarianism etc. Let's take the example of apples: in order to find out how many apples a French person eats on average every day, we take annual production in France, from which we deduct the amount exported, then we add the quantity imported. Then, we divide the result by the number of inhabitants. We do this for every agricultural product. This allows us, using typical menus, to evaluate the quantity of a given pesticide that every French person is likely to ingest every day."

"That's a lot of work! All that to avoid our getting sick from eating?"

"Yes," the toxicologist replied. "But, you know, we, too, are consumers."

"Let's take the example of chlorpyrifos-methyl, which is used as an insecticide on numerous crops. What happens if a consumer reaches the ADI, because she has eaten a few too many treated fruits and vegetables?"

"Yes, I understand your question, but you must also know that the MRL we evaluate is much higher than your actual exposure. According to our monitoring programs, we know that not all the apples you eat have been treated with chlorpyrifos-methyl. It should be understood that our evaluation of consumption really corresponds to the worst theoretical case—that is, to a day in which everything you eat has been treated with the same pesticide. This is very unlikely to happen in the real world. Because, in general, you will have on your plate a mixture of potatoes that have been treated, but also carrots or lettuce that has not been. Thus the probability that you would take in a very high level of chlorpyrifos-methyl residue in one day is extremely low."

"Of course," I said, "but all that isn't really reassuring . . ."

"Of course it is!" Ossendorp replied. "Don't forget that the potential danger of a product has nothing to do with the risk you actually face. Like with salt: if you eat five kilos of salt, you will get sick, but even so, you wouldn't say that salt is very toxic. As Paracelsus said, the dose makes the poison. But, okay, if you really want zero risk, you're right, pesticides shouldn't be used. But that is a political decision. As long as politicians keep saying that they have to be authorized because farmers need them to have plentiful harvests, it's the best we can do."[6]

"The Magician-Chemists of the Postindustrial Era"

I emerged from this long interview quite annoyed. It wasn't that I found Bernadette Ossendorp unpleasant or that I thought she was, to put things plainly, "selling me a story." On the contrary, I got the impression that she was being very sincere, even when she proffered arguments that were hopelessly weak: for example, how can someone dare compare pesticides, which are designed to kill, to table salt? I will note, incidentally, that while suicide by pesticide ingestion is unfortunately relatively common, in contrast, suicide by voluntary table salt poisoning is unheard of! Evidently, the comparison with good old-fashioned sodium chloride is used by all risk evaluation specialists, because I also heard it from Angelika Tritscher, the secretary of the JMPR and the JECFA at the WHO: "The fact that a chemical product or pesticide residue is found in a food does not mean that your health is threatened," she explained. "Like for salt, the question is knowing at what exposure level the danger is revealed. The problem is that food is a very emotional thing. If we add salt to it, we can control the quantity, which is not the case with pesticides. It is this unknown that frightens people, because they feel as if they don't know what's on their plates."

This reasoning shocked James Huff—the carcinogenicity specialist from the National Institute of Environmental Health Sciences (NIEHS) and former director of the International Agency for Research on Cancer (IARC) Monographs Programme—who had previously denounced industry's influence on the WHO agency (see Chapter 10). "I'm surprised that Angelika Tritscher, whose scientific qualifications I am aware of, could use the chemical manufacturers' favorite talking point," he declared indignantly, when I met him one month later. "Salt is a natural substance, one of whose functions is enhancing taste; it is certainly preferable to use it in moderation, but from that to comparing it to pesticides, which are specially designed to have harmful effects on living organisms and which pollute our plates, unbeknownst to us! Honestly, they can't be serious. But this is typical of the mental confusion experts in charge of the evaluation of food pollutants are known for—we're asking them to carry out an impossible task, because deep down, they know very well that ADIs and MRLs are only artifacts and that the only way of truly protecting people is to simply ban a large number of extremely toxic products they persist in evaluating, for better or worse."[7]

This viewpoint echoes German sociologist Ulrich Beck's enlightening analyses in his book *Risk Society*, in which he makes a harsh indictment of the role played by scientists in the health disaster that characterizes "advanced modernity": "As they are constituted—with their overspecified division of labor, their concentration on methodology and theory, their externally determined abstinence from practice—the sciences are entirely incapable of reacting adequately to civilizational risks, since they are prominently involved in the origin and grown of those very risks. Instead—sometimes with the clear conscience of 'pure scientific method,' sometimes with increasing pangs of guilt—the sciences become the legitimating patrons of a global industrial pollution and contamination of air, water, foodstuffs, etc., as well as the related generalized sickness and death of plants, animals and people."[8]

Ulrich Beck devotes a number of pages to risk specialists, scientists whom he implies are "magicians" or "acceptable level jugglers." Since scientists are never entirely clueless,

> they have many words for it, many methods, many figures. A central term for "I don't know either" is "acceptable level." [. . .] Acceptable levels for 'permissible' traces of pollutants and toxins in the air, water and food have a meaning similar to that of the principle of efficiency for the distribution of wealth: they permit the emission of toxins *and* legitimate it to just that limited degree. Whoever *limits* pollution has also *concurred* in it. [. . .] Acceptable values may indeed prevent the very worst from happening, but they are at the same time "blank checks" to poison nature and mankind a bit. [. . .] Acceptable levels in this sense are the retreat lines of a civilization supplying itself in surplus with pollutants and toxic substances. The really rather obvious demand for non-poisoning is rejected as Utopian. [. . .] Acceptable values make possible a *permanent ration of collective standardized poisoning.* [. . .] Acceptable levels certainly fulfill the function of a symbolic detoxification. They are a sort of symbolic tranquilizer pill against the mounting news reports on toxins. They signal that someone is making an effort and paying attention.[9]

Beck concludes with an acerbic commentary on the acceptable limit creators, who are, in his eyes, "late industrial chemical magicians," gifted "seers" with "the ability of the 'third eye'": "It ultimately comes down to how long poisoning will not be called poisoning and when it will begin to be called poisoning. [. . .] No matter how benevolently one looks at it, the whole affair remains a very complicated, verbose and number-intensive way of saying: we do not know either."[10]

Industry Information Is "Confidential"

"I was able to examine the list of studies you supplied to Dow AgroSciences, the producer of chlorpyrifos-methyl. It's very interesting, because they are all 'unpublished' and covered by 'data protection.' Is this always the case?"

Professor Angelo Moretto, an Italian neurotoxicologist overseeing the JMPR session in September 2009, raised his eyebrows at my question. To help him formulate his answer, I held out a sixty-six-page document, published in 2005 by the European Union, which lists the two hundred-odd studies conducted by the American producer on its insecticide.[11] The document includes experiments done on animals to measure the product's toxicity, but also field tests to evaluate residue levels on crops. For example, one measured "Residues of chlorpyrifos- methyl in tomatoes at harvest and processed fractions (canned tomatoes, juice and puree) following multiple applications of Reldan."[12] Another evaluated the "residues . . . in wine grapes at harvest following two applications of [Reldan]."[13] All of these studies are described as "unpublished," while an introductory paragraph stresses that the producer "has claimed data protection." What's more, some of these studies concern chlorpyrifos and not chlorpyrifos-methyl!

After taking a long look at the document, Angelo Moretto finally said, "Yes, that is quite possible. Studies supplied by industry to the JMPR or to national authorities are data that are protected by a confidentiality clause. But, if you consult the documents produced by the JMPR after evaluation meetings or by national authorities, you will find large summaries of the data."

"Summaries, but not the raw data?"

"No, not the raw data, because they belong to the producer. So you have to trust the twenty or so JMPR experts—who come from all over the world and were chosen for their expertise, for the proper analysis and interpretation of the data."

"And there is no reason not to trust you?"

"I should hope not!" the JMPR president answered with a forced smile.[14]

Here, we are touching upon one of the recurring critiques from nongovernmental organizations and civil society representatives with regard to the JMPR or the JECFA, but also to the European Food Safety Authority (EFSA) or any other public agency charged with the evaluation or management of chemical risks. This is because, without batting an eye, they all accept the dictum imposed by manufacturers, who demand their study data be covered by "trade secret."

"The practice of keeping data secret only serves the commercial interests of chemical companies," Erik Millstone, the British professor of science policy (see Chapter 12), told me. "It is completely contrary to the interests of consumers and public health. The WHO and regulatory agencies in no way deserve the public's trust, as long as they don't change their practices. Only the data concerning products' manufacturing process can justify the confidentiality clause, because, in the context of competition, that's what represents sensitive trade information. But all the toxicological data concerning the safety or toxicity of their products should be in the public domain."[15]

I also brought up this delicate question with Angelika Tritscher, the JMPR and JECFA secretary, who, as part of her responsibilities, plays a central role in the organization of the evaluation process. One year before the committee meetings, she publicly announces which substances will be submitted for (re)evaluation, and that "governments, interested organizations, producers of these chemicals, and individuals are invited to submit data. . . . The submitted data may be published or unpublished." In a text she put online in October 2008, in anticipation of the JMPR session in September 2009, she specified, "Unpublished confidential studies that are submitted will be safeguarded and will be used only for evaluation purposes by JMPR."[16]

"Why is raw data not public?" I asked her.

"Honestly, I don't really see what the public could do with all that data. It's thousands of pages," she replied.

"I'm not talking about the public in the broad sense, but, for example, for a consumer or environmental organization that wants to verify the toxicological data for a pesticide. Why is this data covered under trade secrets?"

"It is because of intellectual property protection. These are legal issues. Data are private and belong to the company transmitting them. We do not have the right to communicate them to a third party."

"The fact that data are protected feeds the doubt surrounding their validity and undermines trust based on transparency . . ."

"Of course! I completely understand your remark, because people have the impression we have something to hide," Angelika Tritscher acknowledged with surprising honesty.

"If we take the example of tobacco, the studies supplied by cigarette manufacturers were flawed, even manipulated or falsified, and the WHO had been misled by industry for years," I said.

"I have no comment to make on that."

"But is it true?"

"I have no comment to make on that, even more so because that happened before I came to the organization. I don't know all of the details."

"I know that it was a painful incident here," I continued, "and led to a serious adjustment in 2000 . . ."[17]

"Yes, it is clearly was painful incident. But I am not sure that it can be compared to the situation with pesticides. Nevertheless, data protection is in fact the subject of intense debate here, and we will see where it takes us. You should ask the industry why it holds on to data confidentiality so tightly."[18]

When Manufacturers Evade Embarrassing Questions

I hadn't expected Angelika Tritscher's advice to seek an interview with representatives from the pesticide industry. Naturally, as I was particularly interested in chlorpyrifos-methyl and, in a broader sense, in chlorpyrifos, one of the most controversial insecticides, I contacted the headquarters of Dow AgroSciences, one of its major worldwide producers, in Midland, Michigan, during one of the two long trips I took to the United States to carry out my investigation. On October 2, 2009, Jan Zurvalec, the head of public relations at the multinational company, forwarded my request to Sue Breach, her counterpart at the European branch, based in London. On October 13, she wrote me a very pleasant e-mail asking me to send her the questions I wanted to ask during the filmed interview: "I cannot guarantee direct participation in the program, but we will consider your request and questions with care and duly respond."[19]

To be honest, I harbored no illusions, given that Monsanto had refused all dialogue during my previous investigation recounted in *The World According to Monsanto*; while Dow and Monsanto have always been competitors on the pesticides, plastics, and chemicals markets, they have always stuck together when it comes to defending the interests of the chemical industry. And on October 16, I received a negative response from Sue Breach: "As an organisation we are always open to media interaction regarding our products and activities, particularly in the area of health, safety and the environment. However, while we appreciate the offer to be interviewed, we regretfully have to decline on this occasion as following a review of some of your previous work, we have legitimate concerns as to whether our perspectives would be fairly represented in the proper context." In conclusion, the representative for Dow AgroSciences offered to send me "written feedback" to my questions.

Then something very amusing happened. I decided to contact the organizations that represented the chemical industry in Europe, and very quickly, I observed that their directors were all consulting each other about my "case," exchanging several e-mails, in which a certain Thomas Lyall, from the "European Office of Governmental Affairs" at Dow in Brussels, was actively participating. I realized all this when one of the directors sent me an e-mail and forgot to delete all the exchanges that had preceded it. Eventually, the European Chemical Industrial Council declined my interview offer. As did the European Crop Protection Association (ECPA), an official lobby for large agricultural firms also based in the Belgian capital. On January 28, 2010, I received an e-mail from Phil Newton, its director of public relations, to whom I had sent my questions concerning—on a very "basic" level—the "role played by industry in the process of pesticides evaluation" and "data confidentiality."

"Dear Marie-Monique," he wrote. "It is important to note that all plant protection products used in European agriculture are fully assessed and tested according to relevant EU rules (Directive 91/414/EEC, to be replaced by Regulation 1107/2009/EC). Independent peer reviews of all data are conducted by the European Food Safety Authority. . . . As such, they are the most appropriate source of information on this topic." Finally, on February 1, 2010, Ana Riley, from Croplife International—which presents itself on its website as a "global federation representing the plant science industry" and is financed by the eight main pesticides manufacturers,[20] kindly blew me off, openly citing the refusal of the ECPA, which she admitted contacting.

Which left France, namely the indispensable Crop Protection Industry Association (Union des industries de la protection des plantes, UIPP), which, as we saw in Chapter 2, brings together the "nineteen business marketing and commercializing plant protection products and agricultural services." On January 28, 2010, its press office sent me a terse response to my request to interview its managing director: "We are informing you that we do not wish to proceed with your request to interview Jean-Charles Bocquet." I then called the UIPP headquarters directly, where I came across a very understanding person who clearly had not written the e-mail, as she gave me her director's cell phone number. A long telephone conversation with Jean-Charles Bocquet followed, at the start of which, he admitted: "I saw your film, *The World According to Monsanto*, which I found to be very committed work. I have no problem with that, because you have the right to be committed, but your commitment—generally, against Monsanto—translates, for me, into a very strong commitment against all businesses that produce pesticides, and since I represent them, as it happens, it would be difficult for me to grant your interview request. Also, there are a lot of errors in your film, and I don't think it's right that you didn't try to meet with Monsanto representatives."

"What?" I interrupted. "Listening to you, I'm not sure you really saw my film, because if that was the case, you would have noticed that I went to Saint Louis, which isn't exactly next door, and the company refused to see me, after three months of negotiations. I've always wondered why. Were they afraid of the questions I was going to ask? And by the way, I would like to know what errors I committed in my film."

"Uh . . . Well I assure you I saw it . . . It's true that Monsanto has its own communications policy, but I am generally more open."

"Are you afraid of the questions I might ask you?"

"Not at all! I have no doubt about the questions people ask me, but more about the manner in which my answers are used."

"I work for Arte, which is a high-quality channel, and I can't force you to say the opposite of what you've already said. It is up to you to defend your point of view! For example, how would you justify the fact that the raw data sent by pesticides producers to the JMPR or the EFSA are not public?"

"Because the public is not an expert! The day it is, it will have access. We aren't going to just give information to organizations that aren't responsible for plant product evaluation! When we know how much the numerous studies we have to do cost . . ."

"How much does a toxicological study on the carcinogenic effects of a pesticide cost?"

"Several hundred thousand euros," the UIPP director replied.[21]

In a "written response" finally sent to me on February 24, 2010, Dow AgroSciences claimed that "based on industry surveys the research needed to identify a single new pest control active ingredient and support the government registrations needed to bring it to market takes about eight years to conduct and typically costs more than $180 million." On the subject of chlorpyrifos, the multinational company specified "chlorpyrifos has been marketed since 1965 and is currently registered in about 100 countries worldwide where it is used on more than 50 crops. The aggregate cost of the studies required to support registrations and uses for all of these crops over 45 years would be hard to determine but would certainly be well in excess of $200 million."[22]

Secrecy at the JMPR

Thanks to Angelika Tritscher's help—she made sure I understood the scope of the JMPR experts' task—I obtained exceptional authorization to shoot some footage in the WHO basements. All of the data submitted by producers for the evaluation of their products were stored there. "Placed end-to-end, it takes up several miles of shelves," Marie Villemin, the director of the UN organization's archives, explained to me. "Thankfully the JMPR and JECFA are encouraging manufacturers to send their data digitally now, because otherwise it wouldn't be manageable anymore." Before my eyes stretched rows of shelves, carefully labeled, pesticide by pesticide: there were seven enormous cartons just for "glyphosate," the active ingredient in Monsanto's Roundup. I opened a number of them randomly; they contained studies "on intergenerational and reproductive effects in rats" or field tests on potatoes and carrots. Every study contained hundreds of pages, with thousands of figures, spread out over columns or tables.

"Do the experts really examine all of this data?" I asked Angelika Tritscher.

"Yes, but obviously that doesn't happen during the session, which only lasts nine or ten days. Preparation starts a year ahead of time. The raw data are entrusted to a limited group of experts who summarize it and present that to the panel at the session."

"And who sends the raw data to the limited group in charge of preparation?"

"Producers or sometimes the JMPR offices, it depends."

"So, it is possible that producers might know the names of certain experts on the panel beforehand?"

"Yes . . ."

"Yet, you told me that experts' names were kept confidential until the publication of the JMPR report . . ."

"Yes, that is a WHO rule," the German toxicologist admitted. "This is so we avoid experts being subjected to pressure before work meetings take place, either from industry, by a state that is particularly interested in the subject at hand, or by a consumer organization."

"But this rule has exceptions, since producers might know the identity of certain experts before the sessions?"

"Yes, it's the most practical way to send data . . ."

"How do you choose the experts?"

"We regularly publish what we call a 'call for experts' to make up the JMPR or JECFA panels. Any scientist can apply by sending a detailed curriculum vitae and a list of all of his or her publications. Our choice is based on the competency and expertise of the candidates, but we also have to ensure that all the continents are represented. You should know that the experts selected are not paid—WHO and FAO only cover their travel expenses."

"I consulted your last call to experts. It specifies that they are 'required to declare any potential conflict of interest through completion of a standard form developed by FAO and WHO.'"

As I listened to my interview later, I noted how very delicate my question had been, given that one of the biggest scandals WHO has ever experienced was looming ahead in the fall of 2009: it would involve undeclared conflicts of interest of the experts tasked with advising the organization on the "false pandemic" of the H1N1 flu. Three months before my visit to Geneva, on June 11, Margaret Chan, the director general of the WHO, had stated, in the sober tone the circumstances demanded, "The world is now at the start of the 2009 influenza pandemic," thus triggering the chaos we are all familiar with. One year later, whereas hundreds of thousands of deaths had been anticipated, "swine flu" had created ten times fewer victims than the usual annual epidemic. But the matter was a godsend for the five main vaccine producers—Novartis, GaxoSmithKline, Sanofi-Pasteur, Baxter, and Roche—who split $6 billion in profits. It would be revealed that the "experts" who had advised the WHO were connected to the manufacturers who profited from this pathetic masquerade.[23]

I now better understand why Angelika Tritscher tensed up when I broached the subject of conflicts of interest: "Why are JMPR or JECFA experts' conflicts of interest not published?" I asked her.

"It's a WHO rule," she replied, visibly annoyed. "You have to understand that the expert panels taking place here are not permanent, their composition changes with the cases in question. It would be a tremendous amount of work to publish on our Internet site all of the conflicts of interest of the experts we call upon."

"But the EFSA, for example does [publish the conflict of interest of its experts]."

"That's true, but its expert committees are permanent . . . I understand that this is an important question and, to be completely honest with you, we have discussions about this in our legal department to see how we can develop the system. This does not concern only the publication of conflicts of interest, but also the problem of scientific bias that can be found in certain studies . . ."

"To be frank, I find the functioning of the JMPR and the JECF severely lacking in transparency, because everything is secret here: study data, experts' identities, their conflicts of interest, not to mention the sessions themselves, closed to all outside observers. Yet, it seems to me that the day before I arrived, pesticide producers were speaking before the panel?"

"Yes. We regularly call in manufacturers so they can answer questions concerning their products."

"I can appreciate that it is important to clarify certain points, but why refuse observer status to nongovernmental organizations or academics who wish to obtain it?"

"Work sessions that take place at WHO are private by nature," Tritscher replied. "It is not that they are closed, but to be able to participate, you have to be invited. We also think that seclusion allows the experts to express themselves more easily, free from any influence."

"The Entire System of Food Contaminant Evaluation Should Be Revised"

"It is very difficult to make the system evolve," Ned Groth, a twenty-year veteran biologist for Consumers International, told me with a smile. Yet, at sixty-five, this very respected, charismatic, and well-spoken scientist, who worked at the U.S. National Academy of Sciences, doesn't seem like a trouble-maker. Nevertheless his criticism of the "system" is unflinching. "You mustn't forget," he continued, "that the WHO and the FAO are two enormous bu-reaucracies that depend on the money given by UN member states, but also from private funds, the origins of which are unknown, by the way. They have no interest in clashing with their donors, who follow their activities very closely. And it is clear that the system of chemical evaluation was created *by* and *for* industry."

"Do you really think that the experts do a detailed examination of the thousands of pages of data given by producers?" I asked.

"Of course not!" Ned Groth responded without hesitation. "That's a well-known strategy for manufacturers: they send truckloads of data that no one can verify, unless you spend years on it! That's why it is rare for experts who have no interest in the matter to volunteer for the thankless task, which, on top of everything, isn't paid. If, by some very rare chance, a slightly more vigilant expert panel decides the data are questionable, it's a good thing for industry, too, because it allows them to gain time. The JMPR will ask them to go back over their paper, which will take two years, and during that time, the standards will stay the same."

"What is the expert profile?"

"As the work required is very complicated, the candidates are generally sent by governments that want their point of view to carry weight in the committees' decisions. And those chosen are often people who have retired and have the time, but are not always up to date with the latest scientific advances.

What scientist in the middle of his career would be willing to give up several weeks of his time for an activity where political and commercial interests take precedence over every other consideration? In general, JMPR and JECFA experts are rather mediocre scientists, because the good ones have other things to do."

"Do you think the decisions they make are biased?"

"The problem," Ned Groth sighed, "is that scientists who know enough about the toxicology of pesticides generally work or have worked for industry, as academics or private consultants. And, having rubbed elbows with them often enough, I know that they all come from the same school of thought. They go to the same conferences, speak the same language, and are all convinced that we could not live without pesticides."

"Do you think experts' conflicts of interest could truly influence JMPR or JECFA decisions?"

"Certainly! One of the most characteristic examples is that of the bovine growth hormone, which was evaluated by the JECFA in 1992 and 1998. The committee's work was completely sealed by experts—including the panel's recorder, Margaret Miller—who had worked for Monsanto, the hormone's producer!"[24]

In *The World According to Monsanto*, I devoted two chapters to the exemplary case of the bovine growth hormone, or "rBGH." I had indeed observed the efficiency of the "revolving door" practice: industry representatives take on senior-level posts in governmental or international agencies in order to defend the interests of their favorite employer, to whom they typically return once their mission is complete. The JECFA's work on the transgenic hormone, meant to increase dairy production in cows, had been publicly lambasted by an investigative committee in the Canadian Senate, which had also revealed Monsanto's attempted corruption of experts at Health Canada, the agency in charge of food safety. It was through working on this affair that I had, for the first time, heard of Erik Millstone, professor at the University of Sussex, who had shown in an article how Monsanto had manipulated the interpretation of its data concerning the effects of rBGH on the health of cows.[25]

"The entire system of food contaminant evaluation should be revised," he told me during our meeting in Brighton. "They should get rid of the opacity that characterized both the work and selection of experts. It isn't normal that I can't take part in the JMPR panel because I don't have a PhD in toxicology, even though I've worked on chemical food toxicology for thirty-five years! The lack of transparency is such that they refused to allow me to attend a JMPR session while I was preparing a report for a research organization at the European Union![26] Also, any person working for an industry directly concerned with committee decisions should not be able to hold a high position in their organization chart. This is unfortunately the case with Angelika Tritscher, who worked for Nestlé for several years."

This wasn't new information. Before meeting the German toxicologist, I noted that she had worked at a scientific research center for the Swiss firm, which is a heavy user of food additives, including aspartame (see Chapters 14 and 15). I also noted that she participated in a congress in January 2009 in Tucson, Arizona, organized by the International Life Sciences Institute (ILSI), the "scientific" organization financed by multinational chemical, food industry, and pharmaceutical companies.

"Why should my previous position at Nestlé keep me from working at WHO?" Tritscher was outraged when I questioned her on the subject. "Moreover, the in-house legal department carefully reviewed my CV and decided I had the appropriate background. I have nothing to hide! Know that before being hired by Nestlé, I applied for a position at Greenpeace, in Hamburg, but I wasn't hired. So please don't judge my motives—in life, not everything is black and white."

"Of course," I said, "but could you imagine a former member of Greenpeace in your current job?"

"If he has the necessary scientific qualifications, why not?" Angelika Tritscher replied. "As to my participation in the ILSI congress, it was decided upon with WHO management, who thought I should represent the organization at a round table devoted to risk evaluation. Where's the problem with that?"

I must say that the German toxicologist convinced me of her good faith and her wish to move the system toward more transparency. I will also note that it was thanks to her that I was able to penetrate the very secretive walls of the JMPR, which is a telling sign. My impression was confirmed by Ned Groth's comments: "I know Angelika very well, and I admire her. I don't believe that she could be considered an industry 'mole' inside the JMPR or the JECFA, because she truly does care about public health, and she does everything she can so the committees fulfill their mission as best as possible. There are a lot of excellent scientists who have worked for industry, and a lot of bad scientists who work outside of industry. Beyond the people, it's the system that doesn't work, because it doesn't protect consumers."

"What the JMPR Does Is Not an Exact Science"

To prepare for my trip to Geneva, I reviewed a document published by the FAO and WHO entitled "Principles and Methods for the Risk Assessment of Chemicals in Food." A sentence caught my attention: "JECFA and JMPR determine ADIs based on all the known facts *at the time of the evaluation*."[27] It echoed what René Truhaut had written in one of his retrospective articles, that ADIs are not fixed and unchangeable, and that any new information could lead to their revision.[28] Then I asked myself: if ADIs and MRLs are not definitive values, since they depend on the experts' level of knowledge at the time they set them, how can we claim that they protect us? My doubt as to the efficiency of these infamous standards was reinforced by a document from the EFSA concerning procymidone, a fungicide made by the Japanese company Sumitomo. It was on the list of pesticides submitted for "reevaluation" at the JMPR session in September 2009, for "doubts expressed by the European Union." Actually, the European Union had banned its use in 2008, because it is an extremely strong endocrine disruptor (see Chapter 19), which led the EFSA to reduce its ADI and MRL. While the poison is no longer used in Europe, it is still present in a number of countries that export agricultural products to the EU, hence the necessity of maintaining standards to be used in (potential) inspections.

In the 2009 document, the EFSA explains that it decided to lower procymidone's ADI from 0.025 to 0.0028 mg/kg, and consequently "proposes to change the . . . MRLs in order to reduce the acute and/or consumer exposure to a level *where no negative consumer health effects are expected*."[29] Before analyzing this very troubling passage, it's worth noting that procymidone is used in the cultivation of some forty fruits and vegetables, including pears, apricots, peaches, plums, table grapes, grapevines, strawberries, kiwis, tomatoes, peppers, eggplants, cucumbers, zucchini, melons, lettuce, garlic, onions, etc. If the substance was banned in Europe after more than twenty years of use, it was due to a "multigenerational" study showing that male descendants of rats who had been exposed to 12.5 mg/kg during gestation presented "reduced anogenital distance, hypospadias,[30] testicular atrophy and undescended testes." At a dose five times lower, "increased weight of the testes and decreased weight of the prostate, epididymis[31] and seminal vesicles" were observed. All of this information is revealed in the EFSA document, which specifies that, to establish the new ADI, the experts applied a safety factor of one thousand in relation to the no-observed-adverse-effect level (NOAEL).

I showed the EFSA opinion to Angelo Moretto, the JMPR president, when I met him in Geneva in September 2009: "I have here an EFSA opinion, published January 21, 2009, which concerns procymidone, a fungicide you are set to reevaluate."

"Yes, the European Union expressed doubts about the limits we set," he confirmed, before attentively examining the document I held out to him.

"Did you read the sentence I underlined? It says: 'The EFSA proposes to change the MRLs in order to reduce the acute and/or consumer exposure to a level where no negative consumer health effects are expected.' Does this mean that the ADIs or MRLs set by the EFSA or the JMPR are never definitive?"

"Yes, in life nothing is ever definitive, even science," Moretto eventually replied, after an long, embarrassed silence. "So, if there are new data obliging us to change our previous decisions, we do."[32]

I posed the same question to Angelika Tritscher, who also took a long pause before answering: "I really don't like the way that phrase is formulated," she commented, "because it gives the impression that the previous values were not protecting anyone at all and that consumers were in danger. Which isn't true! People must understand the difference between a chemical's potential danger and the actual risk a consumer faces, because everything depends on the level of exposure. Don't forget that there are significant safety factors . . ."

"Of course, I understand the difference between danger and risk. But let's take the example of lindane, an organochlorine insecticide that's been on the market since 1938. It is a powerful neurotoxin that was classified in 1987 by IARC as possibly carcinogenic to humans and is considered a persistent organic pollutant. It was definitively banned in Europe in 2006. In 1977, the JMPR set an ADI of 0.001 mg per kilo. This standard was completely deceptive, then, even more so because lindane has the ability to accumulate in organisms."

"The problem is that at the time the JMPR did its evaluation, no one was talking about persistent organic pollutants yet," Angelika Tritscher replied. "If you mean to say that what the JMPR does is not an exact science, then I agree with you. Its decisions are based on the scientific knowledge available at the time it carries out its evaluation."

"Excuse me for being a bit trivial, but I find that this whole process seems a bit shoddy."

"The word you just used is very offensive to the experts, who really do everything they can to put out the best scientific judgment," the German toxicologist shot back with a disapproving look.

My remark was certainly a bit harsh, even if it conveyed my sentiments exactly, which were, once again, confirmed by reading *Risk Society*. Save one nuance: contrary to Ulrich Beck, who vilifies the "acceptable level magicians," I think that the main responsibility of the health disaster we are experiencing falls upon politicians, because they are the "risk managers" who we expect will see the big picture and protect our long-term health accordingly. As for the rest, I agree with the German sociologist: "It is not that one *could not* know how the toxic rations affect people individually or in total. *One does not want to know it*. [. . .] Even the already published statistics on such things as diseases or dying forests apparently do not appear eloquent enough to the acceptable level magicians. We are concerned, then, with a permanent large-scale experiment, requiring the involuntary human subjects to report on the accumulating symptoms of toxicity among themselves, with a reversed and elevated burden of proof. Their arguments need not be heeded, because, after all, *there are acceptable levels that were met!*"[33]

January 2010: An Edifying Visit to the EFSA

"I have here an EFSA opinion announcing a decrease in the ADI and MRL for procymidone, because of worries regarding consumer health. Does this mean that the previous ADI, which we thought was protecting us, was in fact not protecting us?" Predictably enough, I posed my perennial question to Herman Fontier, the head of the pesticides unit at the EFSA. There was a palpable silence in the Belgian toxicologist's office. He shot several desperate looks at the three members of the PR team seated behind me. As it happens, they were very nice, and scrupulously recorded the four interviews I completed at the EFSA, on January 19, 2010, in Parma. If they were to listen to their recordings, they would easily see that I am transcribing my interview subject's very muddled answer verbatim.

"It was not protecting for that . . . It did not have the same . . . It did not offer the same protection. Once again, there are safety values that are applied, a value of one hundred in relation to the no-effect level, so there are safeties that are inserted a bit all over in the system. So it is very improbable that the ADI that was set before had health effects . . ."

At first, I was amused by the European civil servant's uneasiness, but then I became quite sad, faced with the extreme fragility of "acceptable level magicians" who are forced to walk a rope so tight that it could snap at the slightest snag: "If I gave you an apple with procymidone and chlorpyrifos residues, would you eat it?" I asked.

"That depends on the residue levels. If they are in compliance with legislation, with a pesticide content below the MRL, yes, I would eat it," he replied, obviously relieved by this new question.

"Even if you know that, in three years, the MRL will be reduced because there will be new data?"

"Yes. We never know what the future has in store for us, but I am confident in the work we do. Absolutely!"[34]

It was precisely in order to "restore and maintain *confidence* in the EU food supply"[35] that the EFSA was created in January 2002, "following a series of food crises in the late 1990s," as its website explains. And it should be noted that in this domain, the EFSA's task is immense. According to a Eurobarometer survey published in February 2006, "40% of people . . . think that their health could be damaged by the food they eat or by other consumer goods. . . . [and] the association of food with health is only made by one person in five."[36] At the top of the list of "external factors" considered particularly "dangerous" by Europeans are "pesticide residues" (71 percent), followed by "residues in meats, like antibiotics and hormones" (68 percent). Finally, the last revelation from the survey: "54% think that their health concerns are taken seriously by the EU. . . . 47% of citizens think that when deciding on priorities, authorities would favor the economic interests of producers over the health of consumers."

Located in Parma, Italy, the EFSA is responsible for evaluating risks connected with the use of chemical products in the food chain. Lacking any regulatory power, it settles for producing "scientific opinions and advice" to "support the European Commission, European Parliament and EU Member States in taking effective and timely risk management decisions." To understand the EFSA's function, it should be viewed as part of the European system of regulation of "plant protection products," governed by directive 91/414 of July 17, 1991. This directive requires that every pesticide be found on a "positive list" of authorized products, the famous "Annex I," before it can be used legally. To secure this listing, the producer must submit a marketing authorization request to one of the Member States of the European Union, considered the "rapporteur Member State," which is tasked with gathering and evaluating the toxicological and ecotoxicological studies on the active substance supplied by the producer. To this end, it calls upon the expertise of the EFSA, which intervenes on two levels.

Firstly, the EFSA gives an opinion on the classification of the molecule as to its potentially carcinogenic, mutagenic, and reprotoxic effects. Adding to the confusion of an already extremely complex subject, the European Union's classification of carcinogenic substances is not the same as that of IARC (see Chapter 10). Category 1 corresponds to the UN agency's group 1, "known to be carcinogenic to humans." Category 2 corresponds to group 2A ("probably carcinogenic to humans"); and category 3 corresponds to group 2B ("possibly carcinogenic").[37] The same principle is used for mutagenic and reprotoxic substances. Secondly, since September 1, 2008, the EFSA has been tasked with proposing ADIs and MRLs for every pesticide submitted for evaluation, which are then enacted by the European Union and are from then on applied/shared by its twenty-seven Member States.

Ultimately, it is the rapporteur Member State that grants the first marketing authorization for a pesticide. Valid for ten years and renewable, it is generally adopted by the other European Member States, following the principle of "mutual recognition," even if each country maintains the ability to "provisionally limit or ban the movement of a product on its territory." By virtue of regulation 1107/2009, which replaced directive 91/414 as of June 14, 2011, the European Commission is now able to "adopt emergency measures in order to restrict or prohibit the use and/or sale of a plant protection product if it is likely to constitute a serious risk to human or animal health or the environment, and that such a risk cannot be contained by the Member State(s) concerned."

"How many active substances in pesticides[38] are currently authorized in Europe?" I asked the head of the EFSA pesticides unit in January 2010.

"You should know that in the 1990s, there were nearly 1,000," he explained. "But today, there are only 300. The European Union led a vast revision program, and a lot of the chemicals did not survive, namely because manufacturers did not defend them, by deciding not to send the data requested of them. In certain cases, the dossier submitted was not complete and inclusion of the products in the new approved list was denied."

"That means that 700 chemicals have been recently banned?"

"Yes, the revision program ended in 2008."

"Does the EFSA take into account the JMPR's work to set ADIs and MRLs?"

"Of course, we follow the JMPR's recommendations very closely. We don't always arrive at the same conclusions, because we sometimes have new studies at our disposal that the JMPR did not have when the evaluation was made. Generally, when there is a difference, our ADI is lower."

"In that case, for the consumer, it would be better to have the EFSA's ADI than the JMPR's?"

"Of course, we think that the EFSA's ADI is the one you should follow!"

Greenpeace's Criticisms of the New European Toxicological Standards

Since September 1, 2008, MRLs have been set by the European Commission, which has led a vast program of harmonization of the existing standards in the twenty-seven Member States of the European Union. In fact, until that date, each country set its own acceptable levels for each agricultural product (vegetables, meats, fruit, milk, eggs, grains, spices, tea, coffee, etc.) and there were close to 170,000 different MRLs for the whole of the European territory! Which was a real headache for the European Commission, who hoped to streamline the process by aligning all the European Union countries with the same standards.

"It was a really good idea, because it allowed all European consumers to be assured the same level of protection," said Manfred Krautter, a chemist who worked for eighteen years for the German branch of Greenpeace, on October 5, 2009, in Hamburg. "Unfortunately, instead of choosing the smallest common denominator, the Commission generally kept the highest MRLs. For Germany and Austria, for example, who had the most ambitious standards, harmonization brought an increase in the authorized residue levels up to 1,000 times higher for 65 percent of pesticides used."[39]

In a report published in March 2008, Greenpeace and Friends of the Earth stressed that "for apples, pears and table grapes, 10% of the set acceptable levels are potentially dangerous for children," who are major consumers of these fruits. As we saw in Chapter 12, toxicological standards are expressed in the quantity of the substance in question in relation to body weight. If an adult consumes quantity X of pesticides residues, he or she will experience fewer effects than a child will. In other words: a child weighing 12 kilograms who eats two apples and a bunch of grapes runs a proportionally higher risk than an adult who weighs 60 kilograms. In their report, the ecological organizations note that "a child who weighs 16.5 kilograms reaches the acceptable level of procymidone by eating just 20 grams of grapes, and that of methomyl (an insecticide) with 40 grams of apples or 50 grams of plums."[40]

"How do you explain that the harmonization led to an increase of numerous MRLs rather than a decrease?" I asked Herman Fontier.

I found his answer rather unconvincing: "First, I'd like to note that the EFSA did away with a certain number of national MRLs that it considered problematic," began the head of the EFSA pesticides unit. "Sometimes, there are indeed differences from one country to another. For example, in State A, the MRL for an agricultural product was, let's say, 1 mg/kg, and in State B, 2 mg/kg. We check whether 2 mg/kg poses any health problems, and if this isn't the case, we decide to take this MRL as a reference, so as to allow country B to continue cultivating the product with the necessary pesticide amount, because obviously agronomic and plant health conditions are not as favorable as in country A. But it should be noted that, in State A, they continue to use the minimum efficient amount, allowing them to stay below 1 mg/kg. This can seem paradoxical, but the increase resulting from harmonization does not lead to a rise in consumer exposure; on the contrary, the fact that we have eliminated certain MRLs has actually increased their safety."[41]

This is what is called "having it both ways." Because, through the simple game of commerce, products from country B arrive in country A, and they contain twice as much residue as those grown at home. So, to claim that an increase in the MRL does not lead to a rise in consumer risk is, at the least, a half-truth and, incidentally, completely contrary to the very principle of acceptable levels. "The increase in a certain number of MRLs means they can embellish the big picture in Europe," Manfred Krautter, the Greenpeace chemist, explained to me, "because the higher the standards are raised, the less chance there is to exceed them! This is what we saw with the publication of the first annual EFSA report on pesticides residues, in which they prided themselves on an observed decrease in the number of times standards were exceeded."

Published on June 10, 2009, the report consisted of a summary of observations made in the twenty-seven Member States of the European Union. In total, 74,305 samples were taken from 350 kinds of food: 354 different pesticides were detected in fruits and vegetables, and 72 in grains. MRLs were exceeded by one or more pesticides in 3.99 percent of the samples, and 26.2 percent of samples contained residues of at least two pesticides (and 1 percent contained more than eight different pesticides). As the report's authors underline, "the percentage of samples of fruit, vegetables and cereals with multiple residues has increased over the years from 15.4% in 1997 to 27.7% in 2006. In 2007, the percentage of samples with multiple residues slightly decreased."[42]

On paper, these results seem more or less reassuring, but it's important to note that the use of a European *average* hides huge disparities from one country to another.[43] In fact, the number of pesticides *sought out* varies from 709 in Germany, which is by far the leader, to 14 in Bulgaria (265 in France and 322 in Italy). The number of pesticides *detected* also varies considerably: 287 in Germany and 5 in Hungary (122 in France and in Spain). Finally, the number of samples analyzed totaled more than 16,000 in Germany, but only a few hundred for Malta or Luxembourg (4,000 for France). "The problem," Manfred Krautter explained, "is that detection of pesticides residues is very expensive, and a number of European countries are not equipped to properly carry out the task. If it was being honest, the EFSA should have specified that the figures they were putting forward fall very short of reality."

In October 2009, I visited the best German laboratory for analyzing pesticides residues and veterinary products, in Stuttgart. Thanks to ultramodern equipment using chromatography and mass spectrometry, the public center can detect more than a thousand molecules (pesticides and their metabolites). "We are one of the rare European labs to have access to this equipment," Eberhard Schüle, its director, explained to me. "And, on average, 5 percent of the food we regularly analyze at the German authorities' request exceeds the standards in effect."

"Do you eat organic food?" I asked, surprising the lab director.

"I could give a personal answer to that question, but as a representative of a public establishment, I prefer to abstain," he replied.[44]

Meanwhile, if several positive clues (I will come back to this in the last chapter of this book) indicate that Europe is setting off on the right track, at the time of writing (in 2010) we still have a long way to go. Going through the EFSA report, I discovered that, among the twelve pesticides that were the most often detected in the samples, two were classified as or suspected of being reprotoxins, one neurotoxin (chlorpyrifos), five carcinogens, and two endocrine disruptors (including procymidone).

"Can carcinogenic pesticides still be found on the market?" I asked Herman Fontier.

"Yes, there are still a few," the EFSA pesticides unit head admitted. "But that will change with the new European regulation 1107/2009, which soon will replace directive 91/414. Because from now on, all category 1 substances classified as mutagenic, carcinogenic, or toxic for reproduction, or suspected of disrupting the endocrine system, will have to be taken off the market."[45]

This is, indeed, good news. However, the EFSA still needs to base its chemical evaluations on high-quality studies, or else the pressure exerted by manufacturers must not completely skew the process. This unfortunately is too often the case, as shown in a sadly exemplary way, and in a field completely different from pesticides, by the unbelievable controversy surrounding an infamous artificial sweetener—aspartame.

14

Aspartame and Regulation:
How Industry Is Pulling the Strings

I recognized my two selves: a crusading idealist and a cold, granitic believer in
the law of the jungle.
— Edgar Monsanto Queeny, Monsanto CEO 1943–1963[1]

"I do want to meet you, as I've heard that you do serious work, but you should
know that I haven't granted an interview on aspartame in fifteen years. The
case is hopeless—it shows that regulatory agencies like the Food and Drug
Administration aren't doing their job, which is to protect consumers before
industry interests." This was my first telephone contact with Dr. John Olney,
a psychiatrist specializing in neuropathology and immunology, who has
worked at Washington University in St. Louis (Missouri) for over forty years.

E 621, E 900, E 951, etc.: The Chemical Food Additives
on Our Plates

At nearly ninety years old, this very respected researcher will be known in
the annals of medicine as the inventor of the term "excitotoxicity," which refers to the capacity of certain amino acids (which are the fundamental building blocks of proteins and peptides), such as glutamic acid or aspartic acid—a
component of aspartame—to excite or overstimulate certain neuronal receptors, and even cause neuron death when present in excess. This neurotoxic
process is associated with neurological diseases like epilepsy and strokes, as
well as neurodegenerative pathologies like Alzheimer's disease, multiple scle-

rosis, and Parkinson's disease. In the book *Neuroscience*, neurologist Dale
Purves and his co-authors write, "the phenomenon of excitotoxicity was discovered in 1957 when D. R. Lucas and J. P. Newhouse [two British ophthal-
mologists] serendipitously discovered that feeding sodium glutamate to infant
mice destroyed neurons in the retina.[2] Roughly a decade later, John Olney
[. . .] extended this discovery by showing that regions of glutamate-induced
neuronal loss can occur throughout the brain."[3]

"My studies clearly demonstrated that glutamate is a neurotoxin that can
create lesions in a region of the brain that is very important in the control of
endocrine functions, thus causing behavioral problems, sexual dysfunction,
and obesity,"[4] explained John Olney in the New Orleans park where we met
in October 2009.[5] At the time, I was attending a conference on endocrine
disruptors (see Chapter 16), and he a symposium on anesthesia and the damage it can cause to children's brains. "But unlike anesthesia, for which we
can assess the benefits vs. the risks (since it's indispensable in operating on
young patients suffering from serious pathologies), glutamate presents nothing but risks. And unfortunately, it's ingested in massive quantities by millions of children and pregnant women," sighed the neurologist.

Beyond its use in Chinese cuisine, glutamate (or monosodium glutamate
[MSG])[6] is one of approximately 300 food additives authorized by the European Union. Designated by the letter E followed by an identification number
(MSG is known as E 621), these infamous "additives" are officially designated as "any substance not normally consumed as a food in itself [. . .] the
intentional addition of which to food for a *technological purpose in the manufacture, processing, preparation, treatment, packaging, transport or storage* of
such food results, or may be reasonably expected to result, in it or its by-
products becoming directly or indirectly a component of such foods," according to the convoluted terms of European directive 89/107, which regulates
their use.[7]

More mundanely, these substances, which are found in a majority of synthetic chemical products, arrived on our plates with the advent of the food-
processing industry that accompanied the "green revolution" after World War
II. Both were eagerly welcomed by manufacturers, as they brought about a
substantial reduction in manufacturing costs.[8] They also filled a number of
"technological" functions, as outlined in great detail by another European directive (95/2): "preservatives," "antioxidants," "acids" or "acidity regulators,"
"anti-caking agents," "emulsifiers," "firming agents," "flavor enhancers" (like

MSG), "foaming agents" or "anti-foaming agents," "gelling agents," "glazing agents," "humectants," "modified starches," "packaging gases," "propellants," "stabilizers," "thickeners" or "sweeteners" (like aspartame).[9]

When a substance is natural, the manufacturers simply use its name, like for the coloring "beetroot red" (also called E 162), but when it is a chemical product with an off-putting and unattractive name—like dimethylpolysiloxane, a silicone derivative used as an anti-foaming agent in fruit juices, jams, wine, and powdered milk—they prefer to indicate a number, say E 900. The majority of food additives contain the acceptable daily intake (ADI) level, which is proof (if it was needed) that they are not harmless. And, as we will see with the example of aspartame, this sacrosanct value is often established from toxicological studies whose quality leaves something to be desired.

The Discovery of Aspartame

Aspartame, or E 951, is an artificial sweetener whose sweetening power is two hundred times greater than that of cane sugar. Present in over 6,000 food products, it is consumed by 200 million people worldwide (including 4 million in France) who ingest it in the form of sweeteners such as Canderel or Equal, breakfast cereals, chewing gum, carbonated beverages (like Diet Coke or other so-called sugarless liquids), yogurts, processed desserts, vitamins, and over 300 medications. Its principal manufacturers are the American companies Merisant and NutraSweet (two former Monsanto subsidiaries) and the Japanese firm Ajinomoto, which collectively produce 16,000 tons of it each year.

The compound was haphazardly discovered by James Schatter, a chemist working for the American firm G.D. Searle on a new drug to fight ulcers. In archival footage, I watched the scientist in his white coat, in one of the Chicago company's laboratories in 1965, describe his surprise at the white powder's sugary flavor after absentmindedly licking traces of it from his hand.[10] He remarked that the substance possessed exactly the same flavor as sugar, minus any caloric content or the metallic aftertaste of saccharin (E 954), the artificial (and highly controversial) sweetener dominating the market at the time.[11] Sensing the substance's potential, Searle launched a series of studies in 1967 before filing a demand for marketing authorization from the U.S. Food and Drug Administration (FDA). And thus began the incredible saga that has today made aspartame, according to some, the "most controversial food additive in history," as stated in the magazine *The Ecologist*,[12] and to others, "the most tested food additive in history," as affirmed by manufacturers and regulatory agencies like the FDA.[13]

To gain a better understanding of the issue, I spent four months reviewing E 951 by consulting nearly a thousand documents—declassified archives, scientific studies, press articles, summaries of American congressional hearings—and interviewing nearly twenty experts. Incidentally, I would like to thank Betty Martini, a particularly tenacious American who gave me access to the basement of her Atlanta home, which she had transformed into the largest private documentation resource on aspartame. For over twenty years, she has been accumulating evidence obtained through the Freedom of Information Act, which allows any citizen to access internal government documents (even if some of them are sometimes "redacted" or truncated).[14] I was gradually able to reconstitute the ongoing saga, which illustrates the aberrations of a "risk society" in which the interests of big business take precedence over the "protection of the public," and the "denials of the responsible parties grow ever higher in volume and weaker substance."[15]

To better understand the stakes at play, it is important to note that aspartame is composed of three molecules: aspartic acid (40 percent), phenylalanine (50 percent) and methanol (10 percent).[16] The first two are amino acids found naturally in certain foods, but with a critical difference: when they are ingested in the form of aspartame, they are not linked to any protein and are therefore "freely" released into the body. In a solution or when heated above 86°F, the two substances often degrade into diketopiperazines or "DKPs," a toxic by-product some researchers suspect to be carcinogenic. Methanol, also known as methyl alcohol or wood alcohol, is also a substance found naturally in fruits and vegetables, except that, unlike aspartame, it is always linked to ethanol (or ethyl alcohol), which counteracts its noxious effects.[17] When it is not neutralized, methanol is metabolized in the liver, where it is transformed into formaldehyde, a substance classified as "carcinogenic to humans" in 2006 (see Chapter 7).

As we will see, it was these three molecules' potentially harmful effects that fed the debate on aspartame for over forty years, not to mention the battle plan developed by Searle in the beginning of the 1970s to impose its sweetener. A very troubling "memorandum confidential" proves that, at the very least, the company was aware that the approval of aspartame was not self-evident. The document—classified as "Trade Secret Information" and revealed during an American congressional hearing (to which I will return later)—was sent by Herbert Helling, one of Searle's top officials, to five company scientists on December 28, 1970:

> These are thoughts on the matter of sweetener strategy. As I see it, our objective is to obtain approval from the Food and Drug Administration of SC-18362 for enough uses to permit consumption (and hence production) at a level that will meet the economic requirements. With that in mind, we have to say what we need to do, know, or accomplish in order to bring about this objective. [. . .] We must decide what factors Food & Drug would be most concerned about and determine which of these food items would present the least serious concerns (after ranking the concerns in order of our difficulty to meet at this time). We should arrange an early informal meeting with Dr. Wodike and Dr. Blumenthal. At this meeting, the basic philosophy of our approach to Food and Drug should be to try to get them to say "v.s." [. . .] We must create an affirmative atmosphere in our dealing with them [. . .] this would also help bring them into a subconscious spirit of participation. My prime concern at this time is with the production of the DKP and our lack of complete toxicological data. [. . .] In effect then, I would first ask for an informal, but not necessarily off the record meeting. As a basis for this meeting, we would present a series of assumptions. These assumptions will be specifically stated and any informal non-binding opinions would be predicated on the basis that we can do the right thing, convince them that the assumptions are true. I would first make the assumption that the material is stable in dry form [. . .] [like] presweetened cereal. I would proceed to the next food category, and take these food categories one at a time to see which we begin to meet resistance. [. . .] I would want to explore the nature of the resistance and what we would have to do to overcome it, particularly in relation to studies that are going on. [. . .] The approach from the meeting standpoint must be made to or thru Virgil Wodicka, Head of the Bureau of Foods, who is from an industrial background [sic].[18]

Searle's "Sloppy" Studies

"As soon as I found out that Searle had requested a marketing authorization for aspartame, I contacted the company to tell them about a study I had conducted in 1970 on aspartic acid, one of the sweetener's components," John Olney explained to me. "It showed that the substance created the same kinds of brain lesions as glutamic acid.[19] Searle's representatives told me they were going to study the issue, and I asked them to send me a sample of aspartame, which they did. I then fed it to infant mice and observed the same brain damage that I had with aspartic acid. In 1974, I read in the *Federal Register* [an official FDA publication that lists all the regulatory texts produced by the agency] that the sweetener's approval was imminent. I immediately requested a meeting with the FDA commissioner and sent him images of the mice's brains. I then contacted attorney James Turner, who played a key role in getting cyclamate banned."

In 1970, cyclamate, another artificial sweetener, was banned in the United States, following a campaign led by James Turner (the lawyer previously mentioned in Chapter 12), one of Ralph Nader's protégés, and with whom he published a best-selling book that same year—*The Chemical Feast*.[20] Relying on a study that showed that cyclamate caused bladder cancers in mice (just like saccharin) when combined with saccharin (at the rate of nine parts to one), Turner pressured the FDA to pull the product from the market, even though it had been sold since 1953.[21] But subsequent events show that Abbott, a cyclamate manufacturer, was not as lucky as Searle, which obtained approval for aspartame in dry products on July 26, 1974. "Right away we teamed up with a consumer association and filed an appeal against the FDA's decision, citing studies conducted by John Olney," James Turner told me. "It sparked a huge controversy because, for the first time in its history, the agency was forced to make public the scientific data on which it had founded its approval. And the least we can say is that the studies Searle provided were sloppy."[22] And the debate was long-lived indeed, thanks to a series of damning facts: for six years, FDA scientists unanimously denounced the many deficiencies and irregularities of Searle's toxicological studies that were used to set the ADI for aspartame (which remains in effect today).

In July 1975, FDA commissioner Alexander Schmidt decided to form a "special task force" charged with determining the validity of twenty-nine Searle studies concerning six drugs, as well as aspartame. The exceptional request was prompted by a review of pharmacological tests that agency scientists had found aberrant. The task force included Adrian Gross, who worked at the FDA from 1964 to 1979. In two letters sent to Senator Howard Metzenbaum in 1987,[23] Gross described in detail discoveries made by the inspectors in Searle's Chicago laboratories, where eleven studies on aspartame were meticulously reviewed, including two regarded as particularly important because they had tested the sweetener's carcinogenic and teratogenic effects.

Gross was one of the signatories of the five-hundred-page report the task force submitted on March 24, 1976, which begins as follows: "At the heart of FDA's regulatory process is its ability to rely upon the integrity of the basic safety data submitted by sponsors of regulated products. Our investigation clearly demonstrates that, in the case of G.D. Searle Company, we have no basis for such reliance now." Over dozens of pages, the report lists the "serious deficiencies in Searle's operations and practices," particularly as concerns studies on aspartame. First, the reviewers note a "lack of concern over the homogeneity, or stability of the ingredient-diet mixture," which means "there is no way in which it can be assured that animals received the intended dosage." They also underline the "numerous errors and inconsistencies amongst observations and findings," "observations being reported for material that never existed," and the lack of training by the "'professional' scientists making observations in teratogenicity studies," as well as the "substantial loss in pathology information due to autolysis, fixation 'in toto,' etc." Finally—and this is probably the most serious observation—they denounce "the excision of tumor masses," which allowed the company to reduce the number of brain cancers observed in the experimental groups (twelve in total.) However, as Gross notes in his letter to Senator Metzenbaum, despite all the discrepancies observed, "the rate of brain tumors amongst the animals exposed to it vastly exceeds that for animals not exposed to it and such excess is very highly significant."

The task force also discovered that Searle had "omitted" the results of two key studies: one was led by Harry Waisman, the director of a laboratory at the University of Wisconsin, and considered to be one of the top phenylalanine specialists. This study, conducted in 1967 on seven young monkeys, ended with the death of one of the animals, whereas five others experienced seizures. The second study was conducted by Ann Reynolds, a zoologist at the University of Illinois, who confirmed the results obtained by John Olney. The lapse was so serious that the task force recommended taking legal action against Searle for "criminal violations of the law." Aspartame's marketing authorization was suspended sine die, even as Alexander Schmidt was publicly denouncing Searle's studies during a Senate hearing in July 1976.

"These are the conclusions of the [task force appointed to that] study. Do you agree with those conclusions?" Senator Edward Kennedy asked the agency director.

"Yes I do," he responded.

"Is this the first time, to your knowledge, that such a problem has been uncovered of this magnitude by the Food and Drug Administration?" asked the Democratic senator.

"It is certainly the first time that such an extensive and detailed examination of this kind has taken place. [. . .] From time to time, we have been aware of isolated problems, but we were not aware of the extent of the problem in one pharmaceutical house."[24]

During his testimony, Alexander Schmidt announced the creation of a new task force charged with investigating a third critical study by Searle concerning the effects of DKP, a metabolite of aspartame. The investigation was led by Jerome Bressler, a reputable FDA scientist after whom the report (published in August 1977) would be named. It lists the irregularities observed by the previous FDA team with, however, some "originalities" that are well worth diving into. "Observation records indicated that animal A23LM was alive at week 88, dead from week 92 through week 104, alive at week 108, and dead at week 112," note the inspectors. The rest of their observations are in the same vein, so I will name only a few, as the list of "anomalies" is quite long: "A tissue mass measuring 1.5 × 1.0 cm was excised from animal B3HF on 2/12/72"; "98 of the 196 animals that died during the study were fixed in toto and autopsied at some later date, in some cases more than one year later"; "a total of 20 animals were excluded from the study due to excessive autolysis"; "animal F6HF, a high dose female, was found dead at 787 days of treatment and the gross pathology sheet reported a tissue mass measuring 5.0×4.5×2.5cm. The submission to FDA reported no tissue mass and the animal was excluded from the study due to marked autolysis"; the discovery of "a polyp of the uterus which was not diagnosed by Searle (animal K9MF). The finding of this additional uterine polyp [. . .] increases the incidence in the midi dose to 5 of 34 (15 percent)."[25]

"In 1979, I was able to consult the Searle studies, thanks to the Freedom of Information Act," explained John Olney with a surprisingly steady voice. "I was horrified by what I discovered . . . I mainly remember a photo taken by a lab technician, where you could see a large chunk of DKP sloppily mixed in with the rats' powdered food. This abnormality was criticized in the Bressler report, because rodents are clever enough to avoid a substance that is particularly foul. I also noted the high number of brain tumors observed in one of the central studies, since I know that this kind of tumor is extremely rare in lab animals. The scientific literature of the time reported an incidence of

0.6 percent, whereas the Searle study reached 3.57 percent, despite its many irregularities. I remember telling myself that given these elements, there was no way the FDA could approve aspartame."[26]

Donald Rumsfeld to Aspartame's Defense

John Olney's hopes would be dashed soon enough, once a formidable new actor entered the scene: Donald Rumsfeld, a congressman from Illinois who later served as secretary of defense under Presidents Gerald Ford and George W. Bush. In March 1977, this "Republican JFK," as he was often called, was named the CEO of Searle. "The company was based in the region Rumsfeld represented when he was elected to Congress," attorney James Turner explained to me. "And since the Searle family was very influential, they supported him throughout his political career. After Jimmy Carter was elected [in November 1976], Rumsfeld found himself in a political dry spell, and Searle needed someone with influence who could save its businesses, which were threatened by revelations about its practices and several ongoing lawsuits. Rumsfeld had the ideal profile, because he was as well connected in Washington, DC, as he was in Chicago."

We will undoubtedly never truly understand the role played by Searle's new CEO in burying the judicial proceedings launched by Richard Merrill, the head of the FDA's legal department who, on January 10, 1977, sued the company for "concealing material facts and making false statements." The case was very significant as it was the first time the agency had requested that a criminal investigation be opened against a manufacturer. Six months later, Samuel Skinner, the U.S. attorney for the northern district of Illinois in charge of the inquiry, was recruited by Sidley Austin, the law firm advising Searle. He was replaced by William Conlon, who would join him in January 1979 after conveniently allowing the statue of limitations on the charges to expire.[27]

In July 1979, the FDA created a Public Board of Inquiry (PBI), supervised by three scientists responsible for synthesizing all the information available on aspartame. In September 1980—even as the Democratic administration's days seemed numbered—John Olney submitted a written deposition to those "judges," in which he highlighted the fundamentals of risk evaluation, which were all the more pertinent when applied to a substance whose utility was far from proven: "A risk-benefit analysis should be performed separately for any population sub-group that might be particularly susceptible to harm (fetuses,

suckling infants, children, phenylketonurics[28]) or that might potentially benefit from the product (e.g., diabetics and the obese) [. . .] as it may lead to
the development of an intelligent plan whereby the product may be made
available to those who stand to benefit, without exposing those for whom it
poses undue hazard."[29]

His suggestions make a great deal of sense, standing in stark contrast to
the evaluation criteria adopted by the regulatory agencies. As concerns the
supposed "benefits" of aspartame, the scientist cites the conclusions of a National Academy of Sciences Forum on Sweeteners held in 1974: "there may
be a psychological benefit for the obese person who, by using low-calorie
sweeteners, is reminded that he is (or should be) on a diet. [. . .] The sweetener *per se* can best be described as a gimmick that serves a reminder function." As for diabetics, the potential benefits are "in terms of pleasure and
convenience rather than health." After underlining the specific risks faced by
children who regularly consume a mixture of MSG and aspartame (like they
do now, by eating a bag of chips with a Diet Coke), John Olney rams his
point home by evoking comments on saccharin prepared by Donald Kennedy, the new FDA commissioner:

(1) no benefits for any consumer group have been demonstrated; (2) children are increasing their consumption of saccharin at an alarming rate, and
(3) the FDA has a special obligation to protect children because they are
not intellectually mature enough to evaluate risk and make sound decisions
for their own protection.[30]

On September 30, 1980, the PBI submitted its report, which seemed to indicate that John Olney and James Turner had won the battle. The three judg-
ing scientists concluded that "approval of aspartame for use in foods should
be withheld at least until the question concerning its possible oncogenic potential has been resolved by further experiments. [. . .] Therefore, it is OR-
DERED that Approval of the food additive petition for aspartame (FAP
3A2885) be, and it is, hereby withdrawn."[31] But five weeks later, Ronald Reagan, the Hollywood cowboy-cum-deregulation apostle, was elected president
of the United States. Donald Rumsfeld, who would remain Searle's CEO
until 1985, joined Reagan's transition team, which was responsible for prep-
ping the new administration before the presidential inauguration on January
20, 1981. Rumsfeld's mission was to clean up the Health Department, the
FDA's parent organization. He recommended Dr. Arthur Hayes, a professor

of medicine at Pennsylvania State University, to head the agency. When Hayes officially took the post on April 3, 1981, the *New York Times* published a prophetic article: "The FDA has responsibility for protecting consumers against impure and unsafe foods, drugs and cosmetics. Its activities, particularly in the area of screening new drugs and food additives seen as potential carcinogens, have been criticized by the food and drug industries."[32] The article added that certain industry representatives considered Hayes to be more open to their points of view than his predecessors. And indeed, everything seemed to indicate several high-placed individuals had asked the new commissioner to wrap up the aspartame case as fast as possible, to signal that the Reagan administration was auguring a new regulatory era.[33] During this period, state intervention in industry affairs would be reduced as much as possible (echoing the doctrines of neoliberalism) and the FDA would become a rubber stamp for industrial products, with its control limited to the bare minimum.

And so, on July 15, 1981, Arthur Hayes granted aspartame market authorization at an ADI of 50 mg/kg. The commissioner justified the decision, which was published in the *Federal Register*, by "reasonable certainty" that (1) aspartame does not "induce brain neoplasms (tumors) in the rat"; (2) it does not "pose a risk of contributing to mental retardation, brain damage, or undesirable effects on neuroendocrine regulatory systems."

This initial approval applied to "dry products" like candy, chewing gum, cereal, and powdered coffee or tea. It would be extended to carbonated beverages and vitamins in 1983, and then progressively to all food categories. In the November 1987 letter he sent to Senator Metzenbaum, Adrian Gross, who was a member of the first FDA task force, bitterly wrote: "It is impossible for anyone to appreciate just how a determination by the FDA that the G.D. Searle & Co. experimental studies with aspartame were of an unacceptable quality in 1976 can be metamorphosed several years later into a view by that same Agency that essentially the *same studies* were sufficiently reliable for anyone to assess that this food additive is 'reasonably certain' to be safe for consumption by humans."[34]

The "Snowball Effect"

"And then there was a snowball effect," said Erik Millstone, a professor of political science at the University of Sussex, with a sad smile. "Reagan's election had repercussions in Geneva, since JECFA [the Joint Food and Agriculture Organization (FAO)/WHO Expert Committee on Food Additives] followed the FDA's example, as did all the European countries! In the United Kingdom, for example, in the mid-1980s, I spoke with a representative of the Ministry of Agriculture, Fisheries and Food, to determine from what scientific basis aspartame had been approved. He replied that there had been several exchanges with the FDA, which had certified that the sweetener hadn't presented any problems, and that's it!"

"What studies did JECFA use to set its ADI at 40 mg/kg?" I asked.

"The same studies used by the FDA—meaning Searle's! Through this case, it's easy to understand why that first authorization is so important to companies: the ideal is to obtain it from the FDA or JECFA, because then the door opens for the rest of the world, which blindly copies their decisions. After that, all they need is some time and no one can remember under what conditions the ADI was set and the product is assured a nice little future."

"How do you explain that the FDA and JECFA did not set the same ADIs, even though they evaluated the same studies?"

"The decision was completely arbitrary, because either way, the studies were absolutely unreliable! It's difficult to find out more because, unfortunately, JECFA reports don't mention anything about the discussions."

Admittedly, the JECFA archives (much like those of Joint FAO/WHO Meeting on Pesticide Residues [JMPR]) are quite tight-lipped. In general, they settle for summarizing the scientific arguments that led to an adopted decision. For aspartame, records indicate that nineteen JECFA experts, including René Truhaut and Blumenthal from the FDA, met on April 14–23, 1975, to evaluate its toxicity. They reviewed the Searle study on the effects of DKP, whose numerous irregularities would be revealed in the Bressler report two years later. "A special problem was posed by the presence of an impurity, 5-benzyl-3, 6-dioxo-2-piperazine (diketopiperazine, DKP)," note the experts. "Lesions seen in long-term feeding studies with DKP with rats were described as uterine polyps. [. . .] The Committee was therefore unable to arrive at an evaluation of this compound. Neither a monograph nor a specification was prepared."[35] The summary issued the following year was even more succinct, although it did echo the concerns being raised about the sweetener on the other side of the Atlantic: "In view of the incompleteness of the information available the Committee decided to defer its consideration of aspartame. Tentative specifications were drawn up but no monograph was prepared."[36] In its 1977 report, JECFA once again evoked the DKP study, namely the "assertion that the data base from which these conclusions were drawn requires validation"; hence the deferral of "its decision pending an assurance that the toxicological data are valid."[37]

JECFA would not address the evaluation of aspartame until 1980, when it summed up the substance in a few very terse sentences in its twenty-fourth report: "The Committee evaluated additional toxicity animal studies and several human studies. The no-adverse-effect level, based on animal studies, was found to be 4 g/kg. An ADI for aspartame was established at 40 mg/kg."[38] Five "studies" were effectively cited in the annex: two were conducted by Iroyuki Ishii, who evaluated the incidence of brain tumors among rats and measured the effects of DKP on behalf of Ajinomoto, the Japanese manufacturer of aspartame. The problem is that the results were dated 1981![39] (Note incidentally that the JECFA secretariat included "Dr. M. Fujinaga, from the Japan Food Additives Association.") The three other studies were provided by Searle and concerned the effects of aspartame on phenylketonurics—the report specifies that they were never published. It is impossible to know more about the scientific data that pushed JECFA to "prepare a monograph," or how it resolved the doubts raised by Searle's toxicological studies during previous meetings. Nevertheless, in 1981, a few months after Ronald Reagan was sworn in as president, the committee definitively confirmed the "ADI allocated in the twenty-fourth report."[40]

Thirty years later, in Geneva, the history behind aspartame's ADI (still in force at the time of writing in 2011) has clearly been lost in limbo. "When JECFA set it in the beginning of the 1980s, it was based on the studies available at the time," explained Angelika Tritscher, WHO joint secretary to JECFA and JMPR. "That standard is still valid because, in the meantime, it's been confirmed by other regulatory agencies." And yet "confirm" is far from the appropriate word in this case, since those agencies did not conduct their own review of the studies submitted by Searle. In fact, they merely borrowed the ADI set by JECFA, as Hugues Kenigswald, head of the Food Additive and Nutrient Division at the European Food Safety Authority (EFSA), explained when we met in Parma in January 2009: "The acceptable daily intake of 40 mg/kg was established by JECFA, then adopted in Europe by the Scientific Commission on Food in 1985."

"Do you know which scientific studies JECFA used to establish its ADI?" I asked.

"The studies financed by Searle, meaning the company that wanted to put aspartame on the market," answered the EFSA expert without hesitation.

"Did you know that the Searle studies were very controversial and judged to be unreliable by many scientists within the FDA?"

"I don't know what to think of those initial studies, because I don't have all the facts needed to make a judgment," admitted Hugues Kenigswald. "Clearly, if there were some doubts about the data's validity, those doubts have been lifted."

"The problem is that Searle has not done any new studies that would allow us to understand why those doubts were lifted, and yet everyone has stuck to that ADI."

"It might be regrettable, but that's often the way it is with decisions made thirty years ago."[41]

And that is how aspartame conquered the world, despite the many health warning signs that regulatory agencies continue to ignore with suspicious unanimity.

15

The Dangers of Aspartame and the Silence of Public Authorities

The wise man doesn't give the right answers, he poses the right questions.
—Claude Lévi Strauss

"Those who attack the safety of aspartame are also attacking the independent determinations of the health and regulatory authorities of the world. The fact is, Senator, that *every single authoritative scientific, medical and regulatory body* in the United States and around the world that has ever examined *the scientific evidence on the safety of aspartame* has each, *independently and separately*, arrived at a single identical conclusion, and that is that aspartame is safe."[1] This strong statement made by Robert Shapiro is particularly intriguing given that aspartame owes its success worldwide to an (unsavory) "herding effect," like the one that led village children to follow the Pied Piper.

In *The World According to Monsanto*, I talked at length about the path taken by the Saint Louis company's ambitious and arrogant director, who wanted to revolutionize the world with genetically modified organisms (GMOs). He began his dazzling career as an attorney at Searle, and in 1983 he was named CEO of NutraSweet, the pharmaceutical company subsidiary tasked with producing aspartame (sold in the United States as "Nutra-Sweet"). His role was further established in 1985 when Searle was bought by none other than Monsanto, whose leadership he would assume in 1995.[2]

1987: Revelations of the Metzenbaum U.S. Senate Commission

In November 1987, Robert Shapiro was called as a witness [see above statement] to a Senate hearing in Washington, DC, organized by Howard

Metzenbaum—a democratic senator from Ohio who had always been openly opposed to aspartame. Aware that an outright ban on the artificial sweetener was out of reach, he fought instead for what he considered to be a measure of public hygiene: obligatory labels indicating the amount of aspartame contained in food products. He had already started down that line of questioning during a congressional session on May 5, 1985: "But with all the concerns raised about the safety of NutraSweet, does it not make sense, is it not logical, for individuals and their physicians to know how much NutraSweet is in the diet soda? What could be so terrible about stating the amount? How else will the user, or the physician, know if the person is exceeding reasonable consumption limits, particularly during the summer months?"[3]

I listened to the five-hour recording from the November 3, 1987, hearing, available on the cable television channel, C-Span.[4] I have to say that I was fascinated by the Americans' capacity to very officially reveal an entire series of highly disturbing facts, even if it ultimately changes nothing. Case in point: nearly a quarter of a century after the hearing, aspartame still has not been banned, or even investigated. I subsequently discovered that the Pentagon had placed the substance on a list of candidate products for the development of chemical weapons. And that no fewer than ten highly placed Food and Drug Administration (FDA) officials—who had worked under Arthur Hayes (the agency head from 1981 to 1983) to obtain approval for aspartame, initially for use in dry foods (1981) then in carbonated beverages (1983)—were all subsequently recruited by Searle or Monsanto. Including Michael Taylor.

In my investigation into Monsanto, I described how Taylor, a lawyer with a consulting firm working for the multinational company, was named to the number two FDA position in 1991 (where he would remain for three years) to direct the (non)regulation of GMOs, before becoming the vice president of Monsanto, the leading company in the field, in 1998. Considered an archetype of the practice of "revolving doors," his oscillating career between the private and public sectors began in the early 1980s when he represented the FDA before a Public Board of Inquiry (PBI) on aspartame. As for Arthur Hayes, who left the agency in November 1983 once his term was completed, he became a consultant at Burson-Marsteller, one of the favored public relations firms of both NutraSweet and Monsanto.[5]

I also discovered that, at the request of Senator Metzenbaum, the Government Accountability Office—considered "the investigative arm of Congress"—had surveyed sixty-seven scientists: "more than half said they had some concerns over [aspartame's] safety"; twelve said they had "major concerns."[6]

I also found that, five years after going on the market, aspartame was the product that had generated the greatest number of spontaneous complaints to the FDA, 3,133 of which related to "neurological problems."

Senator Metzenbaum invited Major Michael Collings to exemplify the (numerous) "side effects"—to which I will return—of the white powder that had "captured the tastebuds of the American consumer," to borrow the senator's expression. Collings, a U.S. Air Force pilot and experienced long-distance runner ("five-to-eight-mile jogs in Nevada's desert heat"), had gotten into the habit of drinking "around a gallon [of Diet Coke and Kool-Aid] per day." Gradually, he began to experience slight tremors in his arms and hands; then, on October 4, 1985, he lost consciousness and had an epileptic seizure. After a medical leave of absence, he set out for the Australian desert where he was deprived of his favorite beverage; his symptoms disappeared. Once home in the United States, he returned to his former habits. And the tremors returned, culminating once again in an epileptic seizure. A physician recommended that he avoid all products containing aspartame. Visibly moved, Collings explained that he had done so, and that all his symptoms had definitively disappeared. He added that ever since then, he hadn't been able to fly, because the Air Force considered him disabled.[7]

Some would call this testimony anecdotal. But certainly not Dr. Richard Wurtman, a leading American neurology expert, who at the time was head of the Clinical Research Center at the renowned Massachusetts Institute of Technology (MIT). During his Senate hearing, he presented his study of two hundred aspartame consumers suffering from epileptic seizures, accompanied by frequent migraines and vertigo, despite having no previous history or detectable physiological cause.[8] With the calm assurance of an implacable specialist, Wurtman explained that the source of the problems could be phenylalanine, an amino acid on which he had been working for fifteen years, and on which his lab had published over four hundred studies. Curtailing the (weak) arguments brought forth by NutraSweet's representatives, who kept repeating that "the amino acid components of aspartame occur naturally in foods," the neurologist affirmed that, on the contrary, consummation of aspartame has nothing to do with that of a normal protein, because phenylalanine is not consumed with other amino acids. That is why it has a far greater effect on blood plasma, which can affect neurotransmitter production and brain function.

"How many clinical studies have been made of NutraSweet, to your knowledge?" asked Senator Metzenbaum.

"On brain diseases? I'm not aware of any," responded Wurtman without hesitation, before going to describe a number of extremely interesting facts.

The Maneuverings of the International Life Sciences Institute

In 1981, Wurtman testified before the PBI in favor of aspartame: the neurologist judged that when included in dry foods, the substance presented only minimal risks, as its consumption would remain limited. At the time, he was working as a consultant for the International Life Sciences Institute (ILSI).

In 1983, Wurtman learned that the company had requested that Nutra-Sweet approval be extended to soda fabrication. He expressed his concerns to ILSI because, knowing Americans' thirst for carbonated beverages, particularly among children, he feared that a massive influx of phenylalanine into the food chain would provoke serious health consequences. He therefore offered to lead a study to measure aspartame's power to modify brain chemistry and favor the onset of epileptic seizures.[9] Upon learning of his project, Gerald Gaull, Searle's vice president, paid Wurtman a visit at his MIT lab and threatened to play the veto card at ILSI to have his funding cut. During his testimony, Wurtman explained that once he "became convinced [industry] was not really interested in exploring the toxicity of aspartame," he decided to go without its financial aid.

Before quitting his "consultant" position, he wrote a letter to Robert Shapiro: "Dear Bob, I know you'll agree that my value to Searle . . . derives in part from my telling the company some things that it would rather not hear . . . and then from helping the company to deal with those things. One such thing is that some consumers may develop significant medical symptoms after consuming very large amounts of aspartame, particularly if they happen, concurrently, to be on a low-calorie, low-protein, weight-reducing diet. . . . If Searle-supported studies are going to contribute to our understanding of these people and their symptoms, then the studies have to include them, and not be restricted to people who have a can or two of soda per day."[10] During the hearing, Wurtman stigmatized industry-supported studies that involved "giving a few doses, two or three doses, for one or two days." "We see symptoms after people have taken the aspartame for weeks so one-day and two-day studies, as far as I am concerned, are of no value," he noted. He then outlined the real problem, which is that there are no public funds to conduct serious studies, and mentioned several colleagues who had submitted projects and whom were told to ask the industry for help. Wurtman added that he continued his work by relying on his own laboratory's funds.

Two other scientists questioned by the senators confirmed this warped system, which allows manufacturers to lock up research on their products. "The questions about phenylalanine effects on human brain function have not been asked," said Louis Elsas, a geneticist at Emory University in Atlanta. "So we have spent millions of dollars through our current system on mostly irrelevant experiments without approaching those particular questions." The pediatrics specialist and researcher was particularly concerned about the effects of amino acids on the fetus. "In the developing fetus such a rise in maternal blood phenylalanine could be magnified four- to six-fold by the concentrative efforts of the placenta and fetal blood brain barrier,"[11] he explained. "The effect of such an increased fetal brain concentration would probably be [. . .] expressed as mental retardation, microcephaly, or potential certain birth defects." He concluded that, through the same mechanism, irreversible brain damage could occur in babies from zero to twelve months old.

"Can you tell the Committee about your own experiences with the International Life Sciences Institute?" asked Senator Metzenbaum.

"Yes, sir. It was not good," answered Elsas. "But I was asked after issuing concerns both privately and then publicly on 'Nightline' to give them a specific protocol for how I would approach these concerns. I did this [. . .] but without ever a written peer review of criticism. And the ideas are now reappearing three years later in other places funded by industry."

William Pardridge, an endocrinologist and professor of medicine at the University of California Los Angeles, had a similar experience at ILSI.[12] Focusing specifically on the blood–brain barrier transport of phenylalanine, he submitted two research projects on the effects of aspartame on children's brains, both of which were turned down.

Confronted with these detailed accusations, ILSI's representatives and collaborators made quite a poor showing. They included John Fernstrom, a psychiatrist at the University of Pittsburgh, who attempted to skirt the issue: "I can't imagine a kid taking that ADI [for aspartame]. Fifty mg/kg is five cans of soda pop and [. . .] there is no way he is going to do that." Then, he launched into a surrealistic discussion on the speed of aspartame breakdown, which is ostensibly "five times faster" in rats than in humans. Visibly exasperated, Senator Metzenbaum cut his stonewalling short by pulling out from behind his lectern—with a mischievous smile, one by one—several dozen common products that contain aspartame: carbonated beverages, chewing gum, cereals, yogurts, medications, vitamins, etc. The extremely theatrical accumulation of products elicited a volley of applause from the audience.

October 2009: The FDA Persists and Declares "the Substance Is Safe"

"I have no qualms in saying that if we are basing the amount of aspartame that we are putting in all these foods today on these [Searle] studies, then it is a disaster." After the muddling of the ILSI scientists, Dr. Jacqueline Verrett's testimony offered surprising clarity, provoking a religious silence in the hearing room. Verrett—appearing very severe in her square glasses and classic, tailored suit—had worked at the FDA as a biochemist and toxicologist from 1957 to 1979. In 1977, she joined Jerome Bressler's team and therefore had access to the raw data for three infamous studies (one on DKP and two others on teratogenicity), which helped set the ADI for aspartame in the United States and Europe (see Chapter 14). With a deadpan tone, she referred to "animals returned to the study" after their tumors were removed, "animals [who] were recorded as dead, but subsequent records indicated the same animal was still alive," and offered the cutting critique that "It is unthinkable that any reputable toxicologist giving a complete, objective evaluation of the data resulting from such a study could conclude anything other than that the study was uninterpretable and worthless and should be repeated." She added, "In a quick scan of [the literature], I do not find studies that repeat any of this research enough to answer the questions that were raised. [. . .] and hence the acceptable daily intake figures remain in question and remain unanswered."

Jacqueline Verrett (who passed away in 1997) co-authored an iconoclastic book in 1974—*Eating May Be Hazardous to Your Health*—in which she described her work at the FDA. Daring to challenge the famous agency's reputation, she unabashedly writes: "Unfortunately, our food is not the safest in the world. [. . .] If some food additives were regulated as drugs they would be forbidden—except by prescription and then forced to carry warnings— especially to pregnant women."[13] She gives the example of Citrus Red 2, a food coloring that causes "stillbirths, fetal deaths and birth defects in animals."[14] The toxicologist also describes the role she played in banning cyclamate (E 952, which is still authorized in Europe). On October 1, 1969, she caused an upheaval by revealing the results of a study she had conducted on

13,000 chick embryos on television (on NBC). After being injected with cyclamate, the chicks were born with "severe birth defects" such as "deformed spines and feet, phocomelia."[15]

Delving into hundreds of food additives authorized by the FDA, the majority of which "have never been tested," she laments the fact that "All of us are involved in a gigantic experiment of which we shall never know the outcome—at least in our lifetime. How dangerous are the food chemicals we are eating? Are they contributing to cancer? To birth defects? To mutations? To liver, brain and heart damage and to a hundred other diseases? We don't know. [. . .] We could at this moment be sowing seeds for a cancer epidemic in the 1980s or 1990s."[16]

After reading this very disheartening book, I contacted the FDA in Washington, DC. It seemed like the perfect moment as President Barack Obama had just named Margaret Hamburg, a physician known for her commitment to community health (a domain largely overlooked by industry), to head the agency in March 2009. Familiar with the necessary procedure (thanks to my investigation into Monsanto), I contacted the press office where I encountered Mike Herndon, a civil servant who, after giving me the runaround, finally sent me the e-mail address of a key figure: James Maryanski, the former FDA biotechnology coordinator. From the way Herndon gently blew me off, it appeared that he had gotten wind of my film *The World According to Monsanto*, in which Maryanski proffered several sensational revelations on the links between the agency and the St. Louis-based company. I then had to write to Joshua Sharfstein, Margaret Hamburg's right-hand man, who quickly intervened on my behalf (proof of a shift in attitudes in America). As a result, poor Mike Herndon found himself obligated to set up a meeting for me with David Hattan, the agency's toxicologist responsible for overseeing food additives. When I entered the senior toxicologist's office on October 19, 2009, I thought I was hallucinating: it was the same man who had sat beside Commissioner Frank Young during the infamous Senate hearing on November 3, 1987. It goes without saying that, at the time, Young had obstinately defended the approval of aspartame, under Hattan's approving gaze.

"I saw you in C-Span's archived footage," I told him, slightly amused.

"Yes . . ."

"Jacqueline Verrett wrote this book here, maybe you know it, *Eating May Be Hazardous to your Health*. Have you read it?" I asked, holding the work out to the visibly tense toxicologist.

"No . . ." he murmured.

"Can you open it to page 96? I would like to have your comment, you have been working here for a long time—she says: 'It's not that government decision-makers are corrupt . . .' That's a good thing?" I interjected, scrutinizing David Hattan, who nodded in agreement with a frozen smile. I continued to read: "'. . . but their sense of duty is constantly eroded by industry contacts and the consideration of short-term effects on industry instead of long-term effects on consumers.' Do you think that's accurate?"

"No, I don't agree with her," responded the toxicologist. "I don't think that any of us in the FDA would feel we're doing our job adequately and appropriately if we didn't put consumer safety ahead of any kind of consideration of industry well-being. That's turning the whole safety assessment paradigm on its head. No, I disagree with that completely."

"You followed the approval process for aspartame very closely, correct? Since you arrived at the FDA when the PBI was being set up?"

"Yes."

"The PBI, like other investigatory groups at the FDA, spoke out against authorizing the artificial sweetener. How do you explain that, several months later, the substance was nonetheless authorized, even though the general opinion within the agency was that the Searle studies were absolutely unreliable?"

"Oh! I would welcome individuals, maybe even challenge individuals, to come and look at the actual administrative records that the FDA has in its files about what the FDA did to resolve that controversy. It took many people and many months and millions of dollars being spent by the sponsor, Searle . . . We are not defending everything that was done. There were some mistakes made and some shortcuts to the way the studies were conducted . . . it actually was before 'good laboratory practices'; the standard of performance of studies was not nearly as rigorous then as it is now . . . Basically, although there were problems with the conduct of some of these studies, none of those problems was serious enough to invalidate the studies' results and none of them changed the studies' results that indicated that the compound was safe."[17]

Ninety-One Side Effects

"The FDA received thousands of complaints about the side effects of aspartame," I continued, as David Hattan glanced repeatedly at Michael Herndon, the press office representative seated behind me. "I have here an internal, declassified document presenting ninety-one symptoms: 'headache, dizziness, vomiting, nausea, abdominal cramps, change in vision, diarrhea, seizures, memory loss, rash, sleep problems, change in menstrual pattern, edema, chronic fatigue, shortness of breath . . .'" The document I handed to the toxicologist (to refresh his memory) made headlines in 1995. It was obtained by Betti Martini, the founder of "Mission Possible," thanks to the Freedom of Information Act (see Chapter 14). It revealed that approximately 10,000 people spontaneously contacted the FDA to report problems they believed to be linked to aspartame.[18] And according to a rule accepted by the agency, only 1 percent of consumers who encounter problems with a substance bother writing, which signifies that 1 million Americans could have suffered from the side effects of aspartame (between 1981 and 1995).

All the symptoms described in the FDA document mirrored those observed by Dr. Hyman Roberts during his long career. The Palm Beach physician, whom I met on October 24, 2009, fortuitously developed an interest in aspartame in 1984.[19] That year, he saw a sixteen-year-old female patient named Tammie who had a seizure in his office. The concerned doctor ordered countless exams, none of which revealed the source of the neurological problems. He concluded that the only possible cause was aspartame contained in the "diet sodas" Tammie had begun to drink in order to limit her sugar intake. Four years later, Roberts published a study on 551 patients who had visited him for at least one of the problems described in the FDA document. "The causative role of aspartame-containing products is supported by (1) the relief of complaints shortly after avoiding such products, and (2) their occurrence within hours or days of reexposure, frequently inadvertent. . . . A brief trial of abstinence might avoid multiple consultations, costly tests, and hospitalization."[20] In 2001, Roberts published a 1,020-page book in which he presents the clinical history of 1,400 patients.[21] He observed an addiction phenomenon, notably among heavy consumers of diet sodas (over two liters per day) or "sugarless" chewing gum (at least one pack a day), which causes cravings during withdrawal.

"Did you contact the FDA?" I asked.

"Of course, but the agency never responded!" replied Roberts. "The industry considers all these cases 'anecdotal,' even though hundreds of thousands of people are concerned."

In a letter he sent to Senator Howard Metzenbaum shortly after the November 1987 hearing, neurologist John Olney ironically noted, "I doubt whether the thousands of lay citizens who have generated these complaints have thought it out ahead of time and conspired to make all of their complaints sound like their central nervous system is being affected."[22]

"Are you familiar with Dr. Roberts' work?" David Hattan raised his eyebrows at the question before, after some hesitation, responding: "In reality, the FDA and the Searle company conducted supplementary clinical studies in order to evaluate those effects, like headaches and seizures. All of that was carefully tested and the result was that, in a controlled environment where we know the exact dose used, the exact moment of ingestion and the individual who consumed it . . . well, we can't reproduce those effects."

"I don't know which studies Mr. Hattan is talking about," said Dr. Ralph Walton coldly, during our meeting in New York on October 30, 2009. "It would be great if he sent me his references, as it's precisely because there are no serious studies investigating the neurological effects of aspartame on humans that I decided to conduct my own research." Walton, who is a professor of clinical psychiatry at Northeastern Ohio Universities, also "stumbled onto aspartame by chance." "In 1985, one of my patients whom I had been monitoring for twelve years for chronic depression began to have epileptic fits and develop manic episodes," he told me. "It was even more odd since she had been doing well for years and her antidepressant treatment hadn't changed. After ruling out bipolar disorder, I did some digging to understand what had changed in her life. And I discovered that she had begin drinking 'Crystal Light' products in order to lose weight and was consuming one to two liters of it a day. As soon as she stopped her intake, the problems definitively disappeared. I submitted a clinical report to a medical journal, and one of the reviewers was Richard Wurtman. He asked me if I knew of other similar cases. As I was head of the medical society in my city, I reached out to my colleagues, who reported dozens of cases. Ultimately, these clinical cases constituted a chapter in a book Dr. Wurtman published on the effects of phenylalanine on cerebral functions."[23]

"Can you explain what kind of study you conducted?" I asked.

"In truth, if I had known the serious reactions that it was going to trigger, I never would have launched the experiment . . . We gave aspartame to volunteers for seven days, in a double-blind trial, meaning that the participants didn't know whether they were receiving the substance or the placebo, nor did the researchers administering the products. A friend and colleague, who was forty years old and a doctor of psychology, experienced a retinal detachment and ocular bleeding and lost vision in one eye permanently. A nurse, who also volunteered, also presented ocular bleeding. The ethical committee supervising the study asked us to stop immediately. But, since thirteen people had followed the entire protocol, we were able to publish it with significant findings. Our conclusion was that people who had already experienced depressive episodes were extremely sensitive to aspartame."[24]

"What dose did you use in your study?"

"Thirty mg/kg, since I wanted to remain below the ADI set by the FDA. That corresponds to about eight cans of Diet Coke a day, but it's a dose that many people consume daily, since aspartame is found almost everywhere."[25]

The Influence of Industry: The "Funding Effect"

On November 18, 1996, in Washington, DC, Ralph Walton, lawyer James Turner, Senator Howard Metzenbaum, and John Olney gave a press conference on an article they had just published: "The article we just published shows an increase of incidence of brain tumors and increased malignancies of brain tumors in human population in the USA, starting about three years after aspartame has been introduced."[26] Olney had reviewed data on brain tumors from the National Cancer Institute collected from 1970 to 1992 in thirteen geographic zones from the United States, which covered 10 percent of the American population. He found that an initial, localized increase in incidence "occurred in the mid 1970's and might be explained primarily by improved diagnostic methods." It was followed by a "second phase [that] occurred abruptly in the mid 1980's, resulting in a 10 percent higher rate of brain tumors which has persisted to the present [through 1992]." He concluded that "the evidence presently available is not adequate to establish whether aspartame does or does not cause brain tumors. Therefore, new studies properly designed to answer this question are urgently needed."

The publication drew a great deal of media attention, and the renowned current affairs show *60 Minutes* decided to dedicate a special episode to aspartame. Lost amidst the massive amount of studies about the artificial sweetener, CBS producers asked Ralph Walton to conduct a systematic review of studies published in peer-reviewed scientific journals. An initial search in different databases, including MedLine, provided 527 references; the psychiatrist kept only those that were "clearly linked to the product's safety for humans."

"First," Ralph Walton told me, "it's important to note that the three fundamental Searle studies, which were used to calculate aspartame's ADI, were never published! Furthermore, of the 166 studies that my team eventually selected, 74 were financed by the industry (Searle, Ajinomoto, or ILSI) and 92 by independent research bodies (universities or the FDA). One hundred percent of the industry-funded studies concluded that aspartame was safe. Of the 74, several had been published several times in different journals, under different names, but it was the same study. Of the 92 independent studies, 85 concluded that the sweetener posed one or more health problems. The remaining seven studies were conducted by the FDA and came to the same conclusions reached by those financed by the industry."

"How can you explain such an incredible result?" I asked.

"Aha! Well, you know, money is very powerful . . ."

The blatant phenomenon Ralph Walton observed is known as the "funding effect." David Michaels describes this worrying trend as follows: "When a scientist is hired by a firm with a financial interest in the outcome, the likelihood that the result of the study will be favorable to that firm is dramatically increased." The new head of the Occupational Safety and Health Administration (OSHA) adds, "Having a financial stake in the outcome changes the way even the most respected scientists approach their research and interpret the results of experiments."[27]

Paula Rochon, a Boston geriatrician, observed the funding effect when she was comparing clinical tests for nonsteroid, anti-inflammatory medications, such as aspirin, naproxen, or ibuprofen (Advil), used to treat arthritis. She showed that the industry-funded tests *always* presented favorable conclusions, even if careful review of the data did not confirm them.[28] Four years later, a team led by the Canadian researcher Henry Thomas Stelfox at the University of Toronto made the same observations for calcium channel blockers—medications prescribed to treat hypertension and suspected of causing cardiac arrests. The researchers reviewed articles published between March 1995 and September 1996 and classified their authors in three groups according to their position in relation to the molecules: "favorable," "neutral" and "critical." The results: 96 percent of "favorable" scientists had a financial link with the calcium antagonists' manufacturers, versus 60 percent of "neutral" authors and 37 percent of "critical" ones.[29] Since then, the phenomenon has also been detected for oral anti-contraceptives, and drugs to treat schizophrenia, Alzheimer's disease, and cancer.[30]

I carefully combed through Ralph Walton's list of seventy-four studies financed by aspartame manufacturers, one of which caught my eye because it illustrates quite clearly the phenomenon of "black boxes" described by Bruno

Latour in his book *Science in Action*. In order for a scientific statement to become an established fact whose origin cannot be reconstituted, it needs to be widely cited in multiple scientific articles. "A statement [is] fact or fiction not by itself but only by what the other sentences made of it later on," explains the philosopher. "To survive or to be turned into fact, a statement needs the next generation of papers."[31] And that's why Searle and company *ensured* the publication of several dozen "studies," which never addressed the essential questions and whose goal was *to dominate the scientific literature*: a *published* study is a study that can be *cited* and thereby contribute to the transformation of a "fiction" to a "fact." It is all the more effective if the industry can simultaneously block the production of independent studies that are quite rightly tackling the essential questions—a job ILSI has fulfilled perfectly.

We have seen the dubious conditions in which the ADI for aspartame was set in 1981. Ten years later, Searle asked two of its scientists, Harriett Butchko and Frank Kotsonis, to publish an article about the conception of the ADI, "using the widely used food additive aspartame" as an example.[32] This was an astute move, since it allowed the report to immediately establish the ADI for aspartame as a "black box," whereas, four years after the Senate hearing, it was still far from being unanimously accepted: "Aspartame has been assigned an ADI of 40 mg/kg/day by the World Health Organization and regulatory authorities in Europe and Canada, and of 50 mg/kg/day by the US Food and Drug Administration," write the authors, who then pepper their article with multiple references (fifty), primarily to studies financed by Searle (though the funding source is not detailed), and including those led by Jack Filer (nine references) who, as we have seen, would become the ILSI director! Who is going to verify that these studies, whose authors claim they were used to set the ADI, all date after 1979? Or that a study by a key figure like Filer—one intended to confirm the innocuousness of aspartame—lasted six hours, during which eight "normal adults" (four men and four women) ingested 10 mg of aspartame every two hours?[33]

"The problem," said Ralph Walton, "is that all these low-quality, even biased, studies are published in peer-reviewed scientific journals. We're still waiting for the 'radical reform' called for by Richard Smith." The director of the prestigious *British Journal of Medicine* caused a sensation by publicly recognizing the limits and weaknesses of the peer review system (see Chapter 9), though it is considered to be indispensable to scientific publications. "We know that it is expensive, slow, prone to bias, open to abuse, possibly anti-innovatory, and unable to detect fraud," he wrote. "We also know that the published papers that emerge from the process are often grossly deficient."[34] In his editorial, which rubbed many the wrong way (namely manufacturers), Richard Smith described an experiment led by Fioda Godlee and two colleagues at the journal: they intentionally inserted eight errors into a study they were going to publish. They then sent the text to 420 potential reviewers, 221 (53 percent) of who responded: the average number of errors highlighted was two, not a single reviewer spotted more than five errors, and 16 percent spotted nothing.

The Ramazzini Institute: "The House and the Forum for Those Who Spend Their Lives in the Name of Truth"

"I've been fighting for twenty years to get the National Toxicology Program [NTP] to conduct a study on aspartame," James Huff explained to me in 2009. Huff is the associate director for chemical carcinogenesis at the National Institute of Environmental Health Sciences (NIEHS) and leads the International Agency for Research on Cancer (IARC) Monographs Programme (see Chapter 10). "Unfortunately, the FDA has always opposed such a study by playing its veto card."[35]

"How do you explain that?"

"I think that the agency is afraid that we will prove the sweetener is carcinogenic,"[36] the scientist answered, before directing me to an article that appeared in November 1996, following the publication of John Olney's study on the increase in brain tumors. It quoted James Huff, as well as David Rall, the former director of the NIEHS who oversaw the NTP for nineteen years (until his retirement in 1990): "It's a wonderful way to ensure that it isn't tested," said Rall. "Discourage the testing group from testing it and then say it's safe."[37]

"And yet, I read that the NTP published the results of a study on aspartame in 2005,"[38] I continued.

"That's true," replied James Huff, "but I was opposed to the study, as were several NIEHS colleagues. It was conducted on transgenic mice in which a gene was inserted that renders them more susceptible to cancer. It's a new experimental model irrelevant to nongenotoxic, chemical products. Like aspartame, which is not genotoxic, meaning it doesn't produce mutations.[39] The results of this study—which cost a lot for nothing—were of course negative and made the industry very happy.[40] I was disgusted, which is why I actively participated in designing studies led by the Ramazzini Institute that actually confirmed the carcinogenic potential of aspartame. In my opinion, they are the best studies thus far conducted on the substance."

The Ramazzini Institute—named in honor of the "father of occupational medicine" (see Chapter 7)—was founded in 1987 by Italian oncologist Dr. Cesare Maltoni, whose work on vinyl chloride spread panic among European and American plastic manufacturers (see Chapter 11). Housed in a magnificent Renaissance-era Bentivoglio castle, twenty miles from Bologna, the environmental oncology center defines its research projects in collaboration with the Collegium Ramazzini, composed of 180 scientists from thirty-two countries. They include some of the scientists cited in this book, such as James Huff, Devra Davis, Peter Infante, Vincent Cogliano, Aaron Blair, and Lennart Hardell. Once a year, this exceptional group gathers in Carpi, the "master's [Bernardino Ramazzini] birthplace." In an article published in 2000, which is both a veritable declaration of faith and a testament, Cesare Maltoni (who died in 2001) described what makes this academic collegium so original and unique. "Our time is characterized by the enormous expansion and the primacy of industry and trade, at the expense of culture (including science) and humanism," he wrote. "The primary and, too often, unique goal of industry and trade is profit. The strategy of industry and trade, in order to meet their objectives even when in conflict with culture and humanism, has been marked by the creation of an alternative pseudoscientific culture, whose major aim is to pollute truth instrumentally, by contrasting culture and science and by muffling the voice of humanists."[41] Cesare Maltoni added that the collegium's raison d'être is "to be the house and the forum for those who spend their lives in the name of truth, and to give solidarity to those who, in pursuing the truth, are attacked and humiliated."

Since its establishment, the institute has tested some two hundred chemical pollutants, such as benzene, vinyl chloride, formaldehyde, and numerous pesticides. These studies have often played a role in decreasing the exposure standards in use thanks to their irrefutable findings. Contrary to the large majority of industrial studies, institute analyses are conducted on megacohorts of thousands of subjects, which of course reinforces their statistical potential.[42] During my visit on February 2, 2010, I was impressed by the laboratory's breadth (thirty thousand square feet). Enormous circular facilities housed nine thousand rats exposed to different levels of electromagnetic waves for an experiment that Dr. Morando Soffritti, who succeeded Cesare Maltoni, described as "top secret" with a knowing smile. "The second characteristic of our institute," he explained, "is that, contrary to the recommendations within the guide to 'good laboratory practice,' our experimental studies don't last two years. Instead, we allow our animals to live until their natural deaths. In fact, 80 percent of malignant tumors detected in humans are found after the age of 60–65 years old. It's therefore absurd to sacrifice experimental animals in the hundred and fourth week, which, when applied to the human species, corresponds to the age of retirement, when the frequency of cancers and neurodegenerative diseases is the highest."[43]

"That's the Ramazzini Institute's greatest strength," confirmed James Huff. "When you arbitrarily interrupt a study after two years, you risk missing the carcinogenic effects of a substance. And several examples prove it. Cadmium is a widely used metal, notably for the fabrication of PVC or chemical fertilizers, which IARC classified in group 1 ['carcinogenic to humans']. And yet, experimental studies showed zero effect. Until the day a researcher decided to let the rats die naturally: he observed that 75 percent developed lung cancer in the last quarter of their life. Likewise, the NTP studied toluene and found no effects after eighty months. On the other hand, the Ramazzini Institute observed several cancers that appeared after eighty months. All researchers should adopt the Ramazzini Institute's study protocols, because the stakes are important: we always glorify extended life spans, but what's so great about living ten or fifteen years longer, if it means living one's retirement stricken by any number of diseases that could have been avoided if our exposure to chemical products was better controlled? That's why the two aspartame studies conducted by the Ramazzini Institute are so troubling."

"Aspartame Is a Multipotential Carcinogenic Agent"

More troubling still is the fact that the European Food Safety Authority (EFSA) and the FDA rejected those two studies, as did all the other national regulatory agencies, including, of course, the French Agency for Food, Environmental, and Occupational Health and Safety (Agence nationale de sécurité sanitaire de l'alimentation, de l'environnement et du travail, ANSES). I must say that, even after reviewing the agencies' arguments at length, I remain unconvinced.

The first study, published in 2006, was conducted on 1,800 rats that ingested daily doses of aspartame between 20 mg/kg and 100 mg/kg, from the age of eight weeks to their natural death. The results: a significant increase, correlated to the dose, in lymphoma, leukemia, and renal tumors among the females, and schwannomas (tumors of the cranial nerves) among the males. "Had we stopped the experiments at 110 weeks of age, we would most likely never have demonstrated the carcinogenicity [of aspartame]," write the study's authors. "The results of this mega-experiment indicate that aspartame is a multipotential carcinogenic agent, even at a daily dose of 20 mg/kg body weight, much less than the current acceptable daily intake."[44]

Curiously enough—given that the FDA generally contents itself with data summaries submitted by manufacturers—the agency insisted in this specific case on receiving the entirety of the Ramazzini Institute's raw study data. At least, that is the official argument it kept brandishing, namely by David Hattan who repeated it without batting an eyelid: "We looked at a small subset of the data. To us, those changes looked like the kinds of changes that you see all the time, the sporadic changes that you see as a result of animal testing. We asked two or three times and each time they said it was the policy of their research institute not to share the primary data with outside parties."

"Why did you refuse to communicate the study's raw data?" I asked Morando Soffritti, the scientific director of the Ramazzini Institute.

"I'm surprised the FDA told you that," he responded with his unfailing lopsided grin. "We've been in contact with the FDA since 2005 and we sent them all the data in our possession."

Nevertheless, in an opinion published on April 20, 2007, the American agency affirmed that "study data [. . .] does not support the conclusion that aspartame is a carcinogen."[45] One year earlier, the EFSA had produced a similar opinion, after a long introduction in which it inevitably invokes the laboriously constructed "black box": "Aspartame has undergone extensive testing in animals and studies in humans, including four animal carcinogenicity studies conducted during the 1970s and early 1980s. These studies, together with studies on genotoxicity, were evaluated by regulatory bodies worldwide and it was concluded that they did not show evidence of genotoxic or carcinogenic potential for aspartame."[46] Then, the European body addresses the Ramazzini Institute study, whose "flaws bring into question the validity of the findings. . . . The most plausible explanation of the findings in this study with respect to lymphomas/leukemias is that they have developed in a colony suffering from chronic respiratory disease." The review concludes "that there was no need to revise [. . .] the previously established Acceptable Daily Intake (ADI) for aspartame, of 40 mg/kg body weight (bw)."

"Why did you reject this study?" I asked Hugues Kenigswald, the chief of the Food Additive and Nutrient Division at EFSA (whom we met in Chapter 14).

"First of all, so it's quite clear—this study was absolutely not rejected. On the contrary, it was studied [*sic*] with the greatest attention. However, what became very clear is that there are a certain number, if not a whole host, of methodological flaws that emerged in this study."

"For example?"

"In particular, the fact that some rats displayed respiratory pathologies."

"What's the relationship between having a respiratory disease and a lymphoma or leukemia?"

"A respiratory disease makes it so that it provokes . . . is the source of tumors and can therefore completely confuse the issue; that's exactly what happened in this study."

Once again, the EFSA's argument amused Morando Soffritti, who, comfortably seated in his chair, replied: "We don't agree, for a number of reasons. First of all, because the inflammatory processes we observed in our animals are very often reliant on the fact that we allow them to die naturally without arbitrarily cutting their lives short. And, like humans in the last phases of their lives, pulmonary and renal complications are very common. What's more, it's never been shown that pulmonary or renal infections that appear at the end of life are capable of causing tumors in such a short period."

"Did the rats in the control group have the same inflammatory problems?"

"Of course. We observed them in both the treated and control groups. The only difference between the two groups was that the experimental groups had ingested aspartame and the control group hadn't."

In 2007, Morando Soffritti's team published a second study, more worrying than the first. This time, four hundred pregnant rats were exposed to daily aspartame doses of 20 mg/kg and 100 mg/kg and their offspring were monitored until they died. "We observed that, when exposure began during fetal life, the risk of developing the tumors observed in the first study increased quite significantly," commented Soffritti. "Add to that the appearance of breast tumors among the female offspring. We believe that these findings should push regulatory agencies to act as quickly as possible, because pregnant women and children are the most frequent consumers of aspartame." In the publication, Soffritti and his colleagues underline that, "At their request, we provided each of these agencies with all available raw data related to the study."[47]

And yet, David Hattan said precisely the opposite: "We didn't review the second study done by the Ramazzini Institute because, unfortunately, we couldn't reach an agreement to obtain the raw data," said the FDA toxicologist.

"That's not true," Soffritti would later retort from his Bentivoglio lab.

"You're saying David Hattan is lying?" I asked.

"You could say he's lying."

In its March 19, 2009, opinion, the EFSA underlines that "data were not provided by the authors," which the Ramazzini Institute director fervently denies. Then, the European agency once again dismisses the leukemia and lymphomas observed, which it obstinately (and shockingly!) categorizes as "characteristic for chronic respiratory disease," before proffering an explanation that outright stunned American scientists James Huff and Peter Infante, who found it "scabrous and unscientific": "The increase in incidence of mammary carcinoma is not considered indicative of a carcinogenic potential of aspartame since the *incidence of mammary tumors* in female rats is rather high and *varies considerably between carcinogenicity studies*," write the EFSA experts. "The Panel also noted that *an increased incidence of mammary carcinomas was not reported in the previous ERF* [European Ramazzini Foundation] *study* with aspartame which used much higher doses of the compound."[48]

"It's incredible that those experts could write that," said James Huff. "It seems like they didn't understand the originality of the study, which was to begin exposure in the womb. What's troubling is precisely the fact that offspring developed breast tumors that the adult rats in the first study did not. We observe exactly the same phenomenon with endocrine disruptors: it's the daughters exposed during fetal life who have breast cancers, not their mothers!"

The EFSA's argument was clearly surprising, and yet it was the only one used by Hugues Kenigswald to justify the decision to ignore the Italian study's results: "The breast tumors described in the second study didn't appear in the first study," he explained, while glancing at the two European civil servants sitting behind me. "Therefore the findings of the two studies are inconsistent."

"How do you explain the EFSA's argument?" I asked Morando Soffritti, who clearly weighed his words before responding.

"Evaluations done by experts from different agencies are often hasty and not entirely thought out," he said. "If they had taken the time to measure the implications of exposure beginning in fetal life, they might not have made a judgment that was so trivial from a scientific point of view." In the meantime, in an April 2009 statement, the International Sweeteners Association (ISA) welcomed the EFSA opinion, which "re-confirms the safety of the low-calorie sweetener aspartame, rejecting claims by the Ramazzini Institute (Italy) alleging that aspartame was unsafe. [. . .] These conclusions also support the previous EFSA opinion on aspartame issued in May 2006." And finally, "These conclusions from EFSA are entirely consistent with the global scientific consensus."[49]

Conflicts of Interest and a Pandora's Box

I have already said this and I will repeat it again: the arguments proffered by the EFSA and FDA are entirely unconvincing. How, then, can we understand why these agencies chose to ignore two studies conducted by an institute regarded as a heavyweight in the field of environmental oncology, while they continue to defend tooth and nail the ADI for aspartame, which is based on studies that, to say the least, present serious "methodological flaws" (to borrow Hugues Kenigswald's words)? Intrigued, I decided to identify the twenty-one experts in the EFSA's Panel on Food Additives and Nutrient Sources Added to Food, known as the ANS panel.

Since 2002, the EFSA experts—who have permanent posts, unlike those at the Joint Food and Agriculture Organization (FAO)/World Health Organization (WHO) Meeting on Pesticide Residues (JMPR) or the Joint FAO/WHO Expert Committee on Food Additives (JECFA)—have been obligated to declare their conflicts of interest, which can be consulted on the EFSA's website. That's where I discovered that John Christian Larsen, the panel president, works for ILSI! As do John Gilbert and Ivonne Rietjens, who also has financial ties to the Flavor and Extract Manufacturers Association (FEMA). Jürgen König has contracts with Dannon, a huge aspartame user. But the grand prize, if I may say so, goes to Dominique Parent-Massin, who sits on scientific committees at Ajinomoto, the Japanese aspartame giant, and Coca-Cola, a long-time user of sweeteners and founding member of ILSI! As director of the Food Toxicology Laboratory at Brest University (France), Parent-Massin even headed the food additives panel at the French Food Safety Agency (Agence Française de sécurité sanitaire des aliments, AFSSA [renamed ANSES in 2010])! The Japanese manufacturer's French "dream team" is rounded out with France Bellisle, a researcher at the National Institute of Agronomic Research (Institut national de la recherche agronomique, INRA) who is a member of the scientific committee at the European Food Information Council (EUFIC)—which is financed by large food processing companies—and Bernard Guy-Grand, head of the Nutrition Department of Hôtel-Dieu Hospital (Paris) and president of Ajinomoto's scientific committee. Note that Dominique Parent-Massin refrained from stating her affiliation when acting as a "health authority" during a hearing to defend aspartame's safety.[50] As such, during the 2006 Bichat conference, she adopted a familiar refrain: "Aspartame is one of the most extensively studied additives in the world."[51]

I of course questioned Catherine Geslain-Lanéelle, the EFSA's executive director, about the conflicts of interest held by certain members of the ANS panel, especially Dominique Parent-Massin. In all honesty, I was very curious to meet this ex-director of the Food Department of the French Agricultural Ministry (and a very zealous one at that), who (as we saw in Chapter 6) had refused to submit the Gaucho market authorization dossier to Judge Louis Ripoll. At the time, he was looking into the Food Department as part of an investigation on the insecticide's toxicity in regards to bees. The very cordial executive director explained to me that the EFSA had begun to "re-evaluate colorants" in 2008, and had recently decided to ban a "colorant used in Europe for thirty years in breakfast products and sausages consumed in Great Britain and Ireland." "A review of the studies showed that it was geno-toxic," she continued, "so we took it off the market, just as we've done previously for certain artificial flavorings."

"That's certainly good news," I said. "As for aspartame, I'm surprised to see someone like Dominique Parent-Massin, who has well-known ties with the principal aspartame manufacturer, on a food additives panel . . ."

"That means that when we conduct an evaluation of aspartame, that expert cannot be a rapporteur, cannot prepare the panel opinion, and cannot participate in deliberations on the subject, because he or she has a conflict of interest."

"So for example, Dominique Parent-Massin didn't participate in the aspartame opinion published in March 2009?"

"No . . . It's important to understand that today public research is often linked to private research and, therefore, it's impossible to find experts who have never had contact with industry—I don't think they exist anymore," Catherine Geslain-Lanéelle admitted. "That's why we established a rule that scientists who have worked or are working directly for a product's manufacturer are disqualified from participating in its evaluation, and that's what happened with Dominique Parent-Massin."[52]

But, in any case, transparency clearly has its limits. Dominique Parent-Massin's declaration of conflicts of interest, which I found on the EFSA site *before* my visit to Parma, disappeared several days later! It was replaced by a new one that omitted any mention of the expert's ties to Ajinomoto and Coca-

Cola. The anecdote (once again) amused Morando Soffritti, who recounted his own: "A senior EFSA official told me once, 'Dr. Soffritti, if we admit that the results of your studies are valid, we'll have to ban aspartame first thing tomorrow. You have to realize that that is impossible.'"

Everything indicates that, even beyond its huge economic stakes, aspartame has become an unassailable fortress, as emphasized by Erik Millstone, the relentless thorn in the regulatory agencies' collective side. "If they admit to making a mistake, that would provoke a loss of confidence. And also, they're undoubtedly afraid that it would open the floodgates," he explained with a very accusatory tone. "People might say: perhaps you haven't made just one error, but many; or maybe the entire process is defective! Aspartame is a Pandora's box—if it opens, the entire system could go up in smoke. That's true for bisphenol A as well, another product that symbolizes the inefficiency of the regulatory system's functioning over the past fifty years."

PART IV

The Shocking Scandal of Endocrine Disruptors

16

"Men in Peril"[1]:
Is the Human Species in Danger?

Treat Nature aggressively with greed and violence and incomprehension: wounded Nature will turn and destroy you.

—Aldous Huxley

"We have to change how we regulate chemical products and protect humans. There are sufficient studies showing that endocrine disruptors cause dysfunctions in the reproductive system, cancer, and behavioral problems. The problem is not scientific—it's political!" It was September 14, 2010, at the French National Assembly. Ana Soto, a cellular biology professor at the medical school at Tufts University in Boston, had just concluded the introductory lecture of a conference on endocrine disruptors sponsored by deputies Gérard Bapt and Bérengère Poletti,[2] and organized by the French Environment Health Network (Réseau environnement santé français, RES). Ostensibly addressing the two elected representatives, the American scientist insisted: "You have to act at the level of the law. Otherwise, what do we do? We wait another hundred years, and then look for which receptor we need to act on to avoid the extinction of the human species!"

Seated in the gallery, André Cicolella, an environmental health researcher and RES spokesman, nodded in agreement. The toxicologist was understandably satisfied: on June 5, 2009, he had organized a similar conference at the Palais-Bourbon in Paris, but the room had been far from full. Fifteen months later, he had to turn away people—proof that the need for a "paradigm shift in the assessment of health and environmental risks" (according to the conference's title) had become a concern far beyond a limited circle of experts.

295

Proof also that persistent efforts to sound the alarm, made for over twenty years by American scientists—including Ana Soto and Carlos Sonnenschein, her longtime partner—were starting to bear fruit, despite industry traps and public agency denials.

"Plastics Are Not Inert Materials"

Everything changed one day in 1987 for the two Tufts University researchers. They were working on breast cancer cells and trying to identify an inhibitor that would block the proliferation of cells typical of tumor development. Two years earlier, they had observed that if they extracted estrogen, a natural female hormone, from a blood serum and applied this "purified" serum to breast cancer cells, they stopped multiplying. On the other hand, if they added the estrogen to cancerous cells, they proliferated at warp speed. "We were trying to identify the inhibitor that, according to our hypothesis, was neutralized by the presence of estrogen," Ana Soto explained to me during my visit to their Tufts University laboratory in October 2009. "To do so, we repeated the same experiment over and over, and always obtained the same results: in the absence of estrogen the breast cancer cells didn't multiply, but in the presence of estrogen, they did multiply. And then, all of a sudden, all the cells began to proliferate indiscriminately, in both experiment groups. We thought that our lab had been contaminated by estrogen and we began to verify each component of the process to understand where the contamination could have come from."[3]

During four (very long) months, the two researchers—who even questioned whether the unusual contamination had been the result of "sabotage"—studied all the materials used, proceeding by elimination: glass pipettes, activated carbon filters that facilitated the extraction of estrogen from the serum, plastic tubes that held the blood cells. But however many times they repeated the experiment after changing the material, the cancerous cells continued to multiply, with or without estrogen!

"We had been using the same plastic tubes made by Corning Inc. for years," explained Carlos Sonnenschein as he showed me an example recognizable by its orange stopper. "In desperation, we decided to change our supplier, choosing the Falcon company instead. And then, to our great surprise, the cancerous cells exposed to the purified serum stopped proliferating! From that, we concluded that there was something leaking from the interior of the Corning tubes and acting like estrogen. We quickly alerted Jean

Mayer, the president of Tufts University and a nutritionist who immediately understood the enormous health implications of our discovery."

A meeting was organized with representatives from Corning on July 12, 1998, at the Hilton Hotel in Boston Logan International Airport. "They informed us that they had recently changed the plastic composition [of their tubes] to make them more stable and less friable, but they hadn't changed their catalog accordingly," Ana Soto told me. "Unfortunately, they refused to tell us the name of the molecule used as an antioxidant, arguing that it was covered by trade secrecy."

"We were very shocked," continued Carlos Sonnenschein, "because we were thinking of the effects this substance could have if present in plastic feeding bottles or food packaging. Even if we aren't chemists, we spent two years extracting from these tubes. And then, finally, the Massachusetts Institute of Technology (MIT) told us that it was nonylphenol."[4]

"It was very worrying," added Ana Soto, "because we learned that this molecule was found in the composition of certain plastics made from vinyl chloride like PVC, or from polystyrene, which could come in contact with food or tap water, or even in spermicides, shampoo, or detergents."

"The manufacturer didn't know that the molecule had an estrogenic function?" I asked.

"No! It's typical of how the industry operates," Sonnenschein answered. "The chemists synthesize new substances that are put on the market and it isn't until much later that we find out the effects they can have. In this case, we fortuitously discovered that, contrary to what was thought, plastics are not inert materials from a biological point of view, and they are made up of synthetic molecules that imitate natural hormones."[5]

"What we call 'endocrine disruptors'?"

"Exactly! This new scientific concept was invented by Theo Colborn, to whom humanity is deeply indebted for exposing a category of pollutants that are behind the majority of modern-day chronic diseases."[6]

The Alarming Discoveries of the Zoologist Theo Colborn

A meeting with Theo Colborn has to be earned. First of all, because at eighty-three years old, the woman often compared to Rachel Carson because of her work's impact has had to limit her activities and carefully filter her many interview and conference invitations. And secondly, because she lives in the middle of nowhere in Colorado, sixty miles from the small Grand

Junction Airport. When I landed on December 10, 2009, over three feet of snow covered the legendary Grand Valley gleaming under the blinding sun. The temperature was −13°F, a brutal change after the 73°F weather in Houston where I had been the previous night. In the car ride to Paonia, the town where Theo Colborn settled down with her family in 1962, I read over my notes about her unusual journey: originally trained as a pharmacist, she decided to raise her four children on a Colorado ranch. Colborn then became involved in a local movement to protect the water quality in the valley from mining and agricultural pollution. She earned a master's degree in freshwater ecology (she was already a grandmother by this point), and then worked toward a doctorate in zoology at the University of Wisconsin, which she obtained in 1985 at the age of fifty-eight! During an interview, she explained that she needed those diplomas in order for her voice to be heard.

Among my notes, I found the last e-mail Theo Colborn had sent me in which she referenced the Rachel Carson Prize that links us. Indeed, in June 2009, I had the incredible honor of receiving the tenth Rachel Carson Prize, awarded by a jury in Stavanger, Norway, to a "woman who has distinguished herself in outstanding work for the environment internationally." Theo Colborn had won the fifth prize ten years earlier. So the "environmental health expert" (according to her business card) began by evoking at length the author of *The Silent Spring* (see Chapter 3) as soon as I walked through her door. "Her book was with me throughout my career," she told me. "First of all, because it opened my eyes to the dangers of pesticides, but also because it illustrated a global vision, by re-creating a link between different living organisms and by peering into the future. For me, the most astonishing part was how she explored the deadly consequences that such a deluge of chemical products could have on generations exposed while still in the womb, and on reproduction, which was completely visionary."

In the chapter "Through a Narrow Window," Rachel Carson cites "medical reports" that reported "oligospermia, or reduced production of spermatozoa, among aviation crop dusters applying DDT," and "atrophy of the testes in experimental mammals," and even the metamorphosis of insects exposed to DDT for several generations into "strange creatures called gynandromorphs—part male and part female."[7] In her sole televised interview, given shortly before her death, she was already worried about the transgenerational effects that chemical products could have. "We have to remember the children born today are exposed to these chemicals from birth, perhaps even before birth," she said. "Now what is going to happen to them in adult life as a result of that exposure? We simply don't know. Because we've never before had this kind of experience."[8]

"Rachel Carson was thinking about cancer in particular," Theo Colborn told me. "A disease from which she herself was suffering and which was the biggest concern at the time. Even I needed a lot of time to move away from the postwar toxicological idea that we should measure a chemical product's toxicity by the number of deaths it has caused in the short or medium term. If I was able to move beyond it, it's partially because I adhered to Rachel Carson's teachings that 'our fate is connected to the animals.'"

"How has your vision changed?"

"It was a long process," answered the zoologist. "In 1987, I was recruited for a joint Canadian and American commission to prepare an assessment of the ecological state of the Great Lakes. I contacted all the biologists who were working on the region. I'll never forget meeting those scientists who had each observed similar phenomena—that is, draconian population declines among certain animal species, reproductive problems wherein the adults had difficulties producing offspring and (when the animals did manage to reproduce) babies born with birth defects that did not survive; they also observed unusual behavioral problems, like females pairing with females, males who stopped defending their territories . . ."

In her 1996 bestseller, *Our Stolen Future*,[9] Theo Colborn describes her colleagues' findings that, little by little, allowed her to put together the "isolated pieces of the puzzle." This included studies conducted by Pierre Béland, an oceanographer who began a "book of the dead" in 1982 in which he recorded the many beluga whale corpses that he found in the Gulf of Saint Lawrence. The autopsies revealed breast, bladder, stomach, esophageal, and intestinal cancers, mouth ulcers, pneumonia, viral infections, thyroid cysts, and also genital defects that had never been seen before. A male beluga— "Booly"—was found with two testicles and two ovaries, a hermaphroditic "phenomenon seldom seen in wildlife and never before reported in a whale."[10] All the corpses were covered with pesticide residues, namely DDT, but also polychlorinated biphenyls (PCBs) and heavy metals. At the same time, Pierre Béland observed that the local dolphin population, which was estimated at five thousand at the start of the twentieth century, had fallen to two thousand at the beginning of the 1960s and to five hundred in 1990.

Theo Colborn also met with Glen Fox, an ornithologist who observed a strange phenomenon in the herring gulls of Lake Ontario and Lake Michigan: beginning in the 1970s, nests contained twice as many eggs as were usually found. This was because two females were nesting together, rather than a male–female couple. "Fox nicknamed them 'gay gulls,'" Colborn told me, "because he had discovered a problem of sexual identity among the males and females due to their contamination by DDT, which acts like an estrogenic hormone, much like PVC." During the same period, biologists Richard Aulerich and Robert Ringer had observed the quasi-extinction of minks, which were feeding primarily on fish that were packed with PCBs.

"Given the gravity of the harmful effects observed, I expanded my research beyond the Great Lakes," Theo Colborn told me. "I discovered studies by Charles Facemire, who had noticed the feminization of male panthers in southern Florida parks, with numerous cases of cryptorchidism (or undescended testicles), reduced concentrations of spermatozoa, or an abnormally high level of estradiol, a feminine hormone, and a consequentially low level of testosterone, the male hormone. Autopsies revealed high concentrations of DDE, a DDT metabolite, and PCB accumulated in the protected species' fat deposits. At the same time, Charles Broley was making similar observations in bald eagles—the national bird of the United States—which had practically disappeared from the Florida coasts. Ultimately, I consulted over a thousand studies conducted in North America as well as in Europe, and I understood that there was no spot in the world that was shielded from the insidious pollution perpetuated by thousands of chemical molecules, chiefly those we now call persistent organic pollutants."

PCBs Are Everywhere

I have already briefly outlined persistent organic pollutants, the infamous "POPs" (see Chapter 2) that were banned by the 2001 Stockholm Convention. Included in those nicknamed the "dirty dozen" are DDT, the postwar "miracle herbicide," dioxin, and PCBs, to which I dedicated a chapter in my book *The World According to Monsanto*. In it, I describe how for five decades the St. Louis company concealed the high toxicity of this chlorinated molecule, which presented remarkable thermic stability and fire resistance, and was used as a cooling liquid in electric transformers and industrial hydraulic devices, as well as a lubricant in applications as diverse as plastics, paint, ink, and paper. I wrote that "PCBs are everywhere," and it was while reading *Our Stolen Future* that I truly understood how they were able to colonize the planet and threaten the survival of numerous animal species, including humans.

In her book, Theo Colborn imagines the journey of a PCB molecule, manufactured in the spring of 1947 in a Monsanto factory in Anniston, Alabama. Named "Aroclor 1254," the PCB molecule was loaded into a train that transported it to a General Electric plant that manufactured electrical transformers in Pittsfield, Massachusetts. Blended with oils—in order to form "Pyranol"—it was poured into an electrical transformer that was then installed in an oil refinery in Texas. In July 1947, a violent storm burnt out the electrical grid and the transformer was abandoned at the public dump where a conscientious worker poured the liquid contents onto the refinery's dirt parking lot where the PCB absorbed the red particles. Four months later, a powerful wind swept up the parking lot dust, and the PCB began a long trek that would lead to the Arctic. Exposed to the sun's heat, the molecule began to float like a vapor, rising very high and following the winds across vast distances. As soon as it encountered cold air, it fell abruptly and haphazardly onto a grassy field, where cows came to graze, and where the lipophilic molecule settled on the milk fat. The PCB could have also landed on the surface of a lake, where it would cling to algae before being snatched up by a water flea, which was then devoured by a crustacean, which would be eaten by a trout, which would end up on the plate of a Sunday fisherman.

By the end of its short ten-day life span, the water flea's PCB concentration would have grown to four hundred times the levels in the water, because the Monsanto molecule is not biodegradable and easily builds up in fatty tissues (and eventually in us, the consumers). If the fisherman missed his catch, the injured trout could have ended up eaten by a female seagull (whose PCB concentration would be 25 million times the levels found in the water), which then flew toward Lake Ontario to mate. It laid two eggs. One hatched six weeks later, but the chick was dead, because PCB (as would DDT or dioxin) penetrated the yolk and killed the embryo. The other egg did not hatch, but it was spotted by another gull that broke it; the yolk fell into the lake and was snatched up by a crayfish that was then eaten by an eel, which headed toward the Atlantic Ocean to spawn, lay eggs, and die. Its body disintegrated in the warm tropical waters of the Bahamas and the liberated PCB molecules resumed their aerial journey, riding the winds, still heading north. The incredible life cycle ended in the fatty rump of a polar bear—"the top predator and largest land carnivore"—whose PCB concentration is three billion times greater than that of its surrounding environment.

In *Our Stolen Future,* Theo Colborn emphasizes that, "Like polar bears, humans share the hazards of feeding at the top of the food web. The persistent synthetic chemicals that have invaded the great bear's world pervade ours as well."[11] She concludes: "Almost half a century later, the PCBs made on that spring day might be found virtually anywhere imaginable: in the sperm of a man tested at a fertility clinic in upstate New York, in the finest caviar, in the fat of a newborn baby in Michigan, in penguins in Antarctica, in the bluefin tuna served at a sushi bar in Tokyo, in the monsoon rains falling on Calcutta, in the milk of a nursing mother in France, in a handsome striped bass landed off Martha's Vineyard on a summer weekend."[12]

"As I was piecing together the effects of PCBs and other POPs on wildlife, I also discovered the first studies carried out on highly exposed humans," said the environmental health expert. "They indicated that Inuit children presented PCB levels seven times greater than those of children in southern Canada or the United States, and that breast milk was heavily contaminated.[13] They also showed that these children were suffering from immune deficiencies, like the St. Lawrence belugas, which led to chronic ear infections and diminished antibody production during vaccinations. Another study conducted on mothers who had consumed fish from Lake Michigan revealed that children exposed to PCBs in the womb suffered from neurological disorders and motor deficiencies.[14] Ten years later, researchers observed that those same children had auditory and visual problems, as well as IQs 6.2 points below the average for their age.[15]

"Today, all that has been largely confirmed, but at the time it was new. So, to understand what was going on, I created enormous spreadsheets with the animal or human species involved on one side, and the observed effects on the other. Finally, after weeks of going around in circles in my office, I understood the connection between all those cases: the endocrine system of living organisms was affected starting from life in the womb, which then caused birth defects, reproductive problems, neurological disorders, and weakened immune systems in offspring. That's why I suggested organizing a meeting between all the researchers who had come against this kind of problem. And it was an unforgettable moment."[16]

July 1991: The Historic Wingspread Declaration

Without a doubt, the "meeting" will go down in the annals of medical history, even if many official medical experts today have never heard of it (or so they claim). But for the twenty-one pioneers who gathered in the Wingspread Conference Center in Racine, Wisconsin, on July 26–28, 1991, it was an

"essential experience," according to Ana Soto, one of the participants. To organize this unprecedented meeting, Theo Colborn solicited the help of John
Peterson Myers—aka "Pete Myers"—a young biologist who had worked on
the declining populations of seabirds migrating from the Arctic to South
America, and who had co-authored *Our Stolen Future*. Entitled "Chemically-
Induced Alterations in Sexual Development: The Wildlife/Human Connection," the conference brought together experts from fifteen disciplines,
including anthropology, ecology, endocrinology, histopathology, immunology,
psychiatry, toxicology, zoology, and even law, to compare their work.

"That meeting represented a turning point in my career," said Louis Guillette, a University of Florida zoologist I met with on October 22, 2009, during
a New Orleans conference. "Basically, I was fighting alone in my corner to
try and describe the disorders I was observing in Florida alligators and, all of
a sudden, everything became clear, thanks to this amazing interdisciplinary
exchange and all of Theo's vast work." And the scientist went on to tell me
his story: in 1988, the Florida state government asked him to gather alligator
eggs in order to create breeding farms. He combed a dozen state lakes and
collected over 50,000 eggs. He placed them in incubators and observed that
only 20 percent of eggs taken from the large Lake Apopka (thirty acres, situated near Orlando and Disney World) had hatched, compared with 70 percent of eggs from other lakes. What's more, 50 percent of the baby alligators
died shortly after birth.

"I remembered that, several years earlier, the lake had been strongly contaminated by an accidental dicofol spill—an insecticide similar to DDT,"
added Louis Guillette. "Curiously enough, there were no traces of pesticide
in the lake water, but everything indicated that it was stored in the sediment,
aquatic life, and alligator fat. When I began to study the alligator population,
I expected to find cancers, but what I saw had nothing to do with tumors: the
females presented ovarian malformations and abnormally elevated estrogen
levels; as for the males, they often had micropenises and extremely low testosterone levels. The only hypothesis that seemed plausible to me, though it
was difficult to explain, was that these deformities were due to an imbalance
that arose during embryo formation, because the eggs were contaminated by
pesticide residues."

"Had you already observed those kinds of abnormalities?"

"Never!" the alligator specialist immediately responded. "At the time, the
scientific literature made no mention of this kind of malformation, which
had never been reported among alligators or any other wildlife. However,

I had read studies on lab animals exposed in utero to distilbene, the drug prescribed to pregnant women during the 1950s and 1960s [see Chapter 17]. They reported deformities in the ovaries or penises. But that only worried me further. I asked myself: since these alligators didn't receive any drugs, nor were they deliberately exposed to a high dose of a synthetic molecule, how could low doses of pesticide present in these organisms cause these effects?"

"What pesticide doses did you measure?"

"They were around 1 ppm—a dose that is generally considered to be biologically inactive and that can be found in our environment, or in what we eat every day."

"How can your findings with alligators be useful for humans?"

"It's important to understand that fauna act as sentinels for human health," responded Louis Guillette. "Wildlife alerts us to the environmental dangers threatening us, especially our children. Mammals are similar to reptiles, they have similar hormones, similar ovarian or testicular structures. In fact, observations I made on crocodiles during the 1980s and 1990s can now be seen among many children almost everywhere in the world."

"Particularly among sons of farmers?"

"Exactly. There are studies that indicate that sons of agriculturalists who used pesticides have a higher rate of micropenises or testicular defects."

"Is Lake Apopka clean now?"

"It's in the process of being restored. The authorities have tried to extract the many pesticides in it, but unfortunately it's not very easy, since many of them—like dicofol and DDT—have been incorporated into the lake's food chain. They're buried in the fats of living organisms and we won't see the end of them until many generations from now."

"Are the alligators healed?"

"No! The females are like us. They reproduce over several decades and we're still seeing the same disorders that we did twenty years ago."

"How did the Wingspread meeting enlighten you?"

"Thanks to discussions with my colleagues, who had made similar observations on other wildlife species, I understood that certain chemical products behaved like hormones, which was truly a revelation," Guillette concluded.[17]

At the end of the conference, its participants signed a manifesto, called the "Wingspread Declaration," which, starting in 1991, drew attention to the harmful effects caused by molecules that, twenty years later, public authorities continue to ignore:

Many compounds introduced into the environment by human activity are capable of disrupting endocrine systems of animals, including fish, wildlife, and humans. The consequences of such disruption can be profound because of the crucial role hormones play in controlling development. Many wildlife populations are already affected by these compounds. [. . .] The pattern for effects vary among species and among compounds. Four general points can nonetheless be made: 1) the chemicals of concern may have entirely different effects on the embryo, fetus, or perinatal organism than on the adult; 2) the effects are most often manifested in offspring, not in the exposed parent; 3) the timing of exposure in the developing organism is crucial in determining its character and future potential; and 4) although critical exposure occurs during embryonic development, obvious manifestations may not occur until maturity.

To conclude, the authors warn that, "Unless the environmental load of synthetic hormone disruptors is abated and controlled, large scale dysfunction at the population level is possible. The scope and potential hazard to wildlife and humans are great because of the probability of repeated and/or constant exposure to numerous synthetic chemicals that are known to be endocrine disruptors."[18]

Endocrine Disruptors: Dangerous "Scramblers"

"Who invented the term 'endocrine disruptor'?" I asked. Quite unexpectedly, the question made Theo Colborn smile. "Ah! That's a long story," she responded. "As the conference progressed, the participants grew increasingly excited—and worried—as they become aware of the gravity of the phenomenon they had just identified. But, when it came down to naming it, we had a hard time. Finally, there was a consensus on the term 'endocrine disruptor,' which I personally find very ugly, but we couldn't come up with anything better!"

"What is an endocrine disruptor?"

"It's a chemical substance that interferes with endocrine system functioning."

" What is the function of the endocrine system? " I continued.

"It coordinates the activity of fifty or so hormones produced by glands within our body, such as the thyroid, the pituitary gland, the adrenal glands, as well as the ovaries or the testicles. These hormones play a critical role because they regulate vital processes, like embryonic development, blood glucose levels, blood pressure, brain and nervous system functioning, or the ability to reproduce. The endocrine system controls all the processes that go into making a baby, from fertilization to birth: every muscle, the programming of the brain or organs, everything depends on it. The problem is that we've invented chemical products that resemble natural hormones and that can slide into the same receptors, turning a certain function on or off. The consequences can be deadly, especially if exposure to these substances happens during intrauterine life."

To better measure the implications of Theo Colborn's statements, it is important to understand precisely how natural hormones operate once glands release them into the blood and fluids surrounding cells. They are often described as "chemical messengers" that circulate within the body in search of "target cells" with compatible "receptors." The other metaphor that is often used is that of a "key" (the hormone) capable of entering a "lock" (the receptor) to open a "door" (a biological reaction). Once a hormone has attached to it, the receptor carries out the instructions it receives, either by modifying the proteins contained in the target cell, or by activating genes to create a new protein that will provoke the appropriate biological reaction. "The problem," explained Colborn, "is that endocrine disruptors have the ability to imitate natural hormones by latching on to receptors and triggering a biological reaction at the wrong time; or, on the contrary, they block natural hormones from acting by taking their place in the receptors. They are equally capable of interacting with hormones by modifying the number of receptors or interfering with the synthesis, secretion, or transport of hormones."

According to André Cicolella and Dorothée Benoît Browaeys, endocrine disruptors are not "toxic in the classic sense," as they "act like decoys, manipulators. They interfere with our most intimate functions, be they digestive, respiratory, reproductive, cerebral, and behave like 'scramblers' carrying false messages. They act at microscopic doses and are, by their chemical nature, very varied."[19] "These chemical substances operate at concentrations of one part per million, or even per billion," Theo Colborn confirmed. "The problem is that a minute shift in hormonal alchemy can cause irreversible effects, notably when it occurs at very sensitive moments during prenatal development, during what we call the 'exposure window.'"

I was particularly troubled by the notion that a fetus had "exposure windows" during pregnancy. As the mother of three teenage girls, I was overcome by a sharp pang of worry, almost visceral in nature, when I learned about the incredible subtlety of organogenesis, or the process of organ formation in a fetus, which largely unfolds during the first thirteen weeks of pregnancy. "There are critical phases during this development," explain Bernard Jégou, Pierre Jouannet, and Alfred Spira, authors of *La Fertilité est-elle en danger?* (Is Fertility in Danger?). "Certain organs or functions start to develop during these periods, which are often very brief and last several hours or several days. Exposure to physical, chemical, and/or biological changes can have different effects, often in a very dramatic fashion, according to the moment of exposure. A variation of a few days in the moment in which an event occurs can translate to radically different effects. [. . .] When maternal, embryo-fetal, and placental mechanisms have to adapt to environmental disruptions, that compensation can also provoke largely negative side effects, which will manifest over the long term."[20]

The three internationally renowned specialists also explain how the endocrine disruptors a mother ingests act like "Trojan horses,"[21] and can disrupt the critical moments of a gestating baby's organogenesis, such as the sexual differentiation that occurs very precisely on the forty-third day, formation of the neural plate that will produce the brain (from the eighteenth to the twentieth day) or that of the heart (forty-sixth and forty-seventh day). Clearly, I did not know any of this when I was pregnant with my daughters in the 1990s. And sadly, mothers-to-be nowadays are not any better informed.

As early as 1996, Theo Colborn and her co-authors had responded definitively to those who claim that synthetic hormones are actually very similar to those naturally produced by plants—an assertion I have read on numerous occasions in literature produced by industry-affiliated scientists and lobbyists. "The body is able to break down and excrete plant-based estrogen, while many of the man-made compounds resist normal breakdown and accumulate in the body, exposing humans and animals to low-level but long-term exposure. This pattern of chronic hormone exposure is unprecedented in our evolutionary experience, and adapting to this new hazard is a matter of millennia not decades."[22]

Human Fertility Reduction and Worrying Reproductive Anomalies

At the same time that the Wingspread pioneers were inventing the term "endocrine disruptors," a Danish scientist, Niels Skakkebaek, was preparing to publish a study that would "drop like a bomb." With his Copenhagen University

Hospital colleagues, Skakkebaek "analyzed sixty-one articles published between 1938 and 1990, regarding a total of 14,947 fertile or healthy men, from
every continent, and which revealed a steady decline in sperm production
over time. While the first studies dating from 1938 reported an average concentration of 113 million spermatozoa per milliliter of sperm, the most recent publications from 1990 observed an average concentration of 66 million
per milliliter."[23] In plain English: the spermatozoa quantity of ejaculate decreased by half in less than fifty years!

The study's results, published in September 1992 in the eminent *British
Medical Journal*,[24] were so surprising that Jacques Auger and Pierre Jouannet
doubted its accuracy. The two French reproductive health specialists—who
founded the Centers for the Study and Conservation of Human Ova and
Sperm (Centres d'étude et de conservation des œufs et du sperme, CECOS),
which were critical to the development of in vitro fertilization (IVF)—
decided to analyze and compare the ejaculate of 1,750 Parisian sperm donors
between 1973 (the year both CECOS and the Kremlin-Bicêtre Hospital
were established) and 1992. Their results confirmed those of the Danish
study: spermatozoa quantity had fallen by 25 percent over two decades, or a
decrease in concentration of approximately 2 percent per year. Men born in
1945 and whose samples were measured in 1975 had on average 102 million
sperm per milliliter, compared to 51 million for those born in 1962 (and measured thirty years later). What's more, the quantitative drop was accompanied by a decrease in the quality of spermatozoa, which presented reduced
mobility and morphological anomalies that caused reduced fertility.[25] In the
book he co-authored with Bernard Jégou and Alfred Spira, Pierre Jouannet
underlines the doubts provoked by the decidedly troubling Danish study:
"The results seemed to run so counter to a commonly fixed fact—the stability of spermatic production—that the prestigious journal that would publish
the article [the *New England Journal of Medicine*] took care to have it evaluated by an external statistician."[26]

But suspicions persisted and, in 2000, American epidemiologist Shanna
Swan decided to repeat Niels Skakkebaek's meta-analysis, adding forty additional publications. She confirmed the Danish team's conclusions—definitively
and on a wider scale—by observing a yearly average decline in sperm density
of 1.5 percent in the United States and 3 percent in Europe and Australia
between 1934 and 1996.[27]

The controversy stirred up by Swan's findings still amuses Niels Skakkebaek (his account was included in Theo Colborn's *Our Stolen Future*). "When

my study came out, everybody focused on the dramatic spermatozoid decrease," he said during our meeting on January 21, 2010, in his Rigshospitalet laboratory in Copenhagen. "But, for me, it contained another piece of information just as troubling, which is the steady increase of testicular cancer rates, notably in Denmark where it multiplied by three between 1940 and 1980. This was all the more worrying because the increase hadn't been observed in neighboring Finland, a barely industrialized country that is essentially covered by forests. I also observed the same disparity for two male genital anomalies, which were four times more common in Denmark than Finland: cryptorchidism and hypospadias."

To better understand the import of the Danish researcher's discovery, it is worth noting that "the descent of the testes into the scrotum is controlled by hormones: the insulin-like factor 3 and testosterone. When the testes haven't descended into the scrotum before three months, we call it cryptorchidism," as the authors of *La Fertilité est-elle en danger?* explain. In the same way, concerning hypospadias, they write that, "the formation of the urethra in the penis is controlled by testosterone. This development can be disrupted. Instead of opening at the glandular level, the urethra will end in a more or less large opening beneath the penis or even at the level of the scrotum."[28]

Troubled by his study's results, Niels Skakkebaek met up with his Scottish colleague Richard Sharpe, who had observed the same reproductive anomalies in the United Kingdom. Together, they combed through the scientific literature and discovered that experiments conducted on rats exposed to distilbene, a synthetic estrogen (see Chapter 17), revealed the same kind of birth defects. "That's how, for the first time, we ventured the hypothesis that reproductive anomalies could be due to heightened exposure to estrogens during prenatal life,"[29] the Danish pediatrician and endocrinologist explained.

"So, you conducted some real detective work?"

"Yes, I think I can say that I did, because at the time this kind of investigation was completely new. I was lucky (if I can say so) because my key research was sustained by my medical practice here at Rigshospitalet [Copenhagen University Hospital], where numerous men with infertility problems came to see me. While examining their testicle biopsies, I discovered that they contained precancerous cells. And then it turned out that many of these men, whom I had monitored for several years, did indeed develop testicular cancer. The other troubling fact was that the precancerous cells present in the testicles of these infertile men were similar to the germ cells found in a fetus. Those cells should not be found in the testicles of an adult male.

Everything indicates that something blocked the development of fetal cells that should have matured and evolved toward sperm production, but they remained at the germ cell stage within the testicles, which means that men were born with immature cells. They remained dormant throughout childhood, but at puberty they began to multiply and eventually developed into cancer."

"How do you explain this phenomenon?"

"The most likely hypothesis is that mothers were exposed to endocrine disruptors during pregnancy, at a critical moment for the development of their babies' genital organs. This prenatal contamination caused a series of issues that are all connected: fertility problems, birth defects like cryptorchidism and hypospadias, and testicular cancer. My colleagues and I named this phenomenon the 'testicular dysgenesis syndrome' because we were encountering multiple symptoms with the same environmental and fetal origin. This also means that men who have problems reproducing should be regularly checked, because the risk of developing testicular cancer before the age of forty is considerably increased."[30]

"How do you respond to those who say that cancer has nothing to do with environmental pollution, but that it's due to an increase in the elderly population?"

"It's not true for testicular cancer, which is characteristic of young men between twenty and forty years old," responded Dr. Skakkebaek. "There is practically no risk for men over fifty-five years old of developing a testicular tumor. It also happens that testicular cancer is one of the cancers that has increased the most over the past thirty years, and the only explanation possible is environmental contamination."

"And how can we protect men from developing these serious problems?"

"The only way to protect them is to protect their mothers! The problem is that endocrine disruptors are everywhere. But there are products that pregnant women should absolutely avoid, like phthalates found in lots of plastic packaging and protective food coatings, objects made with PVC, as well as body care products like shampoos. I recently published a study that shows a correlation between phthalate levels in breast milk and the rate of birth defects, like cryptorchidism, among young boys.[31] Products with bisphenol A, such as hard plastic containers or some food cans [see Chapter 18], should also be avoided, as well as nonstick pots and pans that contain perfluorooctanoic acid (PFOA).[32] I just published a study that shows that men whose bodies were largely permeated by PFOA residues have on average 6.2 million sperm per ejaculation, which nears the threshold of sterility.[33] It's also preferable to eat organic fruits and vegetables since numerous pesticides are endocrine disruptors."

"But concerning bisphenol A or PFOA, regulatory agencies keep repeating that the residues found in our bodies are negligible, as they're far beneath the acceptable daily intake for those products. Are they mistaken?"

"I'm not a toxicologist, but as an endocrinologist, I can tell you that these substances act at minute doses that are much lower than the ADI assigned to them. Everything indicates that the regulatory system is not adapted for endocrine disruptors."

"Do you think the human species is in danger?"

"I think the situation is very serious. In Denmark today, 8 percent of children are conceived via assisted medical procreation techniques like in vitro fertilization (IVF)—that's already a lot, and there are more and more couples coming in with fertility problems. Urgent action is needed."

Dawn Forsythe: A Former Chemical Industry Lobbyist's Devastating Account

"When Theo Colborn's book came out on March 18, 1996, my bosses immediately asked me to buy twenty copies for all the higher-ups, in order to prepare a counter-offensive." It wasn't easy to meet up with Dawn Forsythe, who, until the end of 1996, led the Department of Government Affairs for the American branch of Sandoz Agro—a Swiss pesticide manufacturer that merged with Ciba-Geigy in 1996 to form Novartis. Nonetheless, her account is invaluable since, as we know (see Chapter 13), it is almost impossible to obtain an interview with chemical industry representatives, including former employees. "I'm very well placed to know that communication with multinational chemical companies is completely locked down," explained Dawn Forsythe during our meeting at her Washington, DC, home on October 18, 2009. "As for those who left the 'family,' as I did, they generally prefer to move on to something else and keep a low profile."

"Why did you grant me this interview?" I asked.

"Because Theo Colborn recommended you, and I trust her implicitly."

"And yet she was a real thorn in your former employer's side?"

"Yes . . . The Sandoz executives picked her book apart, especially since we had several pesticides suspected of being endocrine disruptors. I remember a meeting with the vice president who, by way of introduction, told me: 'I just read the chapter on the decrease in sperm. The ecological militants must be happy, since they're in favor of birth control, right?' But more seriously, the pesticide manufacturers were scared that Theo would become another Rachel Carson. So, they started spreading a rumor that she had cancer. They hired media companies to monitor her every move, note every act and gesture. I kept a trunk filled with internal documents, many of which are reports on conferences or public debates in which Theo participated and that were carefully recorded by a 'mole.' My main responsibility was to evaluate them. It's important to point out that the 'hunt' began before the book's publication, as shown by an anonymous report about a lecture Theo gave in Ann Arbor, Michigan, on December 2, 1995."[34]

"What were the stakes for the pesticide manufacturers?"

"They were huge! They had been trying to divert attention from the cancer problem for thirty years. All the tests they were supposed to conduct were based on the principle that 'the dose makes the poison.' They didn't understand anything about the concept of endocrine disruptors and didn't see how they could test the effects of their products on the fetus or on reproduction. At Sandoz, like the rest of the chemical industry, we didn't have a single endocrinologist among our scientific personnel! I have here an unsigned document from March 11, 1996, classified as 'interoffice correspondence,' which nicely summarizes the panic gripping my superiors: 'The best and brightest minds in human history have worked for decades to discover the causes and cure for cancer, and we aren't there yet. It may take decades to decipher the biological processes of endocrine disruptors.'"

"But, within the company, they weren't denying that pesticides could be endocrine disruptors?"

"Not at all! I have another document, dated July 30, 1996, which is the umpteenth draft of the official declaration of the American Crop Protection Association (ACPA), which would eventually be signed by all the pesticide manufacturers. I personally coordinated the preparation of this shared declaration, which was shuttled back and forth between all the signatory companies. This draft was written by nine industry scientists who proposed replacing the term 'endocrine disruptor' with 'endocrine modulation of reproduction,' explaining that the word modulation is less emotionally loaded than 'disruptor.' Then they wrote: 'There is convincing scientific evidence that some organic chemicals, including some pesticides, have caused reproductive effects on local, highly exposed fish and wildlife populations and that these effects are based on modulation of the reproductive endocrine system. Fur-

thermore, current EPA [Environmental Protection Agency]-required laboratory studies generally do not provide sufficient information to evaluate whether
a chemical may cause such effects.'[35] I should clarify that this paragraph
disappeared from the final declaration! Which isn't surprising, given that one
of the arguments I was meant to promote among all my contacts was precisely the opposite! I have here a memorandum from the National Agricultural Chemicals Association, which I widely distributed. It tackles the key
questions about endocrine disruptors and provides preprepared answers. For
example: 'Do current pesticide tests required by EPA detect estrogenic activity?' The answer: 'Yes! The key test which will signal potential estrogenic effects is the two-generation reproduction study.'"

"What was the industry's strategy to counter the impact of *Our Stolen
Future?*"

"Attack the attacker, but not directly! There were many in the industry
who wanted to personally attack Theo Colborn. But others said: if you attack
her, you'll give her more credibility. There was probably nothing better for an
environmental scientist than to be personally attacked by the pesticide industry, which hasn't gotten great press. That's what happened to Rachel Carson, and it was disastrous in terms of image. During the many meetings we
organized—1996 was a very grueling year—we decided to show our good-
will: we created a work group, named the 'Endocrine Issue Coalition,' which
was meant to provide proposals to improve the evaluation of pesticides and
other chemical products. The message I had to circulate was: 'We take all
this very seriously, we're working on it . . .' At the same time, I was responsible for contacting all the 'pro-pesticide groups' that the industry had created
throughout the fifty states."

"'Pro-pesticide groups'? What's that?" I asked, not sure I had clearly understood.

"They're front organizations that we concocted from start to finish and to
whom we directed the press when they requested an interview with an industry representative. Look, I have the list right here: How could anyone distrust
the 'Indiana Coalition for Environmental Protection?' Or the 'Kansas Committee for Environmental Protection and Education?' Or the 'Washington Friends
of Farms and Forests?' We gave them money and information and their role
was to defend our positions, all the while pretending to be independent."

"The goal was to create doubt?"

"Exactly! When journalists asked their opinions on the debate over endocrine disruptors, they would respond: 'Ah! Well you know, we shouldn't get

carried away. We need pesticides in order to produce abundant food cheaply . . . more research is needed . . .' I have a letter here from Terry Witt, the president of 'Oregonians for Food and Shelter,' one of those groups. It's addressed collectively to his contacts at Sandoz, Ciba, DuPont, Monsanto, ACPA, and DowElanco. He asks them to send him 'information and/or names of expert contacts to counter a campaign against organochlorines herbicides led by what he called the 'environmental and anti-technology faction.' I imagine that we gave him the names of several academics we had recruited."

"Academics?"

"Yes! That was another component of my job: create and maintain a network of academic allies whom we could solicit to conduct studies, which paid handsomely, and potentially to speak out publicly to defend our interests . . ."

At this stage of the interview, Dawn Forsythe suddenly stopped talking. After a long silence, she began to speak again, her voice interrupted by sobs: "It was very painful for me, especially during the years after I left, once I understood the role I had played in sabotaging laws meant to protect the population, or to convince people to believe our lies. It was very painful and it still is . . . I'm sorry to have spent part of my life like that. I was a child of the sixties and seventies who wanted to do good, and I sincerely thought that we needed pesticides to feed the world."

"Why did you leave?"

"I attended a lecture by Ana Soto on the link between endocrine disruptors and breast cancer, during which she mentioned several pesticides, including atrazine. At the time, Sandoz was planning to mix the herbicide with a household product, so I brought my concerns to the higher-ups. I very quickly understood that they weren't interested. Little by little, I began to sense that people distrusted me, not only at Sandoz, but in the rest of the industry: one day, during an intercompany meeting, the Dow Chemical representative called me an 'eco-feminist terrorist.' I took advantage of the merger between Sandoz and Ciba-Geigy to leave . . . It wasn't easy afterwards. I was of course discredited in the industry, but also in the environmentalist domain: Who's going to trust a former pesticide lobbyist? Thanks to Theo Colborn's support, I got back on my feet and found a job in civil service. In the meantime, the industry's maneuvering paid off: in August 1996, Congress voted in a law asking the EPA to implement a program to evaluate the potential effects of chemical products on the endocrine system, but thirteen years later it still hadn't been done. What a waste of time!"[36]

Dawn Forsythe was right. As we will see in the two following chapters (on distilbene and bisphenol A), the alarm sounded in 1991 by the Wingspread Declaration scientists had very little effect. But before moving on, I had a final question—one that had been gnawing at me throughout my investigation on the chemical industry: "The people who work for Sandoz or Monsanto are a family: What do they do to protect it?"

"They stick to themselves," answered the former pesticide lobbyist. "Unless there's a merger or mass layoff, it's rare to leave the large family that the chemical industry represents. And, in that universe, chemical risks don't exist. They think like I myself did for many years: they sincerely believe that their company is 'responsible' and that products are seriously tested before being put on the market. In any case, the great majority is convinced of it."

17

Distilbene: The "Perfect Model"?

We have become unwitting guinea pigs in our own vast experiment.
— Theo Colborn

"Distilbene [DES] is truly the chemical product that changed how we think, by showing us about endocrine disruption, and about what we now call the 'fetal origins of adult diseases.'" It is with these words that John McLachlan, the director of the Center for Bioenvironmental Research at Tulane University, began the Ninth Annual Symposium on the Environment and Hormones that took place on October 20–24, 2009, in New Orleans. About sixty international scientists participated, including Ana Soto, Carlos Sonnenschein, and Louis Guillette (see Chapter 16).

"Distilbene was the first synthetic hormone intentionally produced in 1938 by Charlie Dodds," continued John McLachlan, considered one of the world's top specialists in harmful DES. "Dodds had already synthesized bisphenol A in 1936, but seeing that DES had a higher estrogenic power, he put aside bisphenol A. Incidentally, others salvaged it and, since it polymerized easily, they used it to make plastics, which we'll come back to . . . Distilbene was prescribed to millions of women (4 to 8 million) as endocrine support for pregnancies from the end of the 1940s to 1975. You all know what happens next: we observed vaginal cancers and numerous disorders of the female reproductive tract in the daughters of treated women. It's a substance that provokes breast growth in men who have absorbed a miniscule quantity of it . . . To start us off today, I would like to hand it over to representatives of DES Action who have been working closely with my lab for over thirty years . . ."

A "Wonder Drug" Discovered in 1938

Before hearing accounts from the "DES daughters," we should look back on the history of a molecule that, thirty years after being banned, continues to harm families and represents a "model compound for other environmental agents with estrogenic potential."[1] As John McLachlan mentioned, the Englishman Charles Dodds synthesized DES[2] at the very moment when his Swiss colleague Paul Müller was discovering DDT. The inventors of the "wonder drug" and insecticide both received a Nobel Prize in 1948—a record time period for receiving the prestigious honor, which says a lot about the enthusiasm the two molecules aroused. It so happens that they have (at least) two things in common: they are "poisons" that are now banned, and they present a similar chemical structure giving them the ability to imitate estrogen, the female sex hormone. That's what two researchers at the University of Syracuse discovered just as DES was beginning its tragic career in gynecological practices. They observed that when administered to roosters, DDT atrophied the testicles of the poor birds and feminized them.[3]

The "feminizing" power of DES, considered to be a very powerful synthetic estrogen, had been observed in German factories during World War II. Because the molecule had never been patented (as it was synthesized in a publicly funded lab), it was immediately adopted as an anabolic steroid used in farming efforts under the Third Reich: mixed with feed for chickens, cows, and pigs, it "boosted" their development by 15 percent to 25 percent. This gain in time and money was very handy in wartime and would utterly fascinate none other than Robert Kehoe, the inveterate defender of leaded gasoline (see Chapter 8). While traveling in Nazi Germany, where he met with chemists at I.G. Farben to "study incidence of and methods of prevention of bladder tumor among workers in the benzidine plant,"[4] the director of the Kettering Laboratory described with admiration the DES factory that the Zyklon B manufacturers had him visit.

"A drug effect of interest in relation to industrial hygiene is that of DES, in the manufacture of which only female workers are employed, because of the untoward effects induced in males by the absorption of this material in the course of a day's work. Boys develop a mammary swelling with such severe pain the pressure of a shirt cannot be endured. [. . .] On the other hand older males develop some atrophy of the testes and some apparently temporary loss of sexual potency."[5] However, Kehoe, an accredited scientist within the chemical industry, did not say a word about the effects the substance could have on pregnant workers. Nonetheless, if he had consulted the international scientific literature, he would have discovered that Charles Dodds, the inventor of DES, had himself noticed as early as 1938 that the ingestion of estrogen, including DES, at the early stages of pregnancy, caused abortions in rabbits and rats.[6] The same year, two British researchers made similar observations in cows among which DES lowered milk production.[7] In France, Antoine Lacassagne noted that the substance led to mammary cancers in mice.[8] At the same time, American researchers were reporting that female rats exposed to estrogen in utero were born with uterine, vaginal, and ovarian deformities, whereas the males displayed multiple genital anomalies, such as atrophied penises.[9]

Barely one year after the discovery of DES, approximately forty articles highlighted the carcinogenic and teratogenic dangers of natural or synthetic estrogen so emphatically that the *Journal of the American Medical Association* (*JAMA*) sounded the alarm: "The possibility of carcinoma induced by estrogens cannot be ignored. The long continued administration of these proliferating agents to patients with a predisposition to cancer may be hazardous. The idea that estrogens are related in their activity only to sex organs should be abandoned. Other tissues of the body may react in an undesirable manner when the doses are excessive and over too long a period. This point should be firmly established, since it appears likely that in the future the medical profession may be importuned to prescribe to patients large doses of high potency estrogens, such as stilbesterol [also known as distilbene], because of the ease of administration of these preparations."[10]

The *JAMA* editorialist got it right: in 1941, the sale of DES was authorized by the Food and Drug Administration (FDA), shortly followed by most of the countries in Europe. Pharmaceutical companies—Eli Lilly, Abbott, Upjohn, Merck—pounced on the molecule for its cheap (no patent existed) and easy fabrication. The "wonder drug" was massively prescribed in pill form to treat menopausal hot flashes, vaginitis, to suppress lactation, and to treat acne in young girls and control growth; it was even used as "a morning-after pill." In 1947, DES was authorized as a food supplement and as an implant in cow ears or chicken necks to help fatten them up. In 1971, even as the lawyer Ralph Nader was taking up arms against the massive influx of estrogen in the food chain, FDA commissioner Charles Edwards was publicly supporting the drug, using absurd arguments that would shock toxicologist Jacqueline Verrett (see Chapter 15): "A 500-pound animal will reach a marketable weight of 1,050 pounds using 511 pounds less feed and 31 days sooner when fed DES-containing feed."[11]

1962: The Short-Lived Thalidomide Scandal

"The problem," says Stephanie Kanarek, one of the four representatives of the organization DES Action present at the New Orleans symposium, "is that we have a hard time trusting medical authorities. We have serious health problems because of a completely legal drug that was prescribed to our mothers when we were in their wombs and, today, they want to treat us with drugs that are just as legal, but we're very distrustful. This is why we need input from independent scientists—can we believe drug companies and doctors?"

Stephanie Kanarek's consternation is understandable—her parents "sacrificed a lot" in order to strictly follow the "Smith and Smith regime," a very costly treatment recommended in order to have a "fat and healthy" baby, according to the propaganda at the time. George and Olive Smith were, respectively, a gynecologist-obstetrician and an endocrinologist at Harvard Medical School. They were specialists in high-risk pregnancies who, in 1948, published an article recommending the use of DES to prevent miscarriages and gestational diabetes. Relying on very partial observations drawn from a few female volunteers, they recommended using DES from the start of pregnancy, then regularly increasing the dose up until the thirty-fifth week.[12] Widely promoted by pharmaceutical companies, which graciously distributed the "regime" to gynecologists, accompanied by bottles of DES pills, the treatment was rapidly extended to include "all pregnancies" in order to ensure "bigger and stronger babies," as a Grant Chemicals Company advertisement proclaimed in the June 1957 edition of the *American Journal of Obstetrics and Gynecology*. More prosaically, as the sociologist Susan Bell highlights, DES became a "major means of medicalizing pregnancy"[13] and, consequently, made a lot of money, even as warnings continued to accumulate.

In 1953, James Ferguson conducted a study in New Orleans wherein 184 women were treated with DES while 198 received a placebo; the study indicated that the drug had no effect on preventing miscarriages, eclampsia,[14] premature births, or fetal deaths.[15] The same year, William Dieckmann confirmed those results based on a group study of 1,646 women, 840 of which received DES, at the University of Chicago Medical Center.[16] An additional study that reexamined the same group twenty-five years later would reveal that DES actually had the complete opposite effect of what was intended[17]—but I will address this in detail later.

While the FDA and international health authorities haughtily ignored Dieckmann's study, DES continued to be massively prescribed, following an unchanging mechanism that, as I write this, reminds me of the Mediator case.[18] "Drug companies were persuasive in their marketing," writes Pat Cody, founder of DES Action. "Physicians wanted to believe they were helping their patients. They did not have the time to look up the research done on all the medicines they prescribed. They trusted the drug companies. Women trusted their doctors and rarely questioned practice."[19]

But what about health authorities? Wasn't it their role to carry out the scientific monitoring the practitioners couldn't? How can we explain the fact that they did not react to the many studies that, by the end of the 1950s, would herald the disquieting scenario to come, if not by incredible negligence and a shameful indulgence toward drug companies? In 1959, William Gardner (from Yale University) showed that mice exposed to DES in utero developed vaginal and uterine cancers.[20] That same year, a study reported four cases of "masculinization of female infants" whose mothers had followed treatment with DES,[21] whereas another highlighted a case of "hermaphroditism" in a little boy suffering from hypospadias.[22]

In the era of chemistry triumphant, during which we were happy to celebrate insecticides and "wonder drugs," it seems like medical and health authorities were blinded by what Theo Colborn calls the "myth of the placental barrier." Meaning the "belief that the placenta, the complex body of tissue that attaches to the wall of the womb and to the baby through the umbilical cord, acts as an impenetrable shield protecting the developing baby from harmful outside influences. [. . .] According to the thinking of the time, the only thing capable of invading the womb and causing deformities was radiation."[23]

This myth was shattered in 1962, a few weeks before the publication of Rachel Carson's *Silent Spring*, when newspapers around the world splashed images of children suffering from appalling birth defects across their front pages. The majority displayed limb abnormalities, such as the absence of an arm and fingers growing out of a shoulder. This extremely rare disorder was named "phocomelia" after the Greek word for seal because, just like the aquatic mammal, the victims' hands were directly attached to their torso. Their deformities were sometimes accompanied by deafness, blindness, autism, brain damage, and epilepsy. The culprit was thalidomide, a German medication put on the market in 1957 in fifty countries (though not the United States) and prescribed as a tranquilizer, as well as for the treatment of morning sickness in pregnant women. In five years, the drug caused deformities in eight thousand children. Researchers then began to study the striking effects of the substance and discovered that certain exposed infants had been spared, even though their mothers had taken the harmful pill over a long period; conversely, others were horribly deformed, even though their mothers had only taken the medication on one occasion. Scientists realized that the teratogenic impact "appeared to depend on the *timing* of drug use, not on the dose."[24] Mothers who took the drug—be it one or two pills—between the fifth and the eighth week of pregnancy gave birth to children with deformed limbs, because that was precisely the period during which the fetus was forming arms and legs.

"The tragic episode also drove home the lesson that substances and doses tolerated readily by adults can devastate the unborn," write the authors of *Our Stolen Future*. "The principle that 'timing is all' would be demonstrated again and again as scientists explored the power of chemicals to disrupt development. A small dose of a drug or hormone that might have no effect at one point in a baby's development, for example, might be devastating just a few weeks earlier."[25]

While the *New Yorker* was publishing *Silent Spring* in serial form (see Chapter 3), *Life* magazine dedicated its cover to the thalidomide disaster.[26] If, as we will see shortly, the public authorities struggled to make sense of the tragedy, Rachel Carson clearly understood the stakes: "We certainly have had tragic warnings that drugs can cause serious malformations and other defects in generations yet unborn," she said in her sole televised interview on January 1, 1963. "Pesticides may well have the same effect. You don't have to test these on generations of human beings. You can test them on laboratory animals, the same sort of organisms that have been used successfully for many years to determine genetic effect. We must go on to think in terms of other methods of control, of much more scientific, much more accurate and precise methods."[27] The biologist was completely right (see Chapter 19).

The Horrible Tragedy of the "Distilbene Daughters"

"If thalidomide exploded the myth of the inviolable womb forever, the DES experience toppled the notion that birth defects have to be immediate and visible to be important," writes Theo Colborn in *Our Stolen Future*.[28] In

April 1971, the *New England Journal of Medicine* published a "bombshell" of a study,[29] to borrow Jacqueline Verrett's word.[30] In her book co-authored with Jean Carper, *Eating May be Hazardous to Your Health*, the FDA toxicologist notes that at the time that article was making headlines, 30 million head of cattle were being treated with DES every year, and that the secretary of agriculture had been forced to recognize that residues of the substance could be found in meat consumed by Americans. The Harvard-led study cited above presented the case histories of seven young girls, ranging in age from fifteen to twenty-two, who suffered from clear cell adenocarcinoma, a vaginal cancer so rare in this age group that only four cases had been recorded in scientific literature.

Howard Ulfelder, a gynecologist, fortuitously made the "discovery" after being forced to recommend vaginal and uterine ablation to a fifteen-year-old girl. Seeing the specialist's bafflement, the mother asked him if the cause could be the DES she had taken during her pregnancy. The question surprised the gynecologist who, several months later, would see another young girl with the same condition. This time, he asked the mother about DES and was shocked to learn that she had also followed the "Smith and Smith regime." Deeply disturbed, the conscientious doctor contacted one of his colleagues at Harvard, Arthur Herbst, and the epidemiologist David Poskanzer; as a result, five supplementary cases were identified in a single Massachusetts hospital. Six months after their article was published in the *New England Journal of Medicine,* the trio had assembled sixty-two cases of clear cell adenocarcinoma among females under twenty-five.

The case caused such a stir that the FDA was forced to publish a warning indicating that "DES was contraindicated for pregnancy use," but curiously the agency never officially banned the drug.[31] Nonetheless, the curtain had been lifted on DES's harmful effects, which, as the authors of *Our Stolen Future* underline, could have been ignored for quite some time if we had been forced to wait for the public authorities to do their job. "Would doctors have ever linked the medical problems suffered by young women with a drug their mothers had taken decades earlier if it hadn't been for a striking cluster[32] of extremely rare cancers and a chance question posed by a patient's mother?" they quite rightly ask. "Until DES, most scientists thought a drug was safe unless it caused immediate and obvious malformations. They found it hard to believe that something could have a serious long-term impact without causing any outwardly visible birth defects."[33]

"When the study by the Boston doctors was published, I had already undergone my first surgical operation," said Kari Christianson, the program director for DES Action, during our interview at the New Orleans symposium. "I was very young and I will never forget my mother's reaction when she found out from the newspapers that all the disorders ailing me were because of a drug she took while she was carrying me. She completely fell apart . . . She had four miscarriages before my birth and she had always believed that it was the distilbene that allowed her to have me, as well as my younger brother. Incidentally, I was born in perfect health, without any apparent problem."

"What do you suffer from?"

"Cervicovaginal adenosis, a pathology very common among distilbene daughters. It manifests as a mucous membrane on the cervix that can evolve into a cancer."

"How did you discover it?"

"At puberty, like most of us did. Some discovered they had serious problems when they decided to have children."

"That's how it happened for me," added Karen Fernandez, another DES Action activist. "I was a newlywed, I had two extrauterine pregnancies—my babies developed in the Fallopian tubes; and, at the age of twenty-six, I was declared sterile."

"What pathologies are associated with exposure to distilbene?"

"Among girls, there are congenital deformities, like a T-shaped uterus, vaginal or ovarian anomalies often coupled with sterility problems or difficulties carrying a pregnancy to term," Kari Christianson answered. "Uterine or vaginal cancers—such as clear cell adenocarcinoma, which affects one exposed woman in a thousand—have also been observed; the risk of getting breast cancer multiplies by three. And as shown by several epidemiological studies, this risk is applied to our mothers as well. Among boys, there is an increased prevalence of cryptorchidism, hypospadias, testicular cancer, and a weak concentration of spermatozoa. More recently, heightened risks of depression and neurological or behavioral problems have been observed among adults who were exposed in utero. In reality, unlike the mother's natural estrogen, distilbene can reach the fetus's brain, because it has the ability to pass through the placenta. All this has been scientifically established, thanks to the enormous efforts of Pat Cody, the founder of DES Action."

DES Action's Unique Fight

I was very keen on meeting the woman nicknamed the "DES mother," whose work has "attained legendary status in the annals of medicine," according to Susan Bell, a sociologist who dedicated a book to DES Action, the association created in 1978 by Pat Cody.[34] Cody is one of those women whose exemplary commitment to the human community demands our respect. Sadly, the meeting never took place, as she passed away on September 30, 2010, at the age of eighty-seven.

Before starting a movement that, beyond DES, would embody a new way of tackling medical questions, the *Economist* journalist became well known for founding, with her husband, an independent bookstore at Berkeley, where an entire generation of 1960s activists and writers came to change the world. As she recounts in her book *DES Voices: From Anger to Action*, her "entire life" would change on a "Friday in April 1971" when, as she sat "at [her] kitchen table drinking coffee," her heart would "nearly stop" when she read a headline in the *San Francisco Chronicle*.[35] "Drug Passes Cancer to Daughters," read the newspaper presenting the study published by Boston researchers. Pat Cody had taken DES during her pregnancy with Martha, the eldest of her four children. As her heart sank, she recalled the exorbitant treatment that had cost thirty dollars a month (she paid seventy-five dollars for the rent on her house) and calculated that she had ingested ten grams of DES in seven months, or the equivalent of five hundred thousand birth control pills. Filled with regret and worry, she would not say anything to her daughter until she reached adulthood.

Once the initial shock had passed, Martha agreed to have a pap smear: it showed that she had pre-cancerous uterine cells. "She was to come in every six months," said the gynecologist, and, above all, "she wasn't to use birth control [because] estrogen was thought to stimulate the growth of cancer." That was precisely the moment when Pat Cody realized that she had to move beyond her personal tragedy and take action, to inform all the mothers and daughters who had been exposed to DES. And that is where her incredible adventure began, one that became a model for collaboration between women who had an intimate understanding of the harmful effects of DES and the medical and scientific community, as well as legal, political, and health authorities. A rare initiative that might have inspired the French Association of Farmer Victims of Pesticides, whose inaugural general meeting, according to Paul François (see Chapter 1), was scheduled to take place on March 18, 2011, in Ruffec.

Pat Cody and her partners, including Kari Christianson, initially established a network of local committees across the United States, which was used to alert the public via a widely distributed informational letter. Thousands of men and women—distilbene mothers, daughters, and sons—

showed up to share their experiences and concerns. As a result, DES Action was able to create an exceptional database on this modern-day poison, which it then made available to researchers like Arthur Herbst, one of the authors of the 1971 study, who had opened a "Registry for Research on Hormonal Transplacental Carcinogenesis" at Massachusetts General Hospital. The thousands of "anecdotal reports," as the industry likes to (dis)qualify them, were "forward[ed] to the researchers with the hope that a study [could] be done."[36] And it was done, as we will see. At the same time, an awareness campaign targeted physicians and medical facilities so that they could improve their efficiency in terms of both prevention and treatment. DES Action organized workshops and invited practitioners, nurses, teachers, social workers, and scientists, such as John McLachlan, the organizer of the 2009 New Orleans symposium.

Following those efforts, the association very notably provided support for the (numerous) lawsuits filed against manufacturers by DES victims. "If being held accountable for putting poorly tested and ineffective drugs on the market affects their bottom line, perhaps drug companies will be more careful and we can prevent future drug disasters," notes Pat Cody in her book.[37] This very sensible remark reminds me of François Lafforgue, Paul François' lawyer, who encouraged farmers to press charges against pesticide makers during the Ruffec meeting (see Chapter 4).

In 1974, the first plaintiff to bring Eli Lilly (the main manufacturer of DES) to trial was Joyce Bichler, who, at eighteen, was suffering from clear cell adenocarcinoma. The New York court upheld the charges of "negligence." In her plaintiff's brief, Attorney Sybil Shainwald emphasized that "before 1940, there were numerous writings, including eight key articles in medical literature, showing the connection between estrogen, DES and cancer." She went on to enumerate the studies published in the 1940s and 1950s, before concluding: "If the manufacturers knew it could malform a fetus and they knew it was carcinogenic, is it not a fair inference that an available prudent test would be used to see if it was also carcinogenic to offspring? Is the public to be used as a guinea pig until someone discovers in actual use that cancers are being caused?"[38] In 1980, when the verdict was handed down, jurors had to answer seven questions, three of which were crucial: "Should a reasonably prudent drug manufacturer have foreseen that DES might cause cancer in the offspring of pregnant women who took it?" "Yes," the six jurors responded unanimously.

"Would a prudent manufacturer have marketed DES for miscarriage purposes had it known that it caused cancer in the offspring of pregnant mice?" "No," responded the jurors.

"How much do you award Ms. Bichler?" "$500,000."[39]

Eli Lilly appealed, but in vain. The ruling was definitively affirmed in an appellate court. But while Joyce Bichler's victory opened the door, the battle was still far from won for the other plaintiffs, who were facing a seemingly insurmountable difficulty: for a complaint to be admissible, each plaintiff was required to provide evidence of the name of the manufacturer of the product his or her mother had used. An enormous challenge, given that some two hundred companies marketed DES under a wide variety of brands. What's more, Pat Cody underlines, "what mother can be expected to save a bottle of pills from twenty-five years ago?"[40] As for the treating physicians, those who had the courage to testify were rare. Most remained fearful of lawsuits. The DES Action founder ironically notes the incredible number of "floods and fires in doctors' offices" that destroyed their archives. Not to mention the pharmacies that changed hands in the meantime or the practices that closed.

To overcome this obstacle, DES Action rallied to its cause brilliant lawyers who fought so that complaints could be lodged against *all DES manufacturers*, regardless of the brand used by the plaintiffs' mothers. And they won! In March 1980, the California Supreme Court authorized Judith Sindell to file a complaint, even though she did not know which manufacturer made the product her mother had taken, as Pat Cody reports, citing the terms of the historic judgment: "In our contemporary, complex industrialized society, advances in science and technology create . . . goods which may harm consumers and which cannot be traced to any specific producer. The response of the courts can either be to adhere rigidly to prior doctrine or to fashion remedies to meet these changing needs. . . . so should we acknowledge that some adaptation of the rules of causation and liability may be appropriate in these recurring circumstances."[41]

Under the terms of the decision establishing the "market share theory of liability," and given that each company "would be held liable for the proportion of the judgment represented by its share of the market unless it demonstrated that it could not have made the product which caused the plaintiff's injuries," "the plaintiff could name as defendants all the major manufacturers of DES," wrote Nancy Hersh, Judith Sindell's lawyer, in the DES Action newsletter. The California Supreme Court decision set a precedent throughout the United States, notably in Florida, Wisconsin, Washington, and Michigan, where numerous lawsuits were filed against DES manufacturers.

John McLachlan, the "Pivotal Figure" of the Unprecedented Collaboration Between DES Action and Scientists

In her book *DES Voices*, Pat Cody highlights the "importance of lawsuits against the drug companies"—and, I would add, against all poison manufacturers. "First," she writes, "they are important as compensation for the medical costs, pain, and suffering of those who sue. Second, the lawsuits bring media attention to the question of DES exposure. Third, penalties for negligence teach manufacturers lessons about proper testing of drugs, devices, and procedures. [. . .] Fourth, taking action, fighting back, not being a victim but a survivor has a positive effect on the plaintiff and the entire community." Cody also emphasizes that lawsuits are a priority for the association, even if its principal task remains "advocating for research."

In fact, DES Action's great originality was in knowing how to "develop alliances with biomedical scientists in the pursuit of prevention, treatment, [and] research," as Susan Bell notes. After "the activists identified a gap between their intimate, firsthand knowledge of their bodies and the medical literature," DES Action turned to "initiating and conducting their own research."[42] In 1984, the association distributed a detailed medical questionnaire to its members to "help identify whether certain health conditions—beyond those already known—appear more frequently among DES exposed women and men than among nonexposed people." DES Action members picked apart the results of this vast survey in collaboration with Deborah Wingard, an epidemiologist at the University of California, San Diego. Any recurring trends were "discussed with scientists who could follow up with further studies."

One of the "pivotal figures"[43] in the collaborative process between DES Action and critical research was John McLachlan, the organizer of the 2009 New Orleans symposium. Named as head of the Developmental Endocrinology and Pharmacology section of the National Institute of Environmental Health Sciences (NIEHS) in 1976, the biologist led a series of experimental studies that allowed him to verify the "endocrine disruptor hypothesis": "a 'bold and unorthodox insight' developed *in* and *from* DES research," according to Sheldon Krimsky, professor of Urban and Environmental Policy and Planning at Tufts University in Boston.[44] Considered until the mid-1980s as "funky science,"[45] studies by John McLachlan, who became the director of the Center for Bioenvironmental Research at Tulane University in 1985, identified DES as a model for the mechanisms of endocrine disruption. In doing this, McLachlan developed an experimental research protocol that is currently used as a reference by all scientists working on endocrine disruptors, thanks to his constant movement "between the worlds of mouse and human research and between the study of environmental and clinical estrogenic effects."[46] As a result, in 1979, he would create the first symposium on "the environment and hormones," with the participation of Pat Cody. This extremely important scientific event has been held regularly at Tulane University and celebrated its thirtieth anniversary in October 2009.

"What effects did you observe after exposing pregnant mice to distilbene?" I asked the researcher.

"We observed effects on their male and female offspring. Among the females, we noted serious malformations in the genital tract and cancer in their reproductive systems, notably in the vagina; among males, infertility problems, cryptorchidism, and prostate cancer.[47] In fact, everything we observed in mice was verified in humans, and everything that we saw in humans also occurred in the mice. It's actually extremely unsettling: when we conducted our studies on mice twenty-five years ago, we observed that second-generation females experienced menopause more prematurely; and today, we are noticing the same thing among women exposed in utero."

"Why do you consider distilbene a model for understanding the mechanisms of endocrine disruption?"

"It's a perfect model!" the American scientist responded without hesitation. "What we observed with DES has been confirmed today by studies on bisphenol A [which, it is worth recalling, is a synthetic hormone invented by Charles Dodds, before DES]. These two molecules act in the same way from a biological point of view, even when they are used in very weak doses. The distilbene model should be used to anticipate the risks caused by endocrine disruptors found in the environment, because there are very few examples in environmental science where we've accumulated solid data on both animals and on a group of exposed humans who have been monitored for over forty years."[48]

"Endocrine Disruption Is Not a Theoretical Notion . . . It Has a Face"

"I have a twenty-eight-year-old daughter and a twenty-three-year-old son, who are healthy for now," said Cheryl Roth, one of the DES Action representatives attending the New Orleans symposium. "But I'd like to know what the research says about the effects of distilbene on the third generation. What should we tell our members who worry about their grandchildren?"

Retha Newbold, a biologist in the toxicology department of NIEHS for over thirty years, answered the question.

To start, she underlined the relevance of the "DES mouse model," which she helped develop while working with John McLachlan in a NIEHS laboratory. With the aid of a slideshow, Newbold explained that "one percent of female mice exposed to distilbene in utero between the first and fifth day of gestation developed a vaginal adenocarcinoma, which corresponds exactly to the percentage observed among distilbene daughters."[49] "What's more," she added, "the model allowed us to measure the effects of genistein, an isoflavone contained in soy which has a weak estrogenic activity. We observed that females exposed in utero presented ovarian anomalies that could lead to fertility problems. All the studies that we have conducted confirm the fetus's fragility when it is exposed to hormonally active substances at critical moments of its development; this exposure at the very start of the fetal life can trigger diseases in adulthood, which is what we've found in mice and humans."[50]

"Then," continued Retha Newbold, "we wanted to know if susceptibility to tumor formation could be transmitted to subsequent generations. The answer is yes. For example, we observed that 31 percent of mice in the F1 generation, that's to say the mice exposed to DES in utero, developed uterine cancer; among the F2 generation, or the daughters of the F1 generation, we found 11 percent had uterine cancer, compared with 0 percent in the control group.[51] Likewise, among the F2 males, an increased incidence of precancerous lesions and tumors of the reproductive tract was observed.[52] The mechanisms involved in these transgenerational events are still largely unknown, but everything indicates that they are epigenetic. Several laboratories are currently exploring this hypothesis.[53] For humans, several studies have indicated an increased risk of hypospadias among sons of women exposed to distilbene in utero."[54]

The scientist went on to present her latest work on a little-known aspect of the effects that endocrine disruptors can induce. "Aside from cancers and reproductive problems, we observed a link between prenatal exposure to distilbene and obesity, as well as diabetes," she explained. "It's very interesting, especially when we know—as this was recently demonstrated—that adipocytes, the cells present in the adipose tissue, are endocrine organs, in the sense that they have an endocrine function: they can produce and receive signals that interact with the reproductive or immune systems, the liver, or the thyroid. This means that obesity can be considered as an endocrine system disease, which would partially explain the obesity epidemic we are seeing almost everywhere in the world. Obesity is, of course, a complex disease, wherein multiple factors play a role, such as junk food, genetic predisposition, and lack of exercise, but our studies aim to prove that endocrine disruptors, like distilbene, have an 'obesogen' function, to borrow a term invented by our colleague at the University of California Irvine, Bruce Blumberg.[55] Meaning that they can program future obesity in an adult by acting on the fetus in development."

Retha Newbold then presented a series of slides confirming what she calls the "obesogen hypothesis": "You'll see, on the left, the mice who were exposed to DES in utero; and, on the right, those from the control group. Up until the twenty-fourth day, which is the puberty age for these rodents, the exposed mice were slightly thinner than those in the control group. And then, there is a very clear-cut change: in just a few weeks, the exposed mice became obese, and remained so for the rest of their lives, to the point that we had to order larger cages! Other labs have made the same observations with other endocrine disruptors, such as phthalates, flame retardants, PFOA [perfluorooctanoic acid] from nonstick cookware and bisphenol A. It strikes me as a very important field of research because, if confirmed, this means that we can prevent obesity by avoiding exposure to these products, notably among pregnant women."[56]

Then came the time to ask technical questions about the very strictly scientific portion of Retha Newbold's presentation. However, even as the moderator was preparing to introduce the next speaker, Kari Christianson, the program director of DES Action, spoke up: "I wanted to tell you that if we're here, we being the 'DES daughters,' it's so that you, the scientists, will never forget that endocrine disruption is not a theoretical notion, but that it has a face: ours, or those of our children and our grandchildren, for those who have them," she said, visibly very moved. "We don't want the tragedies our families experienced to be tossed aside, or to become a mere footnote in the annals of medicine. We want our suffering to enlighten the future, so that we can avoid similar tragedies. More than ever, we need independent researchers who will work for the good of the community, and we will always be here to remind you of that."

At a time when bisphenol A, another notorious endocrine disruptor, was front and center of the media scene, the DES Action representative's warning rang out like a powerful rejection of regulatory agencies. Because, at the dawn of the 2010s, these agencies are still turning a blind eye to the many warning flags raised throughout the world by dozens of independent researchers.

18

The Case of Bisphenol A:
A Pandora's Box

A new scientific truth does not triumph. Its opponents eventually die.
—Max Planck

"It should be repeated again and again: all the effects of distilbene [DES] observed in humans also occurred in mice and rats," insisted Ana Soto, a biologist at Tufts University (see Chapter 16), during a New Orleans symposium on "the environment and hormones" in October 2009. "And today, with Carlos Sonnenschein, we've obtained the same effects with very small doses of bisphenol A [BPA], similar to what is found in the environment. Nonetheless there is a difference: we obtained our first results in 2007. If we run the comparison with distilbene, we will have to wait until 2032 to verify the effects on humans. Which is going to be very difficult . . . For distilbene, women who were exposed in utero can show their mothers' prescriptions as proof. Whereas the women who will develop a cancer in 2032 won't have any evidence that they were exposed to BPA in utero. I'll leave you with that very worrying thought."

Low Doses with Great Effect

As BPA was making international headlines, the 2009 symposium dedicated an entire day to the molecule synthesized by Charles Dodds in 1936, two years before DES. Considered two thousand times less powerful than natural estrogen, this artificial hormone is largely used in the plastic polymerization process, or as an antioxidant in the composition of certain plasticizers.

With an annual production estimated at 3 million tons, BPA is present in "countless applications" meant to "make our lives easier, healthier, and safer," according to the astonishing website of the manufacturers who produce it.[1] In fact, it is found in a majority (65 percent) of everyday consumer products made of polycarbonate—such as hard plastic containers, water bottles, or baby bottles, prepackaged microwave meals, but also sunglasses, CDs and the thermal paper in cashier receipts—and in much of the epoxy resin coating (35 percent) that lines the interior of food or soda cans, as well as dental cement.[2]

"BPA is one of the most extensively tested materials in use today," according to the manufacturers' propaganda. "Its safety has been studied for more than 40 years. The extensive safety data that exist for BPA show that consumer products made with BPA are safe [. . .] and pose no known risks to human health. The U.S. FDA and international agencies charged to protect public health fully support the use of these materials." So goes the official discourse, which bears a striking resemblance to the one hammered in for three decades by the makers of aspartame.

The acceptable daily intake (ADI) of BPA was set in 2006 at 0.05 mg (or 50 µg) per kg of body weight; although it is also detected in household dust, human exposure occurs primarily through food consumption. In fact, as recognized by the manufacturers themselves, the substance has the capacity to "migrate," that is, to leak from the plastic or resin to penetrate into the foods with which it is in contact. This phenomenon, similar to hydrolysis, is due to the instability of the chemical link between the BPA molecules and polymers, and is augmented by heat (hence the widely publicized controversy about baby bottles heated in microwaves). But, as we will see, the question of baby bottles, as important as it is, is in some ways the tree hiding the forest. In reality, if BPA has captured widespread attention, it is because it symbolizes an issue largely ignored by regulatory agencies—the effects of chemical substances in very low doses, that is, doses that have never been tested because they are considerably smaller than the ADI. Such substances of course include some hormonally active agents, such as endocrine disruptors, which, as we have seen, act in microscopic doses (see Chapter 16), BPA being the most flagrant offender.

But, one might object, isn't the ADI calculated from a dose with no observed adverse effects—the infamous "NOAEL" (see Chapter 12)—to which a "certainty factor," generally of 100, is applied? How can a substance have effects at doses "considerably lower than the ADI"? This question is at the heart of the tug-of-war between European and American regulatory agencies and a growing number of scientists. Unsurprisingly, BPA's manufacturers brush off the "low dose hypothesis" as "nonvalid," stating on their website that the substance "exhibits toxic effects only at very high levels of exposure." And that "the weight of scientific evidence clearly supports the safety of BPA and provides strong reassurance that there is no basis for human health concerns from exposure to low doses of BPA."

The manufacturers' optimism is startling. A study published in 1993 in the journal *Endocrinology* indicates that, on the contrary, such "concerns" are entirely justified.[3] The study concerned a discovery made fortuitously by David Feldman, a researcher at Stanford University who came across the same enigma faced by Ana Soto and Carlos Sonnenschein several years earlier (see Chapter 16). At the time, Feldman was working on a protein present in yeast that he had observed had the power to bind with estrogen. From that observation, he deduced that the yeast contained an estrogen receptor and, therefore, that it likely contained a hormone as well. His team was hunting for the hormone when they noticed that a substance had "squatted" in the estrogen receptor. After long study, Feldman identified the guilty party: BPA, which had migrated from the polycarbonate flasks used to sterilize water for the autoclave experiments. The researcher contacted the manufacturer (GE Plastics), which acknowledged that BPA did migrate toward the contents of flasks and water bottles, particularly when exposed to heat, but also to detergents; to mitigate this imperfection, they had developed a plastic-cleaning regimen that, according to the company, had resolved the problem.

The rest of the story, told in *Our Stolen Future*, is of capital importance because it is at the center of the BPA controversy. David Feldman sent a sample of the contaminated water to GE Plastics, but the company *was unable to detect traces of BPA*, ones that the researcher had nonetheless verified as causing the proliferation of breast cancer cells. The detection limit on the manufacturer's measuring devices was ten parts per billion, whereas the residues detected by the Stanford team ranged from two to five parts per billion. "The Stanford paper shows that bisphenol A prompts an estrogen response in cells in a lab [below the limit of ten parts per billion]," remarked David Feldman. "We don't know enough yet to make this into a public health crisis, but the next logical question is whether it prompts the same response when given in water to an animal."[4]

Several years later, Feldman's colleague Patricia Hunt, a molecular biologist who also witnessed an accidental contamination in her University of Cleveland laboratory in 1998, would follow the Californian endocrinologist's recommendation. At the time, she was conducting experimental studies aimed at understanding why the frequency of pregnancies characterized by *chromosomal abnormalities* increased with the mother's age. Her study consisted of comparing the cellular division of oocytes in mice presenting anomalies to those of "normal" mice. To better understand the significance of her findings, it is worth noting that, among mammals, the production of sex cells (or gametes)—spermatozoa in males and ovules in females—begins during fetal life through a process called meiosis. This means that females form their future ovules while in their mother's womb (oogenesis). In *PLoS Biology*, the scientist described her observation, made while "studying chromosomal alignment in eggs undergoing division before ovulation," that "instead of lining up normally as they should, they were just not lining up at all. [. . .] So we were studying what we thought were the precursors of the chromosomally abnormal egg, which would give rise to, for example, Down syndrome."[5]

Following the example of Ana Soto, Carlos Sonnenschein, and David Feldman, the biologist eventually identified the substance which had profoundly disturbed the ovule formation process: several days earlier, one of the lab employees had cleaned the mice's polycarbonate cages with a powerful detergent that caused the plastic to deteriorate, releasing minute quantities of BPA. The animals were contaminated through dermal exposure. Deeply troubled by the health implications of her fortuitous discovery, Patricia Hunt then decided to deliberately expose gestating mice to a very low dose of BPA, similar to the residue levels observed in the American population. She observed that the developing ovaries of female mice exposed in utero contained an abnormally high number of oocytes presenting chromosome abnormalities associated with miscarriages, congenital defects, and mental retardation in humans. Then, once the mice exposed in utero had reached adulthood, the geneticist had their oocytes fertilized and noted a particularly high level of embryos presenting chromosome abnormalities.[6] "By hitting the mother with a low dose, we increased the likelihood the grandchildren would be abnormal," said Hunt.[7] "Thus, not only is a fetus exposed, but so are eggs that will produce the next generation." And what's more, "These changes to the fetus are permanent and irreversible, whereas impacts of adult exposure are reversible. The fetus is exquisitely sensitive to bisphenol A. One hit during a brief window of time can influence future development."[8]

The Dangers of Fetal Exposure to Bisphenol A

"While it didn't outright deny the validity of our results, the industry did minimize them by arguing that rodents are not humans," Ana Soto explained during my visit to her Tufts University laboratory in Boston. "But what can we do? Deliberately expose pregnant women to bisphenol A to verify that it does produce the same effects as those observed in our experimental studies?" As she was asking this somewhat jaded question, the biologist turned on her computer to show me a series of images relating to the study she conducted with Carlos Sonnenschein on gestating mice exposed to very low doses of BPA. After their fortuitous discovery regarding nonylphenol (see Chapter 16), the two researchers had decided to study the transgenerational effects of BPA. "We thought that it was more useful, because human exposure to bisphenol A is much more significant than to nonylphenol," explained Ana Soto. "That's why, from the beginning, we used doses similar to those found in our environment—that is to say, much lower than the ADI. We even went down to the lowest levels possible, thinking that we wouldn't observe any effects, which unfortunately was not the case."

"What effects did you observe with very low doses of BPA?" I asked.

"Among rats and mice exposed in utero, we observed an increase in the rates of breast and prostate cancer, fertility problems (namely disruptions in menstrual cycles), behavioral issues, such as female mice behaving like males, but also—and this came as quite a surprise—a very pronounced tendency toward obesity. It's very worrying because these pathologies are precisely those rising rapidly among the human population."

"At what dosage did you observe these results?"

"At doses two hundred times lower than the ADI, or 250 nanograms per kilo, like in this experiment," responded Ana Soto while showing me an image captured on an electronic microscope. "You see here the mammary gland of a four-month-old mouse that was not exposed to bisphenol A. We can see the ducts that will eventually drain the milk. There aren't many of them and there's not much branching. Now, I'm going to show you an animal that was exposed to bisphenol A in utero: we observe an exceptional development of the ducts and lateral branching, an abnormal transformation of the terminal end buds, and an increase in progesterone receptors. This is four months after exposure. This would be normal if the mouse was gestating, but that's not the case. Gestation is not a pathology in itself, but if the mammary gland of a female that is not gestating imitates one that is characteristic of gestation, then that is not normal!"

In an article she published in 2001 in *Biology of Reproduction*, Ana Soto wrote "these changes are associated with carcinogenesis in both rodents and humans."[9] To verify these results, the Tufts University team repeated the experiment on rats exposed in utero and observed a significant increase in precancerous breast lesions and, at the highest doses, cancers in situ. "These results are similar to those observed among women exposed to distilbene *in utero* who presented a much greater sensitivity to hormone-dependent cancers," explained Soto.

"But," I insisted, "fetuses are exposed to natural estrogen throughout natal development, which doesn't cause these effects?"

"Natural estrogen appears in the body at the right moment," said Carlos Sonnenschein. "Whereas synthetic hormones can enter anytime and, notably, at the wrong time. The other difference is that the body rapidly metabolizes, and therefore inactivates, the natural hormones, which is not the case with exogenous hormones or hormones that come from outside the body. These have a longer effect because they resist the degradation mechanisms and, what's more, they're lipophilic, meaning they combine with fats."

"Are the effects on the fetus caused by BPA reversible?" I asked.

"Unfortunately, since they occur during organogenesis, or during the organ formation process, it would appear that they are definitive," Soto answered immediately. "The effect of synthetic hormones on developing organs is very different from that on an already formed adult organ."

"Have your results been reproduced in other labs?"

"Of course! And notably by Fred vom Saal, who showed us the way—he was the one who revealed that endocrine disruptors may have no effect at very high doses but very powerful effects at minute doses."

Frederick vom Saal Discovers the Power of Hormones

Studies conducted by Frederick vom Saal, a biologist at Columbia University (Missouri), constituted a "central piece of the puzzle"[10] patiently reconstituted by Theo Colborn throughout a long investigation that led to the discovery of endocrine disruptors (see Chapter 16). In *Our Stolen Future*, the environmental expert describes how this unparalleled researcher, who has an excellent reputation in the United States, was effectively the first to show that "small shifts in hormones before birth can matter a great deal and have consequences that last a lifetime." For this former University of Texas at Austin student, everything began in the 1970s when he was writing his doctoral thesis on the role played by testosterone, the male hormone, in fetal development. He observed that this hormone, indispensable to the develop-ment and smooth functionality of the male reproductive system, also influ-enced a male characteristic: aggression.

This is how he came to spend months observing the behavior of mice from the same gene pool and noticed that certain females with the same mother displayed abnormally high levels of aggression. He hypothesized that this difference in attitudes could be due to where a mouse was situated in the genitor's womb. For these small rodents, a "classic" litter includes on average twelve fetuses, packed in like sardines in a can. Certain females are thus sandwiched between two males. A week before birth, the males' testes begin to secrete testosterone. "The female pups might be bathed in testosterone washing over from the male neighbors," writes Theo Colborn,[11] which could explain why they later adopt a more masculine (read aggressive) behavior. To verify his hypothesis, Frederick vom Saal conducted dozens of Caesarean sections just before the mice's natural birth (which generally occurs on the nineteenth day of gestation). He carefully identified each young mouse in relation to its position in the litter, then observed its behavioral evolution. The results were dramatic: "The aggressive females were the ones who had developed between brothers."[12]

This paramount discovery, confirming the "powerful role of hormones in the development of *both* sexes and the extreme sensitivity of developing mam-mals to slight shifts in hormone levels in the womb," was named the "womb-mate effect"[13] and led to a new scientific concept: the "intra-uterine position (IUP) phenomenon."[14] Frederick vom Saal and the researchers who followed suit, such as Mertice Clark, Peter Karpiuk, and Bennett Galef from McMas-ter University, and John Vandenbergh and Cynthia Huggett from North Car-olina State University, observed that the "intra-uterine position" of females definitively shaped their adult lives. Those who had the "bad luck" to develop between two males—Theo Colborn nicknamed them the "ugly sisters," in contrast with "pretty sisters"—had much less success with male mice, which preferred a "pretty sister" eight times out of ten. "The pretty sisters smell 'sexier' to males because they produce different chemicals than their less at-tractive sisters," Colborn writes. "The prenatal hormone environment leaves a permanent imprint on each sister that is recognized by males for the rest of her life."[15] The researchers also observed that the poor "ugly sisters" experi-enced puberty later and were less fertile than their "pretty sisters," and that, when they did manage to reproduce, their litters were generally made up by a majority of males (60 percent) versus exactly the opposite (60 percent of females) among the "pretty sisters."

But the "intra-uterine position phenomenon" does not exclusively concern the females: Frederick vom Saal and his colleagues observed that young male mice that were sandwiched between two females in the womb were exposed to much higher levels of estrogen than their brothers that were wedged between two males, which produced significant behavioral differences. Nicknamed the "playboys," the first category was characterized by exacerbated levels of aggression that pushed them at times to attack, even kill, infant mice, whereas the second group displayed the irreproachable behavior of "good daddies." What's more, the "playboys" had prostates twice as large as those of their brothers that had not been exposed to their sisters' estrogen, as well as heightened sensitivity to male hormones because they had three times the number of testosterone receptors. Theo Colborn underlines that "although human babies don't usually have to share the womb with siblings, their development can nevertheless be affected by varying hormone levels," which can be caused by "medical problems such as high blood pressure [that] drives up estrogen levels" or the fact that "the mother's body fat contains synthetic chemicals that disrupt hormones."[16] But, the *Our Stolen Future* co-author continues, "it is important to remember that hormones do this without altering genes or causing mutations. They control the 'expression' of genes in the genetic blueprint an individual inherits from its parents. [. . .] The concentrations are typically parts per trillion, one thousand times *lower* than parts per billion."[17]

Among these genes, there is one in particular called "SRY" (sex-determining region of Y gene) whose expression determines sexual differentiation and, more precisely, masculine identity. We know that, among mammals, each female cell has two X chromosomes, whereas a male cell has one X chromosome and one Y chromosome. The mother's eggs therefore all present an X chromosome, whereas the father's spermatozoa present either an X chromosome or a Y chromosome. For a long time, it was thought that the fetus's sex was automatically determined by the presence or absence of a Y chromosome in the father's inseminating spermatozoon; if it contained a Y, the baby would be a boy, and a girl if it contained an X. But since 1990, we know that the process of sexual differentiation is much more complex and that it depends on the activation of one gene, SRY, situated on the Y chromosome.

"Although the sperm delivers the genetic trigger for a male when it penetrates the egg, the developing baby does not commit itself to one course or another for some time," explains Theo Colborn. "Instead, it retains the potential to be either male or female for more than six weeks, developing a pair of unisex gonads that can become either testicles or ovaries and two separate sets of primitive plumbing—one the precursor to the male reproductive tract and the other the making of the fallopian tubes and uterus. These two duct systems, known as the Wolffian and Müllerian ducts, are the only part of the male and female reproductive systems that originate from different tissues. All the other essential equipment—which might seem dramatically different between the two sexes—develops from common tissue found in both boy and girl fetuses. Whether this tissue becomes the penis or the clitoris, the scrotal sack that carries the testicles or the folds of labial flesh around a woman's vagina, or something in between depends on the hormonal cues received during a baby's development."[18] Ultimately, the definitive determination of the fetus's sex depends on the activation of the SRY gene, which triggers a signal sent by the testosterone *at a very specific and unique moment in pregnancy,* as Bernard Jégou, Pierre Jouannet, and Alfred Spira report in their book *La Fertilité est-elle en danger?* (Is Fertility in Danger?).

I am reproducing their description here because it is essential to understanding the subtlety of the sexual differentiation process and the fabrication of genital organs—extremely delicate mechanisms that, as we know, can completely derail if an intruder interferes: "In the seventh week of development, the SRY gene, located in the Y chromosome, sends a signal to the gonad and tells it to transform into a testicle," write the reproductive health specialists.

The differentiation of two sexes depends on the hormonal activity of the fetal testicle that actively secretes two essential hormones. One of the first consequences of the SRY gene is the secretion of the anti-Müllerian hormone (AMH) produced by Sertoli cells and the secretion of testosterone produced by Leydig cells. AMH will provoke the regression of Müllerian ducts while testosterone ensures the presence of Wolffian ducts that develop into the epididymis, the vas deferens and the seminal vesicles. Testosterone and its derivatives, which we still call androgens, also favor the development of the urethra and the prostate, as well as the expansion of the genital tubercle to form the penis and the scrotum in a male. At this stage of development, the testicles are situated in the abdomen. They won't descend into the scrotum until around the seventh or eighth month of pregnancy, or shortly before birth. [. . .] In the female embryo, in the absence of the SRY gene, but also thanks to other genes, the undifferentiated gonad transforms into an ovary. In the absence of testosterone and AMH, the Wolffian ducts regress, while the Müllerian ducts remain and give rise to Fallopian tubes, the uterus and the upper section of the vagina. As for the external genital organs, the urogenital and labioscrotal folds don't merge. They form the small and large vulva lips, respectively, while the genital tubercle forms the clitoris.[19]

A "Time Bomb"

"The chemical industry has put a lot of work into misinforming people that we are not exposed to bisphenol A and that the quantities present in our bodies are not at all worrying," said Frederick vom Saal during the 2009 New Orleans symposium. "The facts indicate the opposite. Just to give you an example: if you insist on having your daily dose of BPA, all you have to do is eat Heinz ketchup or canned tuna in oil. The Centers for Disease Control (CDC) in Atlanta has conducted several studies to measure the level of BPA in the urine of the American population.[20] And, as we can see in that national study, over 95 percent of Americans are contaminated, and the younger you are, the more elevated your bisphenol level.[21] It's important to note that the contamination of premature infants who are placed in an incubator or in an intensive care unit is particularly worrying; it's due to the presence of BPA, as well as of phthalates, in the plastic tubing and intravenous bags.[22] The quantities of bisphenol A measured are identical to those that I've been using in my experimental studies for over ten years."

At the end of the symposium, during which he presented the results of his latest BPA studies, Frederick vom Saal granted me an interview that lasted two hours. It was utterly fascinating to listen to such a brilliant researcher who knows his subject inside out and who talks about it passionately—a dramatic contrast to the vague tepidness of the experts who oversee BPA, as I would shortly observe, within regulatory agencies like the European Food Safety Authority (EFSA) or, the French Food Safety Agency (Agence française de sécurité sanitaire des aliments, AFSSA).[23]

"What effects did you observe when you exposed gestating mice to very low doses of BPA?" I asked.

"Before answering, I want to specify that the doses my team used in our experiments correspond exactly to the levels of BPA found in every impregnation study conducted among American, European, and Japanese populations, meaning that they are much smaller than the acceptable daily intake set by international regulatory agencies. We observed a number of effects: first of all, a decrease in behavioral differences between males and females, as well as loss of gender identity; deformation of the urethra and the bladder, which prevents the animals from urinating correctly at adulthood. Certain deformations are absolutely monstrous, like those seen in these photos where rats exposed in utero to a dose of 20 μg of BPA per kilogram body weight—or a dose two and a half times smaller than the ADI—developed a urethral obstruction leading to dramatic bladder dysfunction. At even smaller doses, BPA provokes insulin secretion, an increase in glucose levels, resistance to insulin, as well as diabetes, cardiovascular diseases, and brain and behavioral disorders. We also observed dysfunctions in the male and female reproductive systems. Among females, ovarian cysts or uterine fibroids. Among males, testicle deformations, a reduction in the quantity of spermatozoa, abnormally low hormone and testosterone levels. Lastly, we observed prostate cancer among the males, and breast cancer among the females. Once we know the extent of human contamination, I sincerely think that we'll be dealing with a veritable time bomb."

"Why is the fetus particularly sensitive to the effects of bisphenol A?" I asked.

"To BPA, but also to endocrine disruptors. The first reason is that, contrary to an adult, the fetus doesn't have a protection system, such as enzymes that would allow it to metabolize chemical substances. Once the substance has penetrated the fetus, it's there to stay. The second reason relates to the unique sensitivity of the fetus, which stems from one sole cell, and then divides into two cells, then four, during the process of cell differentiation. Each cell contains the same genes, be they muscular, adipose, or cerebral, but these genes are programmed differently in order to produce different cells, thanks to action by specific hormones. However, bisphenol A, like all endocrine disruptors, has the power to abnormally disrupt the cellular differentiation process. Once this abnormal path is taken, it's impossible to backtrack. This is what we call 'genetic programming,' which conditions certain organs to function abnormally, in such a way that they develop cancers several decades later."

"So can you confirm the ineffectiveness of the placental barrier in all instances?"

"There is definitely a placental barrier, but, contrary to what we think, it doesn't prevent toxic products from penetrating the placenta and reaching the fetus. It even does the opposite: it acts as a trap, because once the substances have successfully crossed through, the placenta stops them from leaving. My studies have shown that fetal cells are protected in a similar way by natural estrogen present in the mother's body thanks to a system of blood barriers that, nevertheless, is incapable of blocking the intrusion of synthetic hormones into the fetal cells. I made this discovery while working with distilbene, which is in a way the mother of all the endocrine disruptors."[24]

Industry Joins the Fray

In trying to understand how an exogenous estrogen, like DES, could interfere in the key stages of fetal development, Frederick vom Saal was able to confirm what he had already observed in his studies on mice litters: "the extreme power of low hormone doses, which often have an effect much more significant and much more harmful than high doses," as he explained to me during our interview in New Orleans. To better illustrate the stakes of this critical discovery, one regulatory agencies still resist, I will briefly recount the Columbia University researcher's journey.

In 1997, he published an initial study showing that male descendants of mice that had ingested minute quantities of DES during gestation presented an abnormal development in prostate size, similar to that observed among young mice exposed in utero to estradiol, an endogenous hormone that vom Saal understands particularly well.[25] An excessive increase in prostate size and weight is generally considered to be a precursory sign of a potential cancer. After repeating the experiment with BPA, the researchers obtained identical results, which were published the same year: "Our findings show for the first time that fetal exposure to environmentally relevant parts-per-billion (ppb) doses of bisphenol A, in the range currently being consumed by people, can alter the reproductive system in mice," writes vom Saal in an *Environmental Health Perspectives* article that hardly made a stir at the time.[26] But the following year, he published a third study that "immediately caught the attention of the chemical industry—and transformed Fred vom Saal into a tireless crusader against bisphenol A," according to scientific journalist Liza Gross in *PLoS Biology*.[27] For his experiment, the Columbia University researcher fed mice between the eleventh and seventeenth day of gestation with a concentration (dissolved in oil) of BPA of 2 or 20 ng/g body weight—or doses respectively smaller by 25 or 2.5 times the product's ADI. "The 2 ng dose is lower than the amount reported to be swallowed during the first hour after application of a plastic dental sealant," he notes in his article's introduction.[28] However, the observed effects were far from negligible: an increase in size of certain genital organs (preputial glands) or, in contrast, a 20 percent decrease in sperm production relative to the control group. Published shortly after *Our Stolen Future*, which as we have seen elicited a wave of panic among chemical manufacturers (see Chapter 16), the study was similarly devastating, as Frederick vom Saal describes in *PLoS Biology*: "The moment we published something on bisphenol A, the chemical industry went out and hired a number of corporate laboratories to replicate our research. What was stunning about what they did was they hired people who had no idea how to do the work. Each of the members of these groups came to me and said, 'We don't know how to do this, will you teach us?'"[29]

Liza Gross describes the rest of this incredible story: "Vom Saal videotaped his protocols for a group hired by Dow Chemical, and sent one of his students to England to teach AstraZeneca scientists the system. By 1999, a flurry of studies appeared from AstraZeneca, along with a collaborative effort sponsored by the Society of the Plastics Industry (SPI) from the labs of Dow, Shell, General Electric and Bayer, the major bisphenol A producers. (AstraZeneca does not make bisphenol A, but it produces a number of pest-control products that could face similar scrutiny). None of the studies found that low doses of bisphenol A harm the developing prostate."[30]

And yet, another study published the same year by Channda Gupta, a pharmacology professor at the University of Pittsburgh, supported Frederick vom Saal's findings. She fed pregnant mice low doses of BPA and aroclor (the Monsanto pesticide sold under the name Lasso—see Chapter 1), as well as DES, which she used as a "positive control," as did her Columbia University colleague. Liza Gross explains, "if animals fail to respond to DES, whose effects are well understood, it's a sign that the setup is flawed."[31] Channda Gupta observed increased prostate size among males exposed in utero between the sixteenth and the eighteenth day of gestation to a dose of 50 µg of BPA per kilogram of body weight (equivalent to the ADI), as well as enhanced anogenital distance[32] (she obtained similar results with the same dose of aroclor).[33] What's more, when she placed the developing prostates in a culture and treated them with chemical products, she observed the same abnormal organ growth, "indicating that the chemicals targeted the prostate directly."[34]

Industry was quick to react. First, three scientists working for the Chemical Industry Institute of Toxicology—an organization financed by the American Chemistry Council—published a commentary in the *Proceedings of the Society for Experimental Biology and Medicine* vociferously criticizing "the statistical analysis and the resulting conclusions" of Channda Gupta's study.[35] Channda Gupta's response calls to mind a phenomenon that we have already seen in the case of aspartame, namely the "funding effect" (see Chapter 15):

"It is interesting to note that the studies that failed to find an effect of this chemical are funded by the chemical industries, whereas, positive findings are reported by independent academic laboratories. What is also clear, is that scientists who chose to study a chemical of commercial importance are subjected to intense scrutiny by the chemical industry and by the scientists funded by these industries."[36]

Then, relying on its regular poisoning techniques, designed to "pollute the scientific literature," to borrow epidemiologist Peter Infante's expression (see Chapter 9), the industry solicited the "expertise" of a seemingly very respectable organization: the Harvard Center for Risk Analysis (HCRA). On paper, the name looked good and had what was needed to blind naïve experts at the regulatory agencies: Who would suspect an organization affiliated with the famous university of working on behalf of one of the biggest poison manufacturers on the planet? And yet that is the "mission" of the HCRA, which was created in 1989 by a certain John Graham. Documents declassified during the big lawsuits against the tobacco industry revealed that HCRA's first client was Philip Morris,[37] followed by Dow Chemical, DuPont, Monsanto, Exxon, General Electric, and General Motors—companies for whom it conducted long studies that systematically minimized the health risks linked to chemical products. At the same time that the center was solicited by the American Plastics Council to carry out a meta-analysis of studies presenting the effects resulting from low doses of BPA, John Graham joined the George W. Bush administration, named by the president to head the Office of Information and Regulatory Affairs (a key post for chemical product regulation). His nomination provoked a strong outcry, namely from the academic milieu, and on May 9, 2011, fifty-three renowned scientists, including epidemiologist Richard Clapp (see Chapter 11), addressed a letter to the Senate Committee on Governmental Affairs. In it, they denounced "Graham's work [that] has, overall, demonstrated a remarkable congruency with the interests of regulated agencies" and his systematic denial of the "real risks of well-documented pollutants such as dioxin and benzene, and use of extreme and highly-disputed economic assumptions."[38]

After leaving Washington, DC, John Graham was replaced as head of the HCRA by George Gray, a fervent defender of pesticides.[39] He assembled a "panel" of scientists to lead a meta-analysis financed by the plastics industry. It included Lorenz Rhomberg, an "expert" from Gradient Corporation, a consulting firm that worked closely with tobacco companies. In 2006, Rhomberg made a name for himself through his enthusiastic efforts to sabotage a California draft bill (Bill AB 319) that aimed to ban BPA and phthalates in feeding bottles and toys for children less than three years of age. He penned a panoply of arguments that I would soon hear, almost word for word, at the EFSA: "The bill's proponents cite a scientifically unorthodox hypothesis that tiny exposures to bisphenol A—far below those widely considered safe— might harm health. [They] base their claims on unproven speculation that even extremely low levels of bisphenol A somehow could cause harm to children. Many of the studies are of limited or no relevance, however, and those few studies purporting to show bisphenol A effects at tiny doses have not been sustained [by] larger and more rigorous studies. [. . .] No matter how many studies are cited, the sum of weak and inconsistent evidence does not make strong evidence. [. . .] Instead, the safety of bisphenol A-based plastics has been reaffirmed repeatedly. Recent examples include comprehensive government assessments in Japan and Europe, [and] a review by an independent panel of scientific experts organized by the Harvard Center for Risk Analysis."[40]

ADI: Based on "Erroneous Hypotheses That Date from the Sixteenth Century"

Without fail, the meta-analysis published by the HCRA in 2004, thanks to the "support of the American Plastics Council," concluded that there was "no consistent affirmative evidence of low-dose BPA effects for any endpoint."[41] Note that George Gray and the "independent panel of experts" took two years to analyze nineteen of forty-seven studies published in April 2002 and that three panel members ultimately refused to sign the report that recommended, in its conclusion, the "replication of existing studies under carefully controlled conditions."

Just when the plastics industry was broadly circulating its much vaunted report, Frederick vom Saal and Claude Hughes, an endocrinologist who after signing the HCRA opinion ultimately disassociated himself from it, published a new meta-analysis in which they examined not nineteen, but rather 115 studies, a feat that resulted in a 2005 publication on the effects of low doses of BPA.[42] "The results were truly shocking," explained vom Saal during our New Orleans interview. "We noted that more than 90 percent of studies financed by public funds reported significant effects of BPA at low doses—or 94 studies out of 115—but not a single one sponsored by industry did!"

"That's what's called the funding effect," I said.

"Yes . . . And what's more, thirty-one studies conducted on vertebrate or invertebrate animals found significant effects of bisphenol A at a dose lower than the ADI."

"How do you explain the negative results obtained by industry scientists? Did they cheat?"

"Cheating is difficult to prove," answered vom Saal prudently. "However, there are several 'tricks' that can mask potential effects. First off, as Claude Hughes and I wrote in our article, the majority of labs paid by industry used a strain of rats known for being completely insensitive to the effects of estrogenic molecules."

"There are rats like that?" I asked incredulously.

"Yes! This strain, called Sprague-Dawley or CD-SD, was invented, so to speak, fifty years ago by the Charles River company, which selected it for its high fertility and the rapid postnatal growth of its offspring. The result was obese female rats, capable of producing an enormous number of babies, but which were also insensitive to estrogen, like for example ethinyl estradiol (a powerful estrogen found in contraceptive pills): the rats only react to a dose one hundred times greater than the quantity taken daily by women who use an oral contraceptive! Meaning that this strain is completely inappropriate for studying the effects of low doses of synthetic estrogens!"

"And this characteristic of Sprague-Dawley rats isn't known to labs working for the industry?" I asked.

"Apparently not! But curiously enough, all the public labs were informed about it," answered vom Saal with a knowing smile. "The other problem that we encountered with private studies is that they were using technology dating from at least fifty years ago! They're incapable of detecting minute doses of BPA, quite simply because the labs don't have the equipment to do so, or because 'good laboratory practice,' the infamous GLP [see Chapter 12], doesn't require it, which is awfully convenient! It's kind of like an astrologist who wants to study the moon with binoculars when there are telescopes like Hubble in existence! In my lab, we can detect free, or nonmetabolized, bisphenol A residues at a level of 0.2 parts per billion, whereas in the majority of industry studies that we examined, the detection level was fifty to one hundred times higher! In that case, it's easy to conclude that 'exposure to bisphenol A presents no health dangers, because it's been completely eliminated' . . . Finally, the last problem we observed is that scientists in private labs, as well as the majority of regulatory agency experts, generally don't understand endocrinology. They were all trained under the old-fashioned toxicology school of thought, i.e., that 'the dose makes the poison.' However, this principle, which represents the foundation of the acceptable daily intake, is based on erroneous hypotheses that date from the sixteenth century: at the time of Paracelsus, no one knew that chemical products could act like hormones, and that hormones don't follow the rules of toxicology."[43]

"Does that mean that the principle of the 'dose–response' relationship, which is the corollary of the ADI, is also incorrect?"

"Absolutely—it's useless when it comes to endocrine disruptors! It can work for certain traditional toxic substances, but not for hormones, not for any hormone! We know that for certain chemical substances and for natural hormones low doses can stimulate effects while strong doses inhibit them. For hormones, the dose never makes the poison, and effects are not systematically exacerbated, because in endocrinology the linear dose–response curve doesn't exist. I'll give you a concrete example: when a woman has breast cancer, we prescribe her a drug called Tamoxifen. At the treatment's start, the effects are very disagreeable because the molecule begins by stimulating the tumor's progression, and then, once it reaches a certain dose, it blocks the proliferation of cancerous cells. The same phenomenon is observed with Lupron, a drug prescribed to men suffering from prostate cancer. In the two cases, the substance's action is not proportional to the dose, and doesn't follow a linear curve, but rather a curve in the form of an inverted U. In endocrinology, we talk about a two-phase effect: first an upward phase, then a descending one."

"But don't regulatory agencies know about these characteristics?"

"I sincerely believe that their experts should return to medical school and take an introductory course in endocrinology! More seriously, I encourage you to take a look at the consensus declaration recently published by the American Endocrine Society, signed by over a thousand professionals. It makes an official request to the government to take measures to review from top to bottom the manner in which chemical products with a hormonal activity—estimated to be several hundred—are regulated. And the authors of this declaration aren't radical activists demonstrating with huge signs; these are professional endocrinologists who clearly state that, as long as their specialty is not acknowledged by regulatory agencies, consumers and the public will not be protected, because the system will remain ineffective."

So, I read the text published by the Endocrine Society in June 2009 (whose authors included Ana Soto).[44] In a little under fifty pages, it clearly sounded the alarm. "We present the evidence that endocrine disruptors have effects on male and female reproduction," write the authors, as well as on "breast development and cancer, prostate cancer, neuroendocrinology, thyroid, metabolism and obesity, and cardiovascular endocrinology. Results from animal models, human clinical observations, and epidemiological studies converge to implicate endocrine disrupting chemicals (EDCs) as a significant concern to public health." After underlining that endocrine disruptors "represent a broad class of molecules such as organochlorinated pesticides and industrial chemicals, plastics and plasticizers, fuels, and many other chemicals that are present in the environment or are in widespread use," they clarify that "even infinitesimally low levels of exposure—indeed, any level of exposure at all— may cause endocrine or reproductive abnormalities, particularly if exposure occurs during a critical developmental window. Surprisingly, low doses may even exert more potent effects than higher doses. Second, EDCs may exert nontraditional dose-response curves, such as inverted-U or U-shaped curves." In conclusion, they call on "individual and scientific society stakeholders" to "increase understanding of effects of EDCs [. . .], including invoking the precautionary principle [. . .] and communicating and implementing changes in public policy and awareness."

The Study Behind the ADI for BPA is "Ridiculous"

"Do you know which study the EFSA and the FDA used to set the bisphenol A ADI at 50 μg per kilogram of body weight?" I asked Frederick vom Saal, not knowing that I had touched on one of the most incredible elements of this appalling affair.

"The agencies used a study that I will not hesitate to call ridiculous and which should immediately be tossed in the trash bin of scientific history," he responded firmly and seriously, in stark contrast to his cheerful tone at the beginning of our interview. "The study was led by Rochelle Tyl and funded by the Society of the Plastics Industry, Dow Chemical, Bayer, Aristech, Chemical Corp, and GE Plastics, which are the principal manufacturers of bisphenol A. It was published in 2002 and, as its title indicates, it used Sprague-Dawley rats: in other words, it is entirely useless and yet this same study was chosen by the EFSA and the FDA, out of hundreds, to set the ADI!"

On page 32 of its 2006 recommendation,[45] the EFSA notes that the study used to determine the NOAEL for reproductive toxicity was "a comprehensive 3-generation study" by Rochelle Tyl conducted on Sprague-Dawley rats.[46] "When I pointed out in 2005 that Sprague-Dawley rats were insensitive to estrogenic molecules, Tyl's team hastened to conduct a second study with 'Swiss' or 'CD-1' rats," Frederick vom Saal explained. "They were the same ones I used in my lab, but there were still major problems." Effectively, in an elliptic manner at best (if not an enigmatic one), the EFSA's 2006 recommendation evoked the "controversy on possible low-dose effects of BPA in sensitive strains of rodents," before specifying that "a recent two-generation reproductive toxicity evaluation of BPA in mice performed under Good Laboratory Practice (GLP) did not confirm the presence of low dose effects."[47] One can conclude, even if it is not clearly stated, that the ADI of 50 µg set after review of this "ridiculous" study from 2002 was maintained.

"And what were the problems with this second study?" I asked vom Saal.

"There are many!" he exclaimed. "The case is so serious—after all, the ADI for BPA is at stake—that thirty American scientists, including myself, published a long article in 2009 in the journal *Environmental Health Perspectives*[48] to denounce the incredible deficiencies of this study which, like the first one, should be thrown out! Except that it's considered by the EFSA and the FDA as a 'must' for good laboratory practices!"

To better understand the rest of this staggering account, it is worth noting that Rochelle Tyl's team used 280 male mice and as many females, which were divided into three groups: a "control group" (which was not exposed to any substances), a "positive control group" (which was exposed to estradiol as the hormone's effects are very well known), and an "experimental group" (exposed to BPA, via six different doses). Since the study's principal objective was to measure the transgenerational effects of low doses of BPA on the reproductive system, particular attention was paid to the females exposed during gestation and to their male and female descendants.

"The first thing we pointed out in our article," explained Frederick vom Saal, "was that the positive control group mice were extraordinarily insensitive to estradiol. The first effects didn't appear until they were given a dose fifty thousand times greater than that observed in numerous labs, including my own. Everything seems to indicate that Rochelle Tyl's facilities had been contaminated by estrogen. One of the possible explanations could be a fire that ravaged the lab in August 2001, burning twenty or so polycarbonate cages and subsequently releasing quantities of bisphenol A. This hypothesis was recently addressed during a conference in Germany attended by Rochelle Tyl and a FDA representative, where the study's abnormalities were widely discussed.[49] It's incredible that the EFSA and the FDA didn't notice the anomalies that characterized the positive control group—ones that should have quite simply invalidated the study's results since this kind of estrogen contamination makes it impossible to measure the effects of low doses of BPA. The second problem is the highly abnormal prostate weight among males in the control group, which was 75 percent greater than that observed in all similar studies."

In table 3 of her study, Rochelle Tyl noted that the average prostate weight of mice in the control group was greater than 70 mg *at the age of three and a half months*. However, as underlined by the thirty scientists who penned the article with Frederick vom Saal, "this average control weight contrasts sharply with those reported from other laboratories. Specifically, the weight of the prostate in 2- to 3-month-old CD-1 mice is about 40 mg. Several studies have reported that prenatal exposure to very low doses of BPA and positive control estrogens increased prostate size, [. . .] but the enlarged prostate of experimental animals exposed to BPA in these laboratories weighed less than the prostates in the control animals of Tyl."[50] "This exceptional prostate weight can only be explained in one of two ways," explained vom Saal. "Either the dissection techniques were inaccurate, or the animals had diseased prostates. And I have to say that the many justifications provided by Rochelle Tyl for this incongruous size only confirm that the study has no validity."

Admittedly, the industry scientist did stumble on numerous occasions. During a FDA-organized hearing on September 16, 2008, she provided an initial version when questioned publicly by Frederick vom Saal about this obvious anomaly. "The mice weren't three months old, but six," she maintained, "that's why their prostates were larger." The unflappable Columbia University researcher then presented the infamous study, pointing out "two misprints in the paper."[51] Questioned again on the prostates during the conference on BPA held in Germany in April 2009, Rochelle Tyl furnished a third version: this time she said that the animals were actually five months old, prompting some of the fifty-eight scientists in attendance to openly question how such a study could have been chosen as a benchmark by regulatory agencies.[52]

The EFSA's Weak Arguments in Favor of Bisphenol A

But how was this possible? The question kept nagging at me as I drove down the Italian highway linking Bologna to Parma on January 19, 2010. That day, I had a meeting with four EFSA representatives, including Alexandre

Feigenbaum, the head of the Panel on Food Contact Materials, Enzymes, Flavourings and Processing Aids (CEF). Beforehand, I had carefully reviewed the recommendation on BPA published by the European body in November 2006: "The results of the studies reporting low-dose effects are in contrast to the results of studies using protocols [. . .] according to internationally recognized guidelines and performed in compliance with Good Laboratory Practice (GLP). None of these studies [. . .] showed evidence of low-dose effects of BPA in rodents (down to 0.003 mg/kg bw/day by oral exposure)."[53]

Clearly "good laboratory practice" covers all manner of sins. In Chapter 9, I explained how such practices were promoted by the OECD, as well as by the FDA and the EPA, following revelations at the end of the 1970s of behavior that was at best lax, if not fraudulent, on the part of large private laboratories conducting industry studies. Searle studies on aspartame offer a perfect example (see Chapters 14 and 15). However, as stated by Frederick vom Saal and his twenty-nine co-authors in the article cited above ("Why Public Health Agencies Cannot Depend on Good Laboratory Practices as a Criterion for Selecting Data"): "this misconduct was possible because their data usually do not go through the rigorous, multi stage scientific review that is normal for academic data funded by federal agencies and published in the peer-reviewed literature. The lack of these safeguards from academic science had enabled fraud."[54]

In concrete terms, "good laboratory practice" consists of a regulated roadmap according to which scientists conducting studies for regulatory and commercial purposes must scrupulously report all the stages and data of their research, in order to facilitate controls as needed. But all this recording and archiving (on the whole very bureaucratic work) "does not guarantee validity of scientific results," and "specifies nothing about the quality of the research design, the skills of the technicians, the sensitivity of the assays, or whether the methods employed are current or out-of-date."[55] "In our lab, we never have recourse to good laboratory practices because the regular inspections associated with them are too expensive," Ana Soto explained to me. "What's incredible is that the system that was put in place to avoid fraud by private labs has now been turned against university labs—ones already subjected to draconian demands in order to finance their research! And that's why *all* the studies I conducted on bisphenol A, which were *all* published in scientific literature, were *all* rejected by the EFSA!"

"Why did you reject Ana Soto's studies?" I asked Alexandre Feigenbaum at the start of our interview—one recorded by my film crew (as well as by the European body's three representatives looking over my shoulder), as were my other three interviews at the EFSA.

"They simply didn't meet the criteria on study quality," the expert answered. "It's possible that . . . these are isolated effects that were observed: How can you be certain that what you're observing, either in a test tube or on a limited number of animals, has implications for human health? We're obligated to use valid studies that have been accepted by the scientific community. And you know very well that Ana Soto's studies haven't been."

"And what about Frederick vom Saal's studies?" I continued, preferring to ignore the enormity of what I had just heard.

"Mr. vom Saal has been trying to convince the scientific community to take his studies into account for fifteen years. And he hasn't been successful: all the national and international agencies charged with risk assessment, be it the FDA, be it in New Zealand or Japan, the BFR in Germany or the FSA in England, they all agree with our method of risk assessment and with the ADI that we have set."

"How do you explain the fact that the EFSA hasn't taken into account the hundreds of university studies indicating the effects of bisphenol A at doses much smaller than the ADI?" I insisted, growing increasingly discouraged.

"It's clear that we can see effects in the majority of those studies, but we don't know what such effects signify for human health," answered the European expert after a long, incomprehensible monologue that I prefer to spare the reader. "How can an agency responsible for providing recommendations on consumer safety base itself on studies that haven't been validated, or that can't be reproduced?"

"The EFSA used two studies to set the ADI for BPA: those conducted by Rochelle Tyl. What do you think of the fact that she gave three different versions of her mice's age in order to justify their abnormal prostate size?"

"Excuse me? Can you repeat the question?"

"Rochelle Tyl conducted two studies, funded by industry, which were used as a reference by the EFSA to set the ADI of BPA. Yet, in the second study, the prostate size of the control group mice was abnormally large, given the animals' age, which was three months according to what Rochelle Tyl published. To justify this anomaly, she later modified the mice's age, doubling it. Does this comply with 'good laboratory practice?'"

"What you're asking is, does the fact that the mice were six months old rather than three months old completely call into question the study's validity? Is that it?"

"Yes!"

"Can you stop the camera? I want to consult with my colleagues."

"I can't answer your question," Alexandre Feigenbaum finally responded, before changing the subject. "I did some research before you got here: in 2009 alone, there were over one thousand studies on bisphenol A. Some leaned in the direction you mentioned, other agreed entirely with the position endorsed by the EFSA. So, effectively, if you meet scientists who lean in the direction of the low dose effects camp, they might be able to convince you."

"Did you say 'low dose effects camp?' Do you think that the effects of low doses stem from an ideological position and that they aren't supported by anything scientific?"

"They come from schools of thought, yes, but I assure you that they don't represent the majority of the scientific community. Do you think that we can base an opinion that will have a major impact on public health on hypotheses or on unconfirmed data? It's just not possible . . ."

"Ignoring This Data Is Not Reflective of a Scientific Attitude"

"How do you explain that the French AFSSA, or the European EFSA, or even the FDA cling to the ADI of 50 µg/kg even though hundreds of studies on bisphenol A show effects at much lower doses?" Linda Birnbaum, the director of the National Institute of Environmental Health Sciences (NIEHS), smiled at my question during our meeting in her office on October 26, 2009, in Research Triangle Park, North Carolina. Her nomination to the head of the NIEHS by President Barack Obama ten months earlier had been widely applauded in the United States by those who are fighting for public health and the environment to (once again) become a true national concern.

"Why?" she responded, clearly searching for the right words. "Because those agencies haven't studied the new data—that's the problem. Certain regulatory agencies are very slow to adapt to the new science. And yet, throughout the past few years, an enormous quantity of data published in the scientific literature indicates that bisphenol A produces effects on developing organisms at extremely low levels of exposure. I think that ignoring this new data is not reflective of a scientific attitude . . ."

Linda Birnbaum's frankness greatly surprised me, as I was not expecting the director of the largest public research organization in the United States to make waves at the regulatory agencies, even if she does have a reputation as a rigorous scientist who is uncompromising when it comes to professional ethics. A renowned toxicologist—at the time of her nomination to the head of the NIEHS, she was president of the International Union of Toxicology— she led the Experimental Toxicology Division of the Environmental Protection Agency for sixteen years. In 2007, she signed a "consensus statement" on BPA, along with thirty-seven other scientists, which was supported by the NIEHS; this statement summarized three days of work, organized on November 28–29, 2006, in Chapel Hill, North Carolina. Participants were divided into five panels and evaluated seven hundred articles published in scientific literature. Their findings were incontrovertible: "The wide range of adverse effects of low doses of BPA in laboratory animals exposed both during development and in adulthood is a great cause for concern with regard to the potential for similar adverse effects in humans," the authors concluded. "Recent trends in human diseases relate to adverse effects observed in experimental animals exposed to low doses of BPA. Specific examples include: the increase in prostate and breast cancer, urogenital abnormalities in male babies, a decline in semen quality in men, early onset of puberty in girls, metabolic disorders including insulin resistant (type 2) diabetes and obesity, and neurobehavioral problems such as attention deficit hyperactivity disorder (ADHD)."[56]

One year later, Linda Birnbaum participated in the drafting of an extensive report published by the very official National Toxicology Program (NTP) (in a preliminary version in April 2008, then as a monograph in September 2008), which had asked an expert panel to evaluate the toxicity of BPA. Curtailing the industry's waffling, the panel notes that "biomonitoring studies [see Chapter 19] show that human exposure to bisphenol A is widespread" and that "the highest estimated daily intake of bisphenol A in the general population occurs in infants and children." In the conclusion, the authors cautiously acknowledge, "*some* concern for effects on the brain, behavior, and prostate gland in fetuses, infants, and children at current human exposures, [. . .] [and] for effects on the mammary gland and an earlier age for puberty for females."[57]

Admittedly, the tone was very measured, but the Canadian government nonetheless responded immediately: shortly after the NTP's preliminary report was published, it announced the immediate suspension of the sale of any baby bottles containing BPA (note that in Canada, the product's ADI is not 50 µg/kg, but 25 µg/kg). At the same time, Health Canada (the federal department responsible for health) also published a preliminary report on BPA that would drive the point home. It confirmed the widespread contamination of the environment: "Bisphenol A is present in media to which there is no direct release, such as sediment and groundwater. This implies the substance remains sufficiently long in the environment to move from its point of release into other environmental media. [. . .] Bisphenol A is acutely toxic to aquatic organisms. It can also impact the normal development of individual organisms and influence the development of their offspring, with demonstrated adverse effects on reproduction in earthworms, growth in terrestrial plants, and development in mammals and birds. [. . .] As such, it is proposed that bisphenol A be considered as a substance that may constitute a danger in Canada to human life or health."[58]

The Canadian health authority's pioneering decision to de facto ban plastic baby bottles made with BPA was reinforced by the publication in summer 2008 of two studies by Xu-Liang Cao, a Health Canada researcher, which revealed the contamination of canned liquid infant formula, as well as of the contents of baby bottles heated to 70°C.[59] For baby bottles, the substance's migration to the milk varied from 228 to 521 µg/l. On its website, the French Environment Health Network (Réseau environnement santé français, RES) provides a very explicit calculation: "If we assume a volume of 0.5 liters for a one-year-old infant weighing on average 9 kg [20 pounds], the maximum daily dose would be 260 µg/l, or when related to weight: 260/9 = 28.9 µg/kg per day." This "consumption" (*uniquely via a feeding bottle*) certainly falls below Europe's questionable ADI of 50 µg/kg, but is much higher than the risk levels identified by Ana Soto and Carlos Sonnenschein in their study that observed effects on the mammary glands of mice exposed in utero to 250 nanograms of BPA: if regulatory agencies took this study as a reference, they would obtain an ADI less than 2.5 nanograms, or 0.0025 µg per kilogram body weight—in other words, 20,000 times lower than the current ADI.

"Forget all the scientific calculations," Linda Birnbaum told me in a calm, controlled voice, "and let's be practical. I think there is sufficient proof indicating that BPA has the potential to cause harmful effects, particularly during the extremely sensitive period of fetal development. So, if I was a young mother and I was feeding my baby with a bottle, I wouldn't want there to be BPA in that bottle . . ."

BPA-Based Plastic Baby Bottles: Regulatory Agencies' Deceptive Reasoning

"I remind you that the precautionary principle only applies in the absence of a reliable study. As it happens, reliable studies do exist; they conclude that, given the current state of scientific knowledge, baby bottles made with bisphenol A are innocuous [. . .] Those studies were confirmed by all the major health agencies." On March 31, 2009, before the French Parliament, Health Minister Roselyne Bachelot responded to Jean-Christophe Lagarde, the moderate deputy from Seine-Saint-Denis, who had requested that the French government, like in Canada, apply the precautionary principle to—at the least—baby bottles containing bisphenol A. After forcefully affirming that "the precautionary principle is a rational principle and in no case, an emotional one," the imperturbable minister proclaimed: "The Canadian authorities decided on their ban under the pressure of public opinion; however, that decision was in no way founded on a serious scientific study." It's hardly surprising that these regrettable comments would forever tarnish Roselyne Bachelot's image (note that several months later, the minister would throw herself headlong into the H1N1 flu vaccine catastrophe[60]).

In the minister's defense, she was undoubtedly poorly informed by the French and European "experts" responsible for the BPA dossier. In the wake of the Canadian ban, the European Union asked the EFSA to produce new recommendations for BPA use in feeding bottles. The main question was to verify if the breakdown mechanisms of BPA[61] taking place in the bodies of pregnant women, fetuses, and infants (in other words, the most sensitive populations, according to American and Canadian reports) sheltered them from harmful effects. Their results would leave a good number of scientists whom I interviewed on the subject speechless, including Linda Birnbaum, Frederick vom Saal, and toxicologist André Cicolella, spokesman for the RES; the European authority's "CEF Panel" ultimately concluded that "the exposure of a human fetus to BPA would be negligible."[62]

In order to justify their surprising conclusions—which ran counter to the findings of multiple studies conducted on rodents and monkeys, not to mention the Chapel Hill statement and the NTP monograph—the European experts relied on a comparison between BPA and paracetamol, on the basis that the two molecules presented similar structures and, therefore, the detoxification mechanisms in a fetus or newborn must be alike. Although there is no trace of this haphazard reasoning in North American reports or in international scientific literature, the EFSA's argument was blindly adopted by the AFSSA which, in a "recommendation on bisphenol A in polycarbonate feeding bottles likely to be heated in a microwave," published in October 2008, concluded that there was no justification for any "particular usage warnings."[63]

"Fetal exposure via contamination by the mother is negligible," stated Marie Favrot, director of nutrition risk assessment at the AFSSA, during a conference organized on June 5, 2009, at the French National Assembly by RES and Gérard Bapt (president of the environmental health group at the National Assembly). "Of course, these studies weren't conducted with BPA, but we relied on studies conducted with paracetamol, which has structural similarities and notably uses the same detoxification metabolism."

The Need for a Paradigm Shift

"This argument is absolutely ludicrous," said André Cicolella during our meeting in his Paris home on February 11, 2010. "If I had an intern who brought me a report with this kind of hairsplitting, he'd leave with a kick in the backside. It's contrary to everything we're taught and to the very foundations of toxicology. The structures of bisphenol A and paracetamol are clearly very different. They certainly both have a phenolic core, composed of an OH group added to a benzene ring core, but that's all! With this kind of reasoning, any substance with a benzene core should be considered as carcinogenic, which would be completely idiotic and would, quite rightly, provoke outcry from the chemical industry!"

"I remember that you got upset on several occasions during the June 5 conference," I said.

"Yes! There are certain arguments that run in circles and that I just can't listen to anymore, like the one voiced by a plastic industry representative," André Cicolella explained.

Indeed, during the conference, Michel Loubry had cheerfully proposed: "I'm going to take you on a little trip around the world to show you what we, the manufacturers, need—meaning, that is, health agencies' recommendations authorizing us to put different products on the market." Loubry, who is the regional director for Western Europe at Plastics Europe, "the official voice of the European plastics manufacturers," according to the organization's website[64]—added, "The current consensus among health agencies in the United States, Canada, Europe, and Japan is that current levels of BPA exposure in food-related use don't present any risk for human health, including children and babies."

"In the same way, not so long ago, all the agencies agreed that asbestos wasn't a problem," André Cicolella had retorted. "But at the time, everyone said 'where are the victims?' Today, there are three thousand deaths [per year] and there will be tens of thousands twenty years from now . . . So the real question is: Do we wait forty, fifty or sixty years to be sure? Or do we apply the precautionary principle, given the convergence of animal studies, on all the species tested, i.e., mice, rats, and monkeys—data on monkeys are extremely telling . . ."

"The problem," interjected Pascale Briand, director of the AFSSA, "is that we can't protect our fellow citizens correctly on the basis of emotion . . ."

"But how can you talk about 'emotion?'" asked André Cicolella angrily. "How can you talk about emotion, faced with this collection of scientific data?"

"During the conference, Pascale Briand maintained that the AFSSA expert panel was conducting, I quote, an 'independent and impartial scientific evaluation.' What do you think?" I asked the RES spokesman during our meeting in his home.

"The agency's approach is anything but scientific!" he responded. "Otherwise, how can it explain that it based its BPA standard on two very suspect studies, whereas over five hundred serious studies have been published? How can it explain that, on a subject as controversial as bisphenol A, all the recommendations have been unanimous?[65] The root of the problem is the ethics behind these expert assessments. How can a system of expertise that should be constructed in a way to protect public health be perverted to such a degree? Have you read the 'declarations of interest' from the AFSSA or EFSA experts?"

I had read them. Among the experts who submitted the AFSSA's October 2008 recommendation were Jean-François Régnier, who works for Arkema, a polycarbonate panel manufacturer, and Frédéric Hommet and Philippe Saillard, who have contracts with the Technical Center for the Conservation of Agricultural Products (Centre technique de la conservation des produits agricoles, CTCPA). As for the EFSA panel that produced the 2006 and 2008 recommendations, it included at least four members affiliated with the International Life Sciences Institute (ILSI) (see Chapter 12) and one expert, Wolfgang Dekant, who is affiliated with multiple chemical companies, like RCC and Honeywell, etc.

"What's at stake with bisphenol A?" I finally asked André Cicolella.

"If BPA has become an emblematic substance, it's because it embodies the need for a paradigm shift in the assessment of chemical products," he explained. "Current regulation relies on concepts from the 1970s that are completely ineffective for substances like endocrine disruptors. We absolutely have to change our methods or our interpretive framework. Generations of toxicologists have been trained with the notion that 'the dose makes the poison'; and yet, we now see that for numerous substances, it's the period—and sometimes even the day—that makes the poison. For example, testicle formation occurs on the forty-third day of pregnancy: on that day, a pregnant woman is better off avoiding exposure to molecules that have a testicular impact. What's more, the current system is flawed because it doesn't take into account our manifold and constant exposure to hundreds of chemical molecules. Assessments are done independently, molecule by molecule, whereas, in real life, we're subjected to mixtures that can form veritable chemical bombs . . ."

19

The Cocktail Effect

In view of the likely irreversibility of some of the initiated processes, caution is
the better part of bravery and surely a command of responsibility.

—Hans Jonas

Tyrone Hayes arrived at the Ninth Annual Symposium on the Environment
and Hormones that took place on October 20–24, 2009, in New Orleans
wearing a colorful shirt, his dreadlocks pulled back into a ponytail. "I'm
going to tell you about the impact of endocrine disruptors on real life and, of
course, about atrazine and frogs," he said with his legendary cheerfulness,
making the whole audience smile. The well-padded fifty-year-old is one of
the best-known biologists at the University of California, Berkeley, not to
mention a perennial troublemaker for Syngenta, the Swiss chemical, agribusiness, and pesticide giant.[1] With annual turnover of $11 billion (in 2009)
and branches in ninety countries, the company primarily produces the insecticide Cruiser, which is suspected of being partially responsible for the excess mortality rate of bees[2] (see Chapter 6), as well as atrazine, the herbicide
sprayed on my parents' farm when I was born (see Chapter 1).

Atrazine: A "Powerful Chemical Castrator"

During the New Orleans symposium, Tyrone Hayes evoked one of his latest
studies showing that atrazine, an agricultural poison, provoked mechanisms
characteristic of breast and prostate cancers in human cells exposed to doses
similar to those found in the environment.[3] "You've all heard the good news,"
he exclaimed. "The Environmental Protection Agency [EPA] has announced
that it will reexamine atrazine's scientific file! Let's hope it will end up banning it just like Europe did five years ago!" Though the herbicide was banned

by the European Union in 2004,[4] it is still massively used throughout the United States, where approximately forty thousand tons are spread on countless farms growing crops such as corn, sorghum, sugar cane, and wheat every year.[5] Lauded as the "DDT for weeds"[6] when it was put on the market in 1958, today atrazine is the principal contaminant of American surface and ground waters, much like in the majority of European countries (with France in the lead), despite the ban.[7]

Two weeks before the New Orleans symposium, Lisa Jackson, the EPA director nominated by President Barack Obama in January 2009, had effectively announced that the agency would "conduct a new evaluation of the pesticide to assess any possible links between atrazine and cancer, as well as other health problems, such as premature births."[8] "This is a dramatic change," said Linda Birnbaum, director of the National Institute of Environmental Health Sciences (NIEHS) (see Chapter 18). "There is growing evidence that atrazine could be a hazard to human health. This is a strong signal that the world is changing in regards to some of the most widely used chemicals."

If there is one scientist who battled for atrazine's ban in the United States, it is inarguably Tyrone Hayes, even if (as he explained to me during our meeting in his Berkeley laboratory on December 12, 2009) "this battle wasn't a personal decision, but was imposed by events." In 1998, he was contacted by Novartis (the company became Syngenta two years later after its merger with AstraZeneca), which offered him a "handsomely paid" contract to "verify if atrazine [was] an endocrine disruptor," as Theo Colborn and her co-authors note in *Our Stolen Future* (see Chapter 16). For the industry, the matter was quite serious as, seven years earlier, a U.S. Geological Survey report had revealed that "atrazine exceeded drinking water standards in 27 percent of the samples"[9,10] taken from the Missouri, Mississippi, and Ohio Rivers and tributaries. What's more, in the 1980s, two studies conducted on mice[11] and rats[12] had indicated that exposure to the herbicide brought on breast and uterine cancers, lymphomas, and leukemia. The International Agency for Research on Cancer (IARC) judged the results sufficiently convincing and decided to classify atrazine as "possibly carcinogenic to humans" (group 2B) in 1991.[13] As a result, the EPA, leaning on the Safe Drinking Water Act, decreased the atrazine standard to a maximum of 3 μg/l of water, or 3 ppb (parts per billion). In 1994, three studies established a link between rodents' exposure to atrazine and mammary tumors.[14] Then in 1997, one year after the publication of *Our Stolen Future*, an epidemiological study carried out in several rural Kentucky counties found a significant excess of breast cancer among the most exposed women (in correlation with the level of water contamination and proximity of the home to corn cultivations).[15]

Thus began Novartis' great strategic era. Its first tactic was tremendously effective, and brought about IARC's downgrade of atrazine from group 2B to group 3 (not classifiable) in 1999. To justify the surprising decision, the UN agency experts relied on a line of reasoning I described in Chapter 10: "the mechanism by which atrazine increases the incidence of mammary gland tumors in rats is not relevant to humans."[16]

Novartis' second effort revolved around Tyrone Hayes, a brilliant biologist (and the youngest tenured professor at Berkeley) and an amphibian enthusiast who named his daughter Kassina, after an African frog species. "Frogs are my entire life," he explained to me in his laboratory, surrounded by thousands of jars filled with amphibians. "I grew up in the country, in South Carolina, and I was always fascinated by their ability to metamorphose—from an egg to a tadpole, and then to an adult frog."

"Why do frogs provide an interesting model from which to study the effects of endocrine disruptors?" I asked.

"They're a perfect model!" responded the biologist. "First of all, because they're very sensitive to hormones that enable the activation of genes necessary for their various metamorphoses; and then, because they possess exactly the same hormones as humans, such as testosterone, estrogen, or the thyroid hormone."

"How did you go about your study?"

"I should clarify that this process was closely monitored by Novartis, and then Syngenta. Initially, we raised frogs from the *Xenopus laevis* family in water reservoirs to which we had added different doses of atrazine, similar to what's found in field drainage ditches and up to thirty times lower than the existing U.S. standard (3 ppb)—meaning levels that a human being might find in tap water. To give you an idea, that's the equivalent of a grain of salt in a reservoir of water. We observed that atrazine reduced the size of the larynx, which is the voice box in the males. Since they sing to seduce the females, this meant they were sexually handicapped. We also observed very low levels of testosterone among the adult males; some of them were hermaphrodites, which means they had both ovaries and testes. In certain cases, the males became homosexuals and coupled with other males, adopting a feminized behavior; sometimes they had eggs in their testes instead of sperm. Ultimately, atrazine acted as a very powerful chemical castrator that is biologically active at 1 ppb, and even 0.1 ppb."

"Do you know if wild frogs were presenting the same problems?"

"That was actually the second stage of our study: we set out with a refrigerated truck across Utah and Iowa where we collected eight hundred young leopard frogs (*Rana pipiens*) in ditches alongside fields, near golf courses or riverbanks. We dissected them and observed exactly the same dysfunctions that we had seen in the laboratory frogs. It was very upsetting, and that's when I understood that the decline in North American and European frog populations was due to pesticide contamination that affected their reproduction systems."

"How do you explain this phenomenon?"

"Atrazine stimulates an enzyme called 'aromatase,' which transforms the masculine hormone, testosterone, into the female hormone, estrogen. As a result, the estrogen produced by the aromatase leads to the development of female organs, like ovaries or ovules in the testes. However, the levels of aromatase are also linked to the development of breast or prostate cancers. An epidemiological study conducted in a Syngenta atrazine factory in Louisiana, published in 2002, actually indicated a significant excess of prostate cancer among workers."[17]

"How did Syngenta react?"

"Ah!" sighed Tyrone Hayes. "I was very naïve at the time! At first, the company asked me to repeat my study to verify that I would obtain the same results. They offered me 2 million dollars for it and, initially, I accepted . . . Then, I understood their strategy—they wanted to drag things out to gain some time and stop me from publishing. I finally terminated the contract and I published my results in 2002.[18,19] After that, it was war! And I have to say that I never could have imagined that it would be so violent: Syngenta wrote to the dean of UC Berkeley, used the press to discredit me,[20] added a link on its website to junkscience.com, Steven Milloy's site, where I ended up on the list of 'junk scientists' [see Chapter 8]. Today, it makes me laugh because I know that appearing on that list is proof that I was doing good work! They then paid scientists to conduct new studies that, of course, were unable to reproduce my results. Their goal was to create doubt, and it worked, at least in the United States where the EPA ultimately renewed its approval of atrazine in 2007."

In fact, in October 2007, the EPA produced a report that concluded: "Atrazine does not adversely affect amphibian gonadal development: no additional studies are required."[21] Meaning that the unstoppable machine created to destroy any and all uncomfortable truths had, once again, worked marvelously.

At the height of the drama, Tyrone Hayes published an article in *BioScience* in which he deciphered the immutable cogs that I have described throughout this book: manipulations of science, the funding effect, defamation campaigns, the public authorities' complacency, media brainwashing, etc.[22]

Pesticide Mixtures Enhance Individual Effects

"Industry has increased efforts to discredit my work, but my laboratory continues to examine the impacts of atrazine and other pesticides on environmental and public health," writes Tyrone Hayes on his website, ironically named atrazinelovers.com. "My decision to stand up and face the industry giant was not a heroic one. My parents taught me, 'Do not do the right thing because you seek reward . . . and do not avoid the wrong thing because you fear punishment. Do the right thing, because it is the right thing.'"

"My quarrels with Syngenta marked a turning point in my career," explained the Berkeley researcher, "because that's when I began to specialize in a little-explored field: the effects of pesticide mixtures. The leopard frogs that I collected from fields in the Midwest weren't exposed uniquely to atrazine, but rather to a combination of several substances. However, scientific literature is generally only interested in the toxicological effects of pesticides at relatively high doses (in the realm of parts per million), but rarely in low doses and even less so in mixtures of low doses, like those that exist in our everyday environment, namely in tap water and the fruits and vegetables we eat."

This "omission"—on the whole quite surprising, and which also characterizes the regulatory system of chemical products—was similarly highlighted by the U.S. Geological Survey in a 2006 report that is all the more remarkable because it openly describes the pollution of America's surface and ground waters: "Because of the widespread and common occurrence of pesticide mixtures, particularly in streams, the total combined toxicity of pesticides in water or other media often may be greater than that of any single pesticide compound that is present,"[23] writes Robert Gilliom, the principal author. He adds that their findings indicate that the study of mixtures should be an absolute priority.

And so Tyrone Hayes once again hopped aboard his refrigerated truck to cross Nebraska and collect thousands of liters of "chemical brew" flowing through industrial cornfields. Once he returned to Berkeley, he identified nine recurring molecules: four herbicides, including atrazine and alachlor (or

Lasso, which caused Paul François's poisoning; see Chapter 1), three insecticides, and two fungicides.[24] When I met him, he was working on another
mixture composed of five pesticides, including Roundup and chlorpyrifos.
The scientist conducted each study in two ways: he raised frogs in reservoirs
filled with the "brew" from the fields, as well as in the mixture he reconstituted in his laboratory in order to compare the results. And in both cases, the
results were very troubling.

"When we mixed the substances, we noticed effects we hadn't seen with
products taken separately," he explained. "First off, we observed weakened
immune systems in the frogs due to thymus disorders, which meant they
were more susceptible to, for example, meningitis, and that they died of diseases more often than frogs in the control group. That immune weakness
can explain, in part, the population declines. But added to it is the disruption
to reproductive systems, similar to what I observed with atrazine on its own.
Finally, the mixtures had an effect on metamorphosis time and larva size.
And yet, the doses we were using were up to a hundred times lower than the
residue level authorized in water."

"What can we conclude about humans from that?"

"We have no idea!" responded Tyrone Hayes. "But what's incredible is that
the pesticide evaluation system has never taken into account the fact that
substances can interact or accumulate, or even create new molecules. It's
even more surprising given that pharmacists have known for centuries that
it's imperative to avoid mixing certain medications, at the risk of exposing
oneself to serious side effects. For that matter, when the FDA authorizes a
new drug, it always insists that the medicinal contraindications be detailed
in the user instructions. Clearly, this kind of precaution is difficult to implement for pesticides. Imagine the EPA explaining to farmers: you can use pesticide A, as long as your neighbor at the farm next door doesn't use pesticide
B or C! It's impossible! And, if it's impossible, it means that these products
have no business in the fields. In the meantime, knowing the 'chemical body
burden' that characterizes every citizen in industrialized countries, we can
effectively fear the worst."

The "Chemical Body Burden": The "Chemical Brew"
Poisons Us All

I remember very clearly the moment I discovered the expression "chemical
body burden," which paralyzed me. It was in October 2009, on a plane taking

me from New Orleans to Palm Beach, and I was reading *The Body Toxic*, a book by my American colleague Nena Baker that I had bought the previous night. In it, she describes how the concept was created in the beginning of the 2000s by the Centers for Disease Control and Prevention (CDC) in Atlanta, the body in charge of health monitoring in the United States. At the time, the CDC was conducting the first biomonitoring program in the world, aimed specifically at evaluating the "chemical body burden" of the American population. Equipped with an ultramodern laboratory, the CDC measured the residues of twenty-seven chemical products in the blood and urine of 2,400 volunteers, carefully chosen to represent the entire American population (age, sex, ethnic, geographic, and professional origin).

The first report, published in March 2001, was followed by another in 2003 that examined 116 products, then a third in 2005 (148 products) and, finally, a fourth in 2009 (212 products). As soon as I returned to my Palm Beach hotel room, I consulted the fourth report available on the CDC website.[25] I learned that the 212 chemical molecules sought out had *all* been found in the quasi-totality of 2,400 volunteers tested (in their urine and their blood): bisphenol A (BPA) topped the list, followed by polybromo diphenyl (PBDE, a flame retardant), perfluorooctanoic acid (PFOA, used in nonstick pan coating), and a good number of pesticides (or their metabolites), such as alachlor (Lasso), atrazine, chlorpyrifos, as well as organochlorine insecticides from the "dirty dozen"—like DDT (and its metabolite DDE)—which were still present despite being banned.

A true "chemical brew," to borrow Tyrone Hayes's term, revealed thanks to the tenacity of Dr. Richard Jackson, the director of the National Center for Environmental Health at the CDC from 1994 to 2003, which launched the biomonitoring program and published its first report. "I took a fair amount of criticism, but I resisted," he told Nena Baker. "I wanted the larger community and the research community to have it in their hands and use this data the way a doctor would use lab data in making decisions about a patient. The complaint from chemical manufacturers was that the report was just going to scare people. No, you never scare people with real information. You scare them with no information or bad information."[26] Jackson also showed rare courage in the face of what I can easily imagine to be multiple sources of pressure, including from his higher-ups. His observations are true, as evidenced by the fact that the United States is the only country that regularly—every two years—conducts a national biomonitoring program.[27]

No such thing exists in Europe (and especially not in France), where an ostrich-like mentality persists among authorities who choose not to look for anything (so not to find anything), and thereby justify their inaction. The sole initiatives came from nongovernmental organizations (NGOs), like the World Wide Fund for Nature (WWF), which published the results of a far-reaching study in April 2004, named "Detox." The NGO had obtained blood samples from thirty-nine European deputies, fourteen health or environmental ministers, as well as three generations from one family in every European Union state. The findings were on par with those of the CDC: 76 toxic chemical substances were found in the European deputies' blood (of the 101 sought after), which belonged to five broad families: organochlorine pesticides, polychlorinated biphenyl (PCB), brominated flame retardants, phthalates, and perfluorinated compounds (PFCs) like PFOA. On average, each deputy contained a cocktail of 41 toxic products, composed of persistent substances (which do not decompose in nature) or bioaccumulative ones (which accumulate in the body). Ironically, the "grand prize" was attributed to Marie-Anne Isler-Béguin, a European "eco" deputy, who presented a record 51 substances, including a noteworthy quantity of PCB.[28] "It's incredible that it was an NGO that had to carry out this kind of study so that we could have benchmark data," said Isler-Béguin after admitting that she had been devastated to discover her "chemical body burden." "It's up to the authorities and especially the European Commission to carry out such studies."[29]

As for the European ministers, a total of 55 substances were found in their blood, with an average of 37 substances per person (one of which presented up to 43 different residues). Similar results were obtained from the European families (over three generations), like the Mermet family living in the Bretagne region of France: scientists found traces of 34 chemical substances (of the 107 looked for) in blood samples from Liliane Corouge, the grandmother; 26 from the mother, Laurence Mermet; and 31 from the son, Gabriel Mermet.

One year later, in September 2005, WWF and Greenpeace published a new report, this time based on blood taken from forty-seven pregnant or nursing women and the umbilical cords of twenty-two newborns.[30] The findings were sadly predictable: traces of phthalates, BPA, brominated flame retardants (used in furniture, rugs, or electrical equipment), PCB, organochlorine pesticides (DDT, lindane), synthetic musks (present in air fresheners, detergents, beauty products), PFC, and triclosan (used in some toothpastes) were found in the majority of blood samples.[31] "How likely is it that the chemicals found at the concentrations reported are causing adverse effects on the growth and development of the unborn child?" asked the report's authors. "We cannot be sure. [. . .] Additional research is certainly necessary. However, it is already possible to conclude that exposure of the developing fetus to a continuous low dose of a complex mixture of persistent, bioaccumulative and bioactive chemicals is a serious cause for concern. All possible steps should be taken on a precautionary basis to avoid such exposure in the womb. This can only be done by controlling the exposure of the mother to these chemicals— and that means eliminating particularly hazardous substances from the everyday products we use and, ultimately, from the environment in which we live."

A Pesticide Cocktail in the Umbilical Cord

The WWF/Greenpeace recommendation mainly concerned pesticides, traces of which have been discovered in newborns' meconium (the first stools after birth), as demonstrated in a 2001 study conducted in New York by a Columbia University team that detected a cocktail of chlorpyrifos, diazinon (two insecticides known for their effects on the neurological system), and parathion.[32] Two years later, the same team analyzed the umbilical cord plasma of 230 newborns and blood samples taken from their mothers, who were all living in working-class neighborhoods. The researchers observed the presence of twenty-two pesticides, including eight organophosphates such as chlorpyrifos, diazinon, bendiocarb, propoxur, dicloran, folpet, captafol, and captan, present in 48 percent to 83 percent of the samples. They noted a strong correlation between the levels of pesticide residues (and that of their metabolites) in a mother's plasma and that of her newborn and concluded "the pesticides are readily transferred to the developing fetus during pregnancy."[33]

The pesticide saturation of pregnant women appears widespread and it seems to affect urban areas as much as it does rural regions. A study conducted in the 2000s in the Bretagne region of France (on a cohort called "Pélagie") revealed the presence of 52 molecules in the urine of 546 pregnant women, 12 of which belonged to the triazine class (like atrazine), 32 to organophosphates (like chlorpyrifos and chlorpyrifos-methyl), 6 to the amides class, and 2 to carbamates. "The pesticide residues are generally multiple," the authors underlined in 2009. "And their impacts, individual or combined, on the fetus and its development are still uncertain in the epidemiological literature. They will be evaluated in the near future in the Pélagie cohort."[34]

If the "chemical body burden" of pregnant women and babies is particularly troubling, the same goes for children, whose pesticide saturation levels are proportionally much higher than those of adults. Numerous studies have demonstrated this (it is impossible to list them all), including one conducted in Minnesota, a state characterized by intensive farming as well as large urban areas: published in 2001, the study revealed that 93 percent of urine samples collected from ninety children from both urban and rural areas displayed a cocktail of atrazine, malathion, carbaryl, and chlorpyrifos residues.[35] As we have seen, the infamous chlorpyrifos (see Chapter 13) pops up like clockwork in every biomonitoring study: according to the CDC's second report, it is one of the pesticides whose residue levels regularly surpass authorized standards, particularly among children tested. In a document discussing the findings of that report, the Pesticides Action Network emphasized that "it would be difficult to make a case that anyone could be more responsible for the chlorpyrifos in our bodies than Dow Chemical Company. Dow developed and was the first to commercialize the pesticide [. . .] and continues to produce and promote the pesticide in the U.S. and internationally, despite strong evidence of significant public health impacts."[36] The organization also called for accountability from pesticide companies, which pollute our bodies through our plates, and threaten the health of our children in particular.

Based on the findings of a study published by Future Generations (Générations futures, formerly MDRGF, see Chapter 1) in December 2010, the potential culprits are manifold.[37] The organization analyzed the daily food intake of a ten-year-old child, which included three typical meals that followed official recommendations—five fresh fruits and vegetables, three dairy products, and fifty ounces of water per day—and a snack (with sweets). An article published in the French newspaper *Le Monde* noted: "The results were damning. One hundred and twenty-eight residues, eighty-one chemical substances, including forty-two classified as possible or probable carcinogens and five substances classified as certain carcinogens, as well as thirty-seven substances likely to act as endocrine disruptors. [. . .] For breakfast, the butter and tea with milk alone contained more than a dozen possibly carcinogenic residues and three substances proven to be certain carcinogens, as well as nearly twenty residues likely to disrupt the hormonal system. The hamburger meat, the canned tuna, and even the baguette and the chewing gum were stuffed with pesticides and other chemical substances. Analyses of tap water revealed the presence of nitrates and chloroform. But it was the salmon steak at dinner that was revealed to be the 'richest,' with thirty-four detected chemical residues."[38]

As François Veillerette, the founder of Future Generations, explained to the same newspaper, "the contaminated cocktails' probable synergetic effects, induced by ingestion, aren't taken into account, and the ultimate risk for the consumer is probably greatly underestimated. As of right now, we know almost nothing about the impact of chemical cocktails ingested orally."[39]

The New Mixture Math: 0 + 0 + 0 = 60%

"I think that we've been extremely naïve in our research and our regulatory system by focusing on a single chemical product at a time, whereas none of us is exposed to a single substance alone," said Linda Birnbaum during our meeting in her NIEHS office. "I think we completely overlooked the effects that can occur. This is particularly true for natural and synthetic hormones. That's why the challenge we have to tackle now is understanding and evaluating the effects that the chemical mixtures surrounding us can cause. But, unfortunately, there are very few labs working on it."

However, the few laboratories that do (exceptionally) address chemical mixtures within the field of toxicology are European, namely Danish and British. The former is led by Ulla Hass, a toxicologist who works at the Danish Institute for Food and Veterinary Research in Soborg, a suburb of Copenhagen. I met her on a snowy day in January 2010. Before beginning our interview, she had me visit her "menagerie," a clinical white room with cages housing Wistar rats used in her experiments. Thanks to the support of the European Union and in collaboration with the Center for Toxicology at the University of London, Hass conducted a series of studies to test the effects of chemical substance mixtures that had anti-androgen effects on male rats exposed in the womb. The first cocktail tested was composed of two fungicides, vinclozolin and procymidone (see Chapter 13), and flutamide, a drug prescribed to treat prostate cancer.[40]

"What is an anti-androgen?" I asked the Danish toxicologist.

"It's a chemical substance that affects the action of androgens, meaning masculine hormones like testosterone," she replied. "But those masculine hormones are critical for sexual differentiation that, in humans, occurs during the seventh week of pregnancy. They allow the model at the start, which is feminine, to develop into a masculine organism. Therefore, anti-androgens can disrupt the process and prevent the male from developing correctly."

"How did you go about your study?"

"First of all, we observed the effects of each molecule separately to try and find, for each of them, a very low dose that didn't cause any effects. Remember that our objective was to measure the potential effects of mixtures, so it was particularly interesting to see if three molecules that had no effect individually would have one when mixed together. And those are exactly the results we obtained. Take, for example, what we call the 'anogenital distance,' which measures the distance between the anus and the genitalia of the animal. It's twice as long in males as in females, and it is due precisely to the role played by androgens during fetal development. If it's shorter in males, it's an indicator of hypospadias, a serious birth defect in the male reproductive organs. When we tested each product separately, we didn't observe any effects. But, when we exposed the male fetuses to a mixture of three substances, we noticed that 60 percent of them later developed hypospadias, as well as serious defects in their sexual organs. Among the defects we observed, there was notably the presence of a vaginal opening among certain males, who also had testes. In fact, they were sexually in between the two sexes, like hermaphrodites."

Then the toxicologist concluded with this phrase that I will never forget: "We have to learn new math when we work on the toxicology of mixtures, because our results are telling us that $0 + 0 + 0$ equals 60 percent of defects . . ."

"How is that possible?"

"Well, we're seeing a double phenomenon: the effects are adding up and they're synergizing to then increase further," explained Ulla Hass.

"What you're saying is terrifying, especially when we know that every European has what's called a 'chemical body burden.' Can what you've observed in rats also occur in our bodies?"

"In fact, the big problem is that we don't really know," sighed Ulla Hass, echoing the remarks made by her colleague Tyrone Hayes. "It's very difficult to understand why this wasn't taken into account earlier. When you go to the pharmacy to buy a medication, it's written on the user instructions to be careful when taking other drugs, because there could be combination effects. That's why it's not surprising that we have the same phenomenon with chemical pollutants."

"Do you think toxicologists need to completely reevaluate their methods?"

"It's clear that in order to evaluate the toxicity of chemical mixtures, and particularly those of endocrine disruptors, we need to move away from the model we've been taught, which dictates that a low dose has a small effect and a large dose has a big effect, with a linear dose–response curve. It's a simple and reassuring model, but it's useless when it comes to numerous chemical molecules. On the other hand, we should develop tools like those put in place by Andreas Kortenkamp's laboratory in London, with which my lab collaborates. After having entered all the chemical characteristics of the three substances we had tested into a computer system, it could predict, using a specific software program, what the effects of the addition and synergy of the molecules would be. It's a very interesting avenue for the future."

The Breast Cancer Explosion Is Due to Synthetic Hormone Cocktails

I, of course, headed to the United Kingdom to meet Andreas Kortenkamp, who heads the Center for Toxicology at the University of London. In a study he published in 2009 with Ulla Hass and their colleagues, the authors conclude: "Evaluations that ignore the possibility of combination effects may lead to considerable underestimations of risks associated with exposures to chemicals."[41] In his book *Risk Society*, Ulrich Beck says the same thing, but in much more radical terms (which I was ready to adopt as my own once I completed my foray into the chemical world): "How does it help me to know that this or that toxin in this or that concentration is harmful or harmless, if I do not know what the reactions the synergy of these multiple toxins provokes? [. . .] Of necessity, people are threatened in their civilizational risk positions not by individual pollutants, but *holistically*. To respond to their forced questions regarding their *holistic* endangerment with tables of acceptable values for individual substances amounts to collective ridicule with consequences that are no longer only latently murderous. It may be that one could make this mistake in times of a general belief in progress. But to stick to it today in the face of widespread protests and statistical evidence of morbidity and mortality, under the legitimate protection of scientific 'acceptable value rationality,' far exceeds the dimensions of a crisis of faith, and is enough to call for the public prosecutor."[42]

On January 11, 2010—once I was out of my "funk"— I met with Andreas Kortenkamp, a German-born scientist and notably the author of a study on breast cancer that he presented to members of the European Parliament on April 2, 2008.[43] According to him, the steady increase in rates of breast cancer, which strikes one woman out of eight in industrialized countries and represents the leading cause of death by cancer among women aged 34–54, is primarily due to chemical pollution.[44] "The rapid increase in breast cancer in developed countries is very shocking," he explained. "It's due to a range of concordant factors that all relate to the role of estrogen in women's bodies: first off, there's the decision to have children later and, for some, not to breast-feed; there's also, to a lesser extent, the use of birth control pills and, evidently, the use of hormone therapies at menopause. In the United Kingdom, it's esti-mated that the use of hormone replacement therapy has caused an excess of ten thousand cases of breast cancer. We can add the genetic factor, but it shouldn't be overestimated: it's believed to represent one mammary tumor in twenty. Everything indicates that the principal factor is environmental, and that it's tied to the presence of chemical agents capable of imitating the female sexual hormone, and whose effects accumulate at microscopic doses."

"Which products do you blame?" I asked, thinking of all the women, in-cluding several close friends, who are suffering or who have died from breast cancer.

"Unfortunately, the list is long," responded Kortenkamp with a disapprov-ing look. "There are certain food additives like preservatives, anti-UV prod-ucts like sunscreen, parabens, and phthalates found in numerous cosmetic products (shampoos, perfumes, deodorants), alkylphenols used in detergents, paints or plastics, PCBs that continue to pollute the food chain; and then countless pesticides, like the DDT accumulating in our environment, fungi-cides, herbicides, and insecticides that all have an estrogenic activity and are found in residue form in our food.[45] In short, a woman's body is constantly exposed to a cocktail of hormones that can act in conjunction, as shown by a Spanish study.[46] What's more, we know that these hormone combinations are particularly dangerous during fetal development and puberty. That's what the distilbene scandal showed us (see Chapter 17), as did the horrific atomic bomb experiment in Hiroshima: the majority of women who developed can-cers were teenagers at the time of the explosion."

"What studies have you conducted in your lab?" I asked.

"We're testing the synergetic effects of synthetic hormones—either estro-genic or anti-androgenic—on cell lines, meaning in vitro, and not in vivo, as my colleague Ulla Hass does. And our results confirm what she observed in rats: the effects of the xenoestrogens, or environmental estrogens, multiply when they're mixed and, in addition, interact with natural estrogen. We talk a lot about the chemical body burden, but it would be interesting to measure the global hormonal burden of women, which should be a good indicator of the risks of getting breast cancer."

"Do you think that regulatory agencies should revise their evaluation system for chemical products?"

"Certainly!" answered the scientist without hesitation. "They need to change the paradigm to integrate the cocktail effect, which is being completely ignored for the time being. The product-by-product evaluation doesn't make any sense and I've noticed that the European authorities are starting to realize that. In 2004, the European Scientific Committee on Toxicity, Ecotoxicity and the Environment clearly recommended taking into account the cocktail effect of molecules with identical modes of action, like environmental hormones.[47] Likewise, in December 2009, twenty-seven European environmental ministers published a communal declaration asking that the effect of combinations, notably of endocrine disruptors, be integrated into the evaluation system of chemical products. That said, the task is enormous. According to estimates, there are currently between thirty thousand and fifty thousand chemical products on the European market, only 1 percent of which have been tested. If there are around five hundred endocrine disruptors among them, that makes millions of possible combinations . . ."

"Meaning the task is impossible . . ."

"I think we need to approach it pragmatically. River fish provide a good indicator of cocktail effects. We need to determine which substances are affecting them the most, and maybe we'll discover that twenty molecules are responsible for 90 percent of the effects. They could then be pulled from the market, as foreseen by REACH regulations, which are heading in the right direction.[48] But in order to do so, there needs to be a strong political will, because the industry's resistance is tremendous."

"Does the cocktail effect also exist for carcinogenic molecules?"

"Everything seems to indicate that it does. Including Japanese studies that combined pesticides that individually had no carcinogenic effect at the dose used in the mixture, but whose effects were multiplied once they were mixed together."

"Does that mean that Paracelsus's principle that 'the dose makes the poison' should be thrown out, including for products other than endocrine disruptors?"

"Unfortunately, that principle is applied very broadly, but no one really understands what it signifies. Fundamentally, of course, there is a link between a product's toxicity and the dose, but that's not the problem. The flaw in the evaluation system comes from the idea of the no-observed-adverse-effect level (NOAEL). In reality, it's important to understand that surrounding this infamous NOAEL is what statisticians call a 'fog' or grey zone. Meaning we're incapable of knowing what happens at plus or minus 25 percent of the NOAEL. There's no experimental study that can solve this fundamental problem. Of course, we could increase the number of animals tested to reduce the size of the 'fog,' but we can never make it disappear completely. According to the official discourse, applying uncertainty or security factors solves the problem, but again, that's completely arbitrary because, once again, we can't be certain of that. And this is particularly true for the toxicology of mixtures, wherein the combined effects of very low doses of products, which appear harmless when taken alone, are impossible to predict with certainty, apart from applying extremely high security factors, which would considerably limit the use of such products."

"Do you think the current system puts children's lives in particular danger?"

"It's clear that fetuses and young children are particularly sensitive to cocktails of chemical products and, notably, endocrine disruptors. That's what the development of childhood pathologies indicates."

A "Silent Pandemic": Children Are the First Victims

It is impossible to conclude this book without talking about the fate of our children, who are the first victims of generalized environmental pollution, to the extent that Philippe Grandjean, professor of environmental health at Harvard University, and his colleague Philippe Landrigan from the Mount Sinai School of Medicine in New York, speak of a "silent pandemic."[49] While their observations relate to the many neurological problems that affect children—autism, attention disorders, hyperactivity, mental retardation— they can also be applied to all the other diseases affecting hundreds of thousands of children born in so-called developed countries, on account of their exposure to chemical poisons filling their environment, including their mothers' wombs.

And yet the "acceptable level magicians"[50] seem determined to ignore the fact that, in the field of toxicology, and contrary to popular belief, "children are not little adults," as emphasized in a well-documented study carried out at the request of the European Parliament.[51] This is an inarguable assertion, given that the cost of air and water pollution, and lead poisoning of children and youths under twenty comes to a hundred thousand deaths per year in Europe (or 34 percent of deaths for each age group).[52] After mentioning the "specific vulnerability" of pregnant women and fetuses, the European deputies underline the "the physiological and behaviour characteristics of infants and children that increase their vulnerability to negative health impacts from pesticides."[53] This is due to "the fact that their bodies are still developing, and that the chemical-based signaling systems used to steer development are vulnerable to disruption when exposed to environmental toxicants."[54]

What's more, "the blood brain barrier is not fully developed until an infant reaches six months, leaving the developing brain far less protected than for older children and adults.[55] Due to less developed detoxification pathways, a child's metabolism is less able to metabolize and eliminate toxicants.[56] In addition, children eat and drink more per kilogram of body weight than adults. [. . .] A specific dose of a pesticide will have a greater impact on a child than on an adult."[57] The authors add, "Children exhibit hands-to-mouth behavior, have shorter statures, play close to the ground and spend increased time outdoors. Children often have diets rich in fruit and vegetables, so increasing their exposure to pesticide residues. In addition, the processing undertaken to produce infant foods tends to result in higher concentrations of pesticide residues. Nursing infants can ingest pesticide residues through breast milk."[58] The European deputies conclude: "Despite the evidence of increased vulnerability of infants and children and the disabling and chronic nature of the resulting health effects, substance-specific data on postnatal developmental toxicity are lacking for many of the currently used pesticides."

All the characteristics described by the European study are also mentioned on the website of the EPA, which in 1996 had to add a security factor of ten to the habitual factor of one hundred (see Chapter 12) used to calculate the acceptable daily intake (ADI) for pesticides, following the congressional decision to amend the federal law on fungicides, insecticides, and rodenticides in order to better protect children. The agency created an office especially dedicated to children's environmental health, and on its website (under the section "Pesticides and Food"), it clearly explains, "why infants and children may be especially sensitive to health risks posed by pesticides" and the diseases they can develop following exposure to chemical products. The first is, of course, cancer, which "is the second leading cause of death among children ages 1 to 14 years of age, with unintentional injuries being the leading cause." The EPA specifies that "Leukemia is the most common cancer in children under 15, accounting for 30 percent of all childhood cancers, followed by brain and other nervous system cancers."

The occurrence of childhood leukemia is all the more unfortunate since

everything seems to indicate that it could be considerably reduced if pregnant women were informed of the role played by pesticides, and especially by
insecticides, in its etiology. In fact, "a dozen recent epidemiological studies
indicate that the use of indoor insecticides during pregnancy doubled, at the
very least, the probability that the unborn child would develop leukemia or a
non-Hodgkin lymphoma," explained Jacqueline Clavel, director of the Environmental Epidemiology of Cancers team at the French Institute of Health
and Medical Research (Institut national de la sante et de la recherche medicale, INSERM),[59] which led one of those studies.[60] In 2009, a team from the
University of Ottawa conducted a meta-analysis of thirty-one epidemiological studies published between 1950 and 2009 that investigated links between
childhood leukemia and parental exposure pesticides. The results were irrefutable: prenatal maternal exposure to household or agricultural insecticides
multiplied a child's risk of leukemia by 2.7, and that risk was then multiplied
by 3.7 following maternal occupational pesticide exposure.[61]

Children Deformed by Pesticides

During filming, there are certain moments that are particularly intense, and
which leave haunting memories that do not fade with time. I often think back
on my visit, in December 2006, to the Tû Dû hospital, in Ho Chi Minh City
(Vietnam), where Dr. Nguyen Thi Ngoc Phuong kept dozens of jars filled
with preserved fetuses deformed by Monsanto and Dow Chemical's Agent
Orange.[62]

Similarly, I will never forget my trip to Fargo, the North Dakota town that
gave its name to one of the darkest Cohen brothers films. When I arrived on
Halloween 2009, it was freezing cold in the nearby Red River Valley, which
looked ready to welcome snow for the long winter ahead, after which the intensive farming of wheat, corn, beets, potatoes, and transgenic soybeans could
start up again. In this region straddling North Dakota and Minnesota, pesticides are generally distributed by plane, as the average farm size surpasses
several hundred acres.

I was to meet Professor Vincent Garry, from the University of Minneapo-
lis, who participated in the Wingspread conference (see Chapter 16) and conducted three studies on the link between exposure to agricultural poisons
and birth defects.[63] They showed a heightened and very significant risk of
anomalies—cardiovascular, respiratory, urogenital (hypospadias, cryptorchidism), and musculoskeletal (deformities of the limbs, or number of digits)—in

farming families in the Red River Valley, *but also* among residents near the rivers. Compared with that of urban populations in North Dakota or Minnesota, the risk was multiplied by two to four, depending on the type of anomaly. When he studied the farming families more closely, Vincent Garry observed that birth defects and miscarriages were more frequent when the child's conception occurred during spring, meaning the period when pesticides are applied (namely Monsanto's Roundup, which Garry demonstrated was an endocrine disruptor). The researcher also noted a shortage of males among children of pesticide users. Together, we visited David and his family, including his young brother who suffered from serious birth defects and cognitive disabilities. I will never forget the family's emotional attentiveness and discomforted silence around the kitchen table when Vincent Garry presented his study's results—findings they had never heard before.

Ten months later, I set off for Chile (see Chapter 3) where, after meeting with seasonal workers who had suffered from acute poisoning, I spoke with Dr. Victoria Mella, an obstetrician gynecologist at the Regional Hospital of Rancagua. In this central province of the Andean country, intensive farming aimed toward exportation has been practiced since the 1980s, relying in large part on pesticides. Mella had observed a dramatic increase in serious birth defects among children born in her hospital throughout the 1980s. In 1990, she drafted a report, based on ten thousand births, in which she described the many anomalies primarily affecting children of seasonal workers who were exposed to pesticides during their pregnancies: "hydrocephalus," "congenital cardiopathy," "upper and lower limb deficiencies," "defects of the urinary system or neural tube," "oral clefts," "spina bifida," "fetal deaths."[64] Deeply troubled by what she was seeing on a daily basis in her exam room, the gynecologist decided to film the babies' tormented bodies in order to have proof she could submit to the public authorities. I will never forget those horrible images of children deformed by man's chemical folly . . .

CONCLUSION

A Paradigm Shift

"We need a new way of thinking if mankind is to survive," said Albert Einstein over fifty years ago. Today, his words resonate like an alarm bell. Everything indicates that we are at a crossroads and that we urgently need "a paradigm shift in how we manage public health," as André Cicolella told me. "There is a global ecological crisis," continued the RES spokesman, "which concerns four critical domains for the future of humanity: biodiversity, energy, climate, and health. To fix it, there needs to be a veritable revolution in public health similar to the one in the nineteenth century that allowed us to fight against infectious diseases through education and improvements in water quality and hygiene. This new revolution should be based on what I call 'expology,' or the taking into account of all the chemical exposures we undergo in our environment. We can't wait to act any longer, because all the warning meters are at red."

In early 2011, two articles in the French newspaper *Le Monde*, published ten days apart, highlighted two of those "warning meters." On January 27, the first reported that "the life expectancy of Americans had regressed"[1] for the first time in the country's history; the second observed, "The number of people suffering from obesity worldwide has doubled in thirty years."[2] Curiously enough, the publication didn't mention the role of chemical pollutants in this double evolution even though, as we have seen, obesity is a chronic disease whose etiology stems partially (largely?) from environmental origins. Proof that in order to affect a "paradigm shift," we need to stay assiduously informed, because knowledge truly is power.

Anti-Cancer Foods

In the research world there are two schools ignoring each another, even though it would appear that they are complementary: on one side, researchers working exclusively on the environmental origins of chronic diseases—that is to say, the effects of chemical pollution explored in this book; on the other, scientists interested only in "lifestyles" and namely the consequences of a junk food diet composed of an overabundance of fats and sugars (including white flour) and a deficiency in vegetables. In the course of my long investigation, I learned that these two points of view represent two faces of the same coin: the "green revolution" mirrors the "food processing revolution" that profoundly changed what we eat. In addition, as agronomist Pierre Weill underlines in his book *Tout gros demain?* (All Fat Tomorrow?), "our genes are 'old.' They don't change with each generation. The frequency of a gene's spontaneous mutation is about once every hundred thousand years."[3] In his excellent (and courageous) work *Anti-Cancer: Prevent and Fight Through Our Natural Defenses*, David Servan-Schreiber explains, for example, how the fact that cows no longer eat grass or flax rich in omega-3 fatty acids, but corn and soy rich in omega-6, has repercussions on our physiology: "Omega-6s help stock fats and promote rigidity in cells as well as coagulation and inflammation in response to outside aggression. [. . .] Omega-3s are involved in developing the nervous system, making cell membranes more flexible, and reducing inflammation. They also limit the production of adipose cells. Our physiological balance depends very much on the balance between omega-3s and omega-6s in our body, and therefore in our diet. It turns out that it is this dietary balance that has changed the most in the last fifty years."[4] Effectively, it's gone from 1/1 to 1/25, even 1/40. Which, as oncologists can attest, is far from insignificant: inflammation paves the way for cancer.

Behind every tumor, there is a cell aggressed by an exterior agent that could be a virus, radiation, or a chemical product. If the body is in good health, the damaged cell is detected by natural killer (NK) lymphocytes that force it to "commit suicide." This phenomenon is called "apoptosis." But once the immune system is weakened by a chronic inflammation or steady aggression from chemical agents, apoptosis fails and the defective cell begins to multiply; this is the beginning of the tumor, which needs to be supplied by blood vessels in order to develop. This phenomenon is called "angiogenesis." At full term, angiogenesis leads to the creation of metastases, or colonization of the body by cancerous cells.

"Cancer is like a weed," explained Professor Richard Béliveau to me. "To initiate, it needs a seed, which needs to feed off promoting agents in order to develop. When we eat industrial, processed food that uses, for example, hydrogenated oils or trans fats rich in omega-6, we metabolically and physiologically go into proinflammatory mode, and facilitate the seed's growth. On the other hand, if we eat a large quantity of vegetables, we block the weed's development." Richard Béliveau holds the Chair in Cancer Prevention and Treatment at the University of Quebec at Montreal, where he leads a team of thirty researchers studying the anticarcinogenic potential of fruits and vegetables. He has penned over 230 publications in international medical journals.

"What the research has shown over the past twenty years is that certain vegetables contain molecules that have the same pharmacological effects as certain chemotherapy drugs, thanks to their phytochemical components," he said during our meeting in his Montreal laboratory on December 7, 2009.[5] Some of those molecules are cytotoxic: they destroy the cancerous cells. Others are proapoptotic: they prompt cancerous cells to destroy themselves. Others still are anti-inflammatory: they block the inflammation the cancerous cell needs to develop. When the cancer is in its early stages and slowly trying to implant itself, we can create a hostile environment—by consuming these molecules—that prevents the clonal selection of the cancer initiating cells, or the ones that, in twenty, thirty, or forty years, will produce cancer. We can therefore prevent the promotion of cancer through food. This arsenal of anticancerous cells is present in the cruciferous family (*brassica*): cabbage, cauliflower, Brussels sprouts, or, best of all, broccoli,[6] whose glucosinolates favor apoptosis.[7] There's also the allium family: garlic, onions, leeks, or shallots, whose sulfur compounds offer excellent protection against cancer, notably of the prostate.[8] There's also the small berries family: blueberries, blackberries, blackcurrants, strawberries, and especially raspberries, which contain ellagic acids that work to block angiogenesis.[9] We shouldn't forget green tea, whose polyphenols and catechines block the start of angiogenesis. I myself tested its effects on lines of cancerous cells and I observed that it slowed the growth of breast, prostate, kidney, skin, and mouth cancer cells, as well as leukemia cells.[10] There's also dark chocolate,[11] citrus fruits, or red wine, which contains resveratrol."[12]

"Why isn't all this better known?"

"Because there's no money to be made from the results of my studies! I have to constantly battle to obtain funding. Take the example of curcumin, the main component of turmeric: numerous studies have showed that it's a powerful anti-inflammatory, which acts at all stages of cancer. But turmeric can't be patented since it's been used in Indian cooking since time immemorial!"

The Land of Turmeric Threatened by Chronic Diseases

Shortly before Christmas 2009, I spent several days in India, aka the land of turmeric—a spice with a slightly acidic taste that gives curry its yellow color and whose therapeutic virtues have been described in Ayurvedic medicine for at least three thousand years. I participated in the Third International Symposium on Translational Cancer Research in Bhubaneswar, the capital of Orissa (in southeast India). One of the organizers was Bharat Aggarwal, the chief of the Cytokine Research Section at the MD Anderson Cancer Center, who met with me the previous week in Houston, Texas. He presented his remarkable work on curcumin, which is capable of increasing the apoptotic power of gemcitabine (a classic treatment for pancreatic cancer) on cancerous human cells. Then, he showed me images of a pancreatic tumor induced in mice wherein we could see how the curcumin was gradually draining the vessels nourishing it, in order to make it completely disappear.[13] "Curcumin has the capacity to suppress the activation of proteins in transcription factor NF-kappa B, which plays a key role in inflammatory processes," explained the Indian-American researcher. "That's why it acts on apoptosis, angiogenesis, and metastases. John Mendelsohn, president of the MD Anderson Cancer Center, and I are currently running clinical trials on patients, which look very promising."

During the Bhubaneswar conference, the scientists unfailingly spoke at length about curcumin, transcription factor NF-kappa B, the inflammatory mechanisms of cancer, but also about the "privilege of India"—which is in the midst of disappearing, as underlined by Professor Arvind Chaturvedi, director of the Rajiv Gandhi Cancer Institute and Research Centre in New Delhi. Using PowerPoint slides, he projected statistics compiled by the International Agency for Research on Cancer (IARC) in Lyon, France (see Chapter 10). In 2001, the incidence rate of the twenty principal cancers was three to thirty times lower (depending on location) in India than in the United States. The difference was particularly marked for breast and prostate cancer. "Unfortunately, the situation is in the process of changing," Chaturvedi cautioned. "In Punjabi (northern India), cradle of the green revolution, where we use lots of pesticides on high-yield wheat crops, certain cancers have clearly progressed. We're observing the same thing in large cities, where breast and prostate cancers are sharply increasing as a result of the change in lifestyle and eating habits."

"If we don't learn from others' lessons, we'll end up paying a very high price," said the Indian professor in an interview he granted me during the conference. "The answer is simple: no chemical pollutants, no processed food, but rather a healthy lifestyle with physical exercise, no or little red meat, no alcohol or smoking or chewing tobacco, and, of course, with organic food . . ."[14]

Eating "Organic"

"What can we do to avoid chemical pollutants?" I've been asked this question countless times during the debates that followed the many screenings of my film *The World According to Monsanto*. I will undoubtedly hear it many times more after this book/film comes out. And without fail I will give the same response: "We have to eat 'organic' food, as much as possible." I will not go into the cost of organic food at this time as it is not the appropriate venue to do so (I will be launching a new investigation on the subject shortly), but I do want to mention recent studies that show that organic food effectively protects children from the dangers of pesticides (at low levels).

The first of these studies was published in 2003 by researchers from the Universities of Washington and Seattle. They analyzed the urine of eighteen children between the ages of two and five who were fed exclusively with organically produced food, and twenty-one children in the same age range whose parents shopped at conventional supermarkets. The scientists searched for the presence of five organophosphorus pesticides (and their metabolites) and observed that children from the second group presented average residue levels six times greater than those in the first group. They concluded, "Consumption of organic produce represents a relatively simple means for parents to reduce their children's pesticide exposure."[15]

Another study published three years later showed that a change in diet very rapidly prompted the disappearance of pesticide residues measured in the urine of children fed with chemically processed food. It was led by researchers from the same two universities, in collaboration with the Centers for Disease Control and Prevention (CDC) in Atlanta, and observed twenty-three elementary school children put on an organic diet for five days. Their results: levels of organophosphorus pesticide residues, including malathion and chlorpyrifos, fell to a practically undetectable level after ten days. The authors also concluded, "these children were most likely exposed to these organophosphorus pesticides exclusively through their diet."[16] These results were confirmed two years later by another study that was conducted over four consecutive seasons, during which twenty-three children between the ages of three and eleven changed diets several times. Each time, and regardless of the season, the level of pesticides measured in their urine disappeared less than ten days after switching to organic food.[17]

A Pesticide Ban Would Save Lots of Money

"This system produces disease because political, economic, regulatory and ideological norms prioritize values of wealth and profit over human health and environmental well-being," wrote David Egilman and Susanna Rankin Bohme, professors of occupational and environmental health at Brown University, Rhode Island, in 2005. "Corporations 'largely ignore social and environmental costs,' chiefly through externalizing them, or shifting costs to governments, neighbors, or workers." They further underlined the supreme paradox of the "system" (citing economist Robert Monks) by noting that a corporation "'tends to be more profitable to the extent it can make other people pay the bills for its impact on society.'"[18]

It suffices to go through the annual bulletins (available online) from the French National Insurance Fund (Caisse nationale d'assurance maladie, CNAM) to measure at what point "the cost of nontransmittable diseases for individuals, societies and health systems isn't tenable," to borrow the words of WHO.[19] In France, the number of people insured under the general plan who are declared as suffering from a "long duration disease" (LDD) went from 3.7 million in 1994 (11.9 percent of employees) to 8.6 million as of December 31, 2009 (or one person in seven).[20] While this phenomenon doubled in fifteen years, it has accelerated considerably since 2004: between 2006 and 2007, the increase totaled +4.2 percent. LDDs are estimated to represent 60 percent of CNAM's spending (a total of 42 billion euros). As underlined by the organization itself in a "monthly update" dated April 5, 2006, between 1994 and 2004, "the number of people covered for long duration diseases increased significantly (+73.5% since 1994, +53.3% if we apply this number to the evolution of the general population during this period)."

This explosion of LDDs—which appears to stem primarily from environmental origins—sheds new light on the infamous "social security deficit," whose breadth has been bemoaned by successive French administrations. In fact, as André Cicolella very rightly points out, "a simple rule of three shows that, if we had had the same proportion of LDDs in 2004 as existed ten years earlier, health spending would have been reduced by sums largely superior to the deficit of recent years, even including the increase of life expectancy of people with LDDs (and the consequential increase in overall costs of treatment)."[21] He concludes, "The economic gains produced by the nonoccurrence of health pathologies are hardly ever taken into account by the overall economic calculation."[22] The exception to the rule: in 2001, a team of Ontario researchers set out to calculate the cost of four pathologies with suspected links to the environment in the United States and Canada: diabetes, Parkinson's disease, the neurodevelopmental effects of hypothyroidism, and mental impairment. According to their estimates, the "non-occurrence" of these diseases would *save* $57 to $397 million per year, depending on the importance of environmental factors attributed to their etiologies.[23]

This same perspective can be found in the conclusion of an important (and rather radical) report made public by the European Parliament in 2008: "Most of the potential health benefits from restricting the use of certain pesticides would accrue through avoiding the costs of health impacts associated with pesticide exposure. These costs could include health service costs, the value of an individual's lost quality of life, the value of a statistical life lost due to a pesticide-related death, or loss of productivity (days of work lost) due to a pesticide-related poisoning, whether acute or chronic."[24] This lengthy document affirms that "active substances classified as carcinogen category 1 or 2, mutagen category 1 or 2, or toxic for reproduction category 1 or 2 ('CMR 1 & 2'), or those considered to have endocrine disrupting ('ED') properties [. . .] shall not be approved." The authors then cite a series of studies showing the considerable financial gain that the outright ban of CMR pesticides and endocrine disruptors would bring about.

The first study, conducted in 1992, made the very conservative estimate that the health costs of pesticide exposure amount to $787 million per year in the United States.[25] A similar study conducted in Europe fifteen years later—based only on the cost of cancer deaths—estimated the potential savings of a ban of the most dangerous pesticides at $26 billion per year.[26] In 2003, the European Commission evaluated the health benefits that would result from restricting use of chemical products through application of the REACH program (which only concerns pesticides; see Chapter 19): 50 million euros over the next thirty years, 99 percent of which would come from a reduction in cancer deaths.[27]

Regardless of the adopted angle—several reports exclusively addressed the health costs of the dramatic rise in autism[28]—all the studies cited by the European report confirm that, contrary to industry propaganda, applying the precautionary principle would not provoke an economic catastrophe, but would actually *save* lots of money. But, as epidemiologist Richard Clapp explained to me during our Boston meeting, "the logic behind the precautionary principle runs counter to the private interests of the pharmaceutical industry, for whom cancer is the 'crab with golden claws.'" He added, with a knowing smile: "And those who sell us drugs to treat our chronic diseases are the same people who polluted us, and continue to pollute us. They're winning on all fronts . . ."

The Precautionary Principle: The Necessary Democratization of Risk Evaluation Processes

"In order to protect the environment, the precautionary approach shall be widely applied by States according to their capabilities. Where there are threats of serious or irreversible damage, lack of full scientific certainty shall not be used as a reason for postponing cost-effective measures to prevent environmental degradation." It was with these terms that the "precautionary principle" was defined for the first time during a UN conference on development and the environment held in Rio de Janeiro in June 1992. Six years later, the European Commission provided its own definition, which was subsequently adopted by most European countries: "The Precautionary Principle [is] an option open to risk managers when [. . .] scientific information concerning the risk is inconclusive or incomplete in some way. [It] is relevant in those circumstances where risk managers have identified that there are reasonable grounds for concern that an unacceptable level of risk to health exists but the supporting information and data may not be sufficiently complete."[29]

To understand the stakes of the debate surrounding the precautionary principle, it is important to correctly distinguish between *precaution* (management of uncertainty) and *prevention* (management of an identified risk). The asbestos case illustrates this difference perfectly, as French sociologists Michel Callon, Pierre Lascoumes, and Yannick Barthes explain in their book *Acting in an Uncertain World.*[30] Indeed, the dangers of asbestos to humans have been known since (at least) the beginning of the 1930s, and "the risk of pulmonary diseases was sufficiently known from 1975 on for real preventive measures to be taken, the most radical being prohibition," which was enacted by different industrialized countries, except for France, who waited until 1997. "Before 1975, the measures that could have been taken would have come under precaution in the face of identified, but poorly defined dangers."[31]

In this context, "precautionary practice requires a preliminary evaluation of the overflows and associated dangers." It also requires that "exploration weave together the dispersed and heterogeneous information in order to construct 'bundles of convergent indices.' The objective is not to find one consolidated and replicable proof, but the gradual construction of hypotheses combining theoretical data with empirical observations, objective and subjective data."[32]

However, this new way of looking at chemical risk evaluation presupposes the comprehensive revision of the relationship between science and politics, as well as that between science and society. Finished, in effect, would be the absolute truths handed down by what Michel Callon calls "secluded research,"[33] or "laboratory research [that] has distanced itself from the world in order to increase its productivity" and asserted itself as the sole party authorized to provide recommendations on the risks people face. The well-reasoned application of the precautionary principle implies, on the contrary, collaboration with research "in the wild" practiced by laypersons or locals who have acquired expertise through concrete experience with a situation they believe poses an environmental or health risk.

The closed doors of regulatory agencies would be gone as well, along with data shielded by far-reaching "trade secrets," denials of "minority segments of the scientific community," or the valuable work of "whistleblowers." The precautionary approach relies on a "democratization of . . . democracy"—based on dialogue and not on authority arguments—wherein the "'acceptability' of risk is a social process and not a predeterminable objective."[34] Because, as FDA toxicologist Jacqueline Verrett wrote in 1974: "Regulatory agencies [must] immediately stop the practice of vesting chemicals with rights. Chemicals do not have rights: people do."[35]

Notes

Introduction: Knowledge Is Power

1. Marie-Monique Robin, *The World According to Monsanto: Pollution, Corruption, and the Control of the World's Food Supply*, trans. George Holoch (New York: The New Press, 2010).

2. Marie-Monique Robin, *Les Pirates du vivant* and *Blé: chronique d'une mort annoncée?*, Arte, November 15, 2005.

3. Marie-Monique Robin, *Argentine: le soja de la faim*, Arte, October 18, 2005. Along with *Les Pirates du vivant*, it is available on DVD in the "Alerte verte" series.

4. The DVD is distributed by Icarus Films in the United States: www.icarusfilms.com/new2011/pois.html.

1: The Ruffec Appeal and the Battle of Paul François

1. The MDRGF was renamed Générations futures in November 2010.

2. Joël Robin, *Au nom de la terre: La foi d'un paysan* (Paris: Presses de la Renaissance, 2001).

3. Marie-Monique Robin, *Le Suicide des paysans*, TF1 (winner of the documentary prize at the Festival International du Scoop of Angers).

4. Marie-Monique Robin, *The World According to Monsanto: Pollution, Corruption, and the Control of the World's Food Supply*, trans. George Holoch (New York: The New Press, 2010).

5. François Veillerette, *Pesticides, le piège se referme* (2002; Mens: Terre vivante, 2007); see also Fabrice Nicolino and François Veillerette, *Pesticides, révélations sur un scandale français* (Paris: Fayard, 2007).

6. In July 2010, AFSSA merged with the French Agency for Environmental and Occupational Health Safety (Agence française de sécruité sanitaire de l'environnement et du travail, AFSSET) to form the French Agency for Food, Environmental, and Occupational Health and Safety (Agence nationale chargée de la sécurité sanitaire de l'alimentation, de l'environnement et du travail, ANSES).

7. www.victimes-pesticides.org. See also "Un nouveau réseau pour défendre les victimes des pesticides," *Le Monde.fr*, June 18, 2009.

8. The Confédération paysanne is a left-leaning minority farmers' union which campaigns for a sustainable agricultural model that is more familial, less industrial.

9. "Malade des pesticides, je brise la loi du silence," *Ouest France*, March 27, 2009.

10. Every pesticide contains an active ingredient—for Lasso, alachlor—and many additives, also known as inert substances, such as solvents, dispersants, emulsifiers, and surfactants, whose purpose is to enhance the physicochemical properties and the biological effectiveness of the active ingredients; they have no effect as pesticides themselves.

11. "Alachlor," WHO/FAO Data Sheets on Pesticides, no. 86, www.inchem.org, July 1996.

12. The European Union decided not to include alachlor in appendix 1 of directive 91/414/CEE. Notified under the number C(2006) 6567, this decision specifies that "exposure resulting from the handling of the substance and its application at the rates, i.e., the intended doses per hectare, proposed by the notifier, would . . . lead to an unacceptable risk for the operators."

13. "Maïs: le désherbage en prélevée est recommandé," *Le Syndicat agricole*, April 19, 2007.

14. The analysis also showed that Lasso includes 6.1 percent butanol and 0.7 percent isobutanol.

15. "Un agriculteur contre le géant de l'agrochimie," www.viva.presse.fr, April 2, 2009.

16. Jean-François Barré, "Paul, agriculteur, 'gazé' au désherbant!," *La Charente libre*, July 17, 2008.

17. www.medichem2004.org/schedule.pdf (no longer accessible).

18. Monsanto was found liable in February 2012, and the company is considering an appeal.

2: Chemical Weapons Recycled for Agriculture

1. Geneviève Barbier and Armand Farrachi, *La Société cancérigène* (Paris: Seuil, 2007), 51.

2. Ibid., 58.

3. Pesticide Action Network UK, *Pesticides on a Plate: A Consumer Guide to Pesticide Issues in the Food Chain* (London: PAN, 2007).

4. "Safe Use of Pesticides," Public Service Announcement, 1964 (see my film *Notre poison quotidien*, Arte, 2011).

5. Emphasis added.

6. In December 2010 the address was agriculture.gouv.fr. (It remains the same.) The official title of the institution was then (poetically) "Ministry of Agriculture, Food, Fishing, Rural Development, and Land-Use Planning."

7. "Its aims: to better define the acute and sub-acute effects of these products to increase individual protection taking into account actual work and improve collective prevention by the transmission of information to the authorities and manufacturers."

8. "Pesticides et santé des agriculteurs," references-sante-securite.msa.fr, April 26, 2010. Emphasis added.

9. Julie Marc, *Effets toxiques d'herbicides à base de glyphosate sur la régulation du cycle cellulaire et le développement précoce en utilisant l'embryon d'oursin*, doctoral thesis, Université de Biologie de Rennes, September 10, 2004.

10. See Marie-Monique Robin, *Les Pirates du vivant*, Arte, November 15, 2005.

11. The ban was the result of a study carried out by the MSA, which asked that arsenic be listed in the table of occupational diseases for agriculture because of its carcinogenic effects.

12. Paris green also served as a pigment, widely used by the Impressionist painters. Its toxicity is said to be the cause of Cézanne's diabetes and Monet's blindness, as well as Van Gogh's neurological disorders.

13. Guano is the excrement of marine birds; used as biological fertilizer, its nutrient and organic matter content has never been equaled, especially because, unlike chemical fertilizers, an overdose does not affect the environment or soil quality (see my report, *L'Or noir du Pérou*, "Thalassa," FR3, 1992).

14. Arthur Hurst, "Gas Poisoning," in *Medical Diseases of the War* (London: Edward Arnold, 1918), 311–12, quoted in Paul Blanc, *How Everyday Products Make People Sick: Toxins at Home and in the Workplace* (Berkeley: University of California Press, 2007), 116.

15. Hanspeter Witschi, "The Story of the Man Who Gave Us Haber's Law," *Inhalation Toxicology* 9, no. 3 (1997): 201–9.

16. Ibid., 203.

17. David Gaylor, "The Use of Haber's Law in Standard Setting and Risk Assessment," *Toxicology* 149, no. 1 (August 14, 2000): 17–19.

18. WHO/United Nations Environment Programme, *Sound Management of Pesticides and Diagnosis and Treatment of Pesticide Poisoning: A Resource Tool* (2006), 18.

19. Karl Winnacker and Ernst Weingaertner, *Chemische Technologie: Organische Technologie II* (Munich: Carl Hanser Verlag, 1954), 1005–6.

20. See the specifications for Zyklon B in the list of "withdrawn phytosanitary products" on the French Ministry of Agriculture's website, e-phy.agriculkture.gouv.fr.

21. Ibid.

22. Witschi, "Story of the Man Who Gave Us Haber's Law."

23. Rachel Carson, *Silent Spring* (1962; New York: Houghton Mifflin 2002), 8, 16.

24. Among the pesticides in the dirty dozen are a fungicide, hexachlorobenzene, and eight insecticides: aldrin, chlordane, dieldrin, endrin, heptachlore, mirex, toxaphene, and DDT.

25. See the chapter "PCBs: White Collar Crime," in Marie-Monique Robin, *The World According to Monsanto*, trans. George Holoch (New York: The New Press, 2010), 9–29.

26. In 2001 Geigy was taken over by Syngenta, a leading Swiss multinational in the pesticide market.

27. William Buckingham Jr., *Operation Ranch Hand: The Air Force and Herbicides in Southeast Asia 1961–1971* (Washington, DC: Office of Air Force History, 1982), iii.

28. Georganne Chapin and Robert Wasserstrom, "Agricultural Production and Malaria Resurgence in Central America and India," *Nature* 293 (September 17, 1981): 181–85.

29. International Programme on Chemical Safety, "DDT and Its Derivatives" (Geneva: WHO, 1979), www.inchem.org.

30. Carson, *Silent Spring*, 21.

31. The use of DDT was finally limited to campaigns for the eradication of mosquitoes carrying malaria, although they remain highly controversial. Recent studies pointing to a link between DDT and certain cancers might lead the WHO to declare a definitive ban of the insecticide (see Agathe Duparc, "L'OMS pourrait recommander l'interdiction du DDT," *Le Monde*, December 1, 2010).

32. An organophosphate is an organic compound including at least one phosphorus atom directly linked to a carbon atom.

33. The Monitoring Authority for Pesticide Residues has been administered since 2010 by the French Agency for Food, Environmental, and Occupational Health and Safety (Agence nationale chargée de la sécurité sanitaire de l'alimentation, de l'environnement et du travail, ANSES). See its website: www.observatoire-pesticides .gouv.fr.

34. Sarin gas was used in a terrorist attack in the Tokyo subway on March 20, 1995, killing ten and injuring several thousand. The gas was also made in Chile in the 1970s by General Augusto Pinochet's secret police to assassinate his opponents (see Marie-Monique Robin, *Escadrons de la mort, l'école française* (Paris: La Découverte, 2004).

35. This agricultural "revolution" was later called "green" because it was supposed to block the "red revolution" in "underdeveloped" countries, particularly in Asia, where Mao Zedong's seizure of power in China in 1949 risked producing imitators (see Marie-Monique Robin, *Blé: chronique d'une mort annoncée*, Arte, November 15, 2005).

36. "In the beginning: the multiple discovery of the first hormone herbicides," *Weed Science* 49 (2001): 290–97.

37. Chlorophenols are organic compounds made up of a phenolic nucleus in which one or more hydrogen atoms are replaced by one or more chlorine atoms. There are nineteen varieties of chlorophenol, the toxic effects of which are in direct proportion to their degree of chlorination.

38. See Robin, *World According to Monsanto*, 30–68.

39. 2,3,7,8-tetrachlorodibenzo-*p*-dioxin, or TCDD, known as the Seveso poison, was discovered by Wilhelm Sandermann, who was working at the Wood Industry Institute.

40. Jean-Claude Pomonti, "Viêt-nam, les oubliés de la dioxine," *Le Monde*, April 26, 2005.

41. The most reliable estimates were published by Jane Mager Stellman, "The Extent and Patterns of Usage of Agent Orange and Other Herbicides in Vietnam," *Nature* 422 (April 17, 2003): 681–87.

42. Chloracne is the disease that disfigured the Ukrainian president Viktor Yushchenko in 2004, following a poisoning attributed to the country's secret services.

43. Established in 1847, the American Medical Association claims to have 250,000 members. Its journal, *JAMA*, is the most widely read medical weekly in the world.

44. Blanc, *How Everyday Products Make People Sick*, 233.

45. Ibid., 234.

46. Carson, *Silent Spring*, 155.

3: "Elixirs of Death"

1. Rachel Carson, *Le Printemps silencieux* (Paris: Plon, 1963). A *normalien* and chemical engineer who became a renowned mycologist and ardent defender of natural resources, Roger Heim was the author of *Destruction et Protection de la nature* (Paris: Armand Colin, 1952).

2. Rachel Carson, *Silent Spring* (1962; New York: Houghton Mifflin, 2002), 15.

3. Ibid., xi. See also Linda Lear, *Rachel Carson: Witness for Nature* (New York: Henry Holt, 1997).

4. See Gerald Leblanc, "Are Environmental Sentinels Signaling?," *Environmental Health Perspectives* 103, no. 10 (October 1995): 888–90.

5. Quotation from "Rachel Carson Dies of Cancer; 'Silent Spring' Author Was 56," *New York Times*, April 15, 1964, http://www.nytimes.com/learning/general/on thisday/bday/0527.html. See also the rare BBC interview "Rachel Carson Talks About Effects of Pesticides on Children and Future Generations," www.bbcmotion gallery.com, January 1, 1963.

6. Quoted by Dorothy McLaughlin, "*Silent Spring* Revisited," www.pbs.org.

7. "The Desolate Year," *Monsanto Magazine*, October 1962, 5.

8. *Time*, September 28, 1962, 45–46.

9. "The *Time* 100: Rachel Carson," *Time*, March 29, 1999.

10. Close to industry and a virulent anticommunist, Ezra Taft Benson, agriculture secretary from 1953 to 1961 during Eisenhower's presidency, was a major figure in the Church of Jesus Christ of the Latter-Day Saints, and its president from 1985 to 1994.

11. Lear, *Rachel Carson*, 429.

12. President's Science Advisory Committee, "Use of Pesticides," May 15, 1963.

13. David Greenberg, "Pesticides: White House Advisory Body Issues Report Recommending Steps to Reduce Hazard to Public," *Science* 140, no. 3569 (May 24, 1963): 878–79.

14. EPA, "DDT Ban Takes Effect," www.epa.gov, December 31, 1972.

15. Carson, *Silent Spring*, 99.

16. "Indien: die chemische Apokalypse," *Der Spiegel*, December 10, 1984.

17. Ibid.

18. Marie-Monique Robin, *Les Pirates du vivant*, Arte, November 15, 2005. The patent was finally revoked by the European Patent Office after a legal battle lasting more than ten years.

19. WHO, *Public Health Impact of Pesticides Used in Agriculture* (Geneva: WHO, 1990).

20. See Marie Monique Robin, *The World According to Monsanto*, trans. George Holoch (New York: The New Press, 2010), 291, where I recount the funeral of an Indian peasant who had committed suicide by swallowing a pesticide, because he was in debt and his transgenic cotton crop had been a fiasco. See also Ashish Goel and Praveen Aggarwal, "Pesticide Poisoning," *National Medical Journal of India* 20, no. 4 (2007): 182–91.

21. Jerry Jeyaratnam et al., "Survey of Pesticide Poisoning in Sri Lanka," *Bulletin of the World Health Organization* 60 (1982): 615–19. All the studies cited in this section are part of the corpus of the WHO document.

22. Ania Wasilewski, "Pesticide Poisoning in Asia," *IDRC Report*, January 1987. See also Jerry Jeyaratnam et al., "Survey of Acute Poisoning Among Agricultural Workers in Four Asian Countries," *Bulletin of the World Health Organization* 65 (1987): 521–27; Robert Levine, "Assessment of Mortality and Morbidity Due to Unintentional Pesticide Poisonings," unpublished WHO document, WHO/VBC/86.929. See also the seminal book by Mohamed Larbi Bouguerra, *Les Poisons du tiers monde* (Paris: La Découverte, 1985).

23. Edward Baker et al., "Epidemic Malathion Poisoning in Pakistan Malaria Workers," *Lancet* 311, no. 8054 (January 7, 1978): 31–34.

24. WHO/United Nations Environment Programme, *Sound Management of Pesticides and Diagnosis and Treatment of Pesticide Poisoning: A Resource Tool* (Geneva: WHO, 2006).

25. 2,4,5-T, the other Agent Orange ingredient, was banned at the end of the Vietnam War.

26. Ataxia is a neuromuscular disease causing loss of coordination.

27. Emphasis added.

28. The presence of my film crew was noted on the lycée site: www.bonne-terre.fr.

29. Pesticide Action Network Europe and MDRGF, *"Message dans une bouteille": Étude sur la presence de résidus de pesticides dans le vin*, March 26, 2008.

30. AFSSET, "L'Afsset recommandé de renforcer l'évaluation des combinaisons de protection des travailleurs contre les produits chimiques liquids," January 15, 2010.

31. Author interview with Jean-Luc Dupupet, Pézenas, February 9, 2010.

4: Ill from Pesticides

1. "Le métier d'Odalis: relier les fournisseurs aux distributeurs et agriculteurs," www.terrena.fr.

2. The brochure was financed by "the agricultural chambers of Brittany, the regional council of Brittany, the state, Europe."

3. "Maladie professionelle liée aux fungicides: première victoire," www.Nouvelobs.com, May 26, 2005; see also Santé et Travail 30 (January 2000): 52.

4. Brigitte Bègue, "Les pesticides sur la selette," Viva, August 14, 2003.

5. Author's interview of Jean-Luc Dupupet, Pézenas, February 9, 2010.

6. Rachel Carson, Silent Spring (1962; New York: Houghton Mifflin 2002), 188

7. Michel Gérin, Pierre Gosselin, Sylvaine Cordier, Claude Viau, Philippe Quénel, and Éric Dewailly, Environnement et santé publique: Fondements et pratiques (Montreal: Edisem, 2003), 74.

8. Fabrice Nicolino and François Veillerette, Pesticides, révélations sur un scandale français (Paris: Fayard, 2007), 289.

9. According to the National Institute of Agronomic Research (Institut national de la recherche agronomique, INRA), the average annual number of treatments is 6.6 for wheat, 3.7 for corn, and 6.7 for rapeseed (Pesticides et Environnement: Réduire l'utilisation des pesticides et en limiter les impacts environnementaux, joint expert report prepared by the INRA and the National Centre of Agricultural Machinery, Agricultural Engineering, and Water and Forests (Centre d'etude du machinisme agricole du génie rural des eaux et forêts, CEMAGREF) at the request of the Ministry of Agriculture and Fisheries and the Ministry of Ecology and Sustainable Development, December 2005).

10. INRS, Tableaux de maladies professionnelles: Guide d'accès et commentaires, www.inrs-mp.fr/mp/cgi-bin/mppage.pl?, 216–18.

11. Alice Hamilton, "Lead Poisoning in Illinois," in American Association for Labor Legislation (AALL), First National Conference on Industrial Diseases: Chicago, June 10, 1910 (New York: AALL, 1910).

12. Article L.461-6 of the Social Security Code requires every doctor to declare the diseases which seem to him or her to be likely of occupational origin. The number of individuals recognized as victims of occupational disease in France rose from 4,032 in 1989 to 45,000 in 2008.

13. INRS, Tableaux de maladies professionnelles, 299.

14. "A New Domestic Poison," The Lancet 1 (1862): 105.

15. "Chronic Exposure to Benzene," Journal of Industrial Hygiene and Toxicology 21 (1939): 321–77.

16. Estelle Saget, "Le cancer des pesticides," *L'Express*, January 5, 2007; see also Estelle Saget, "Ces agriculteurs malades des pesticides," *L'Express*, October 25, 2004.

17. This letter is in Dominique Marchal's file, which I was able to consult.

18. The Medical Insurance Association for Farmers (Groupement des assureurs maladie des exploitants agricoles, GAMEX) paid the costs of the analysis.

19. On January 15, 2011, almost a year to the day after the Ruffec meeting, Yannick Chenet died.

20. David Michaels, *Doubt Is Their Product: How Industry's Assault on Science Threatens Your Health* (New York: Oxford University Press, 2008), 64.

21. Geneviève Barbier and Armand Farrachi, *La Société cancérigène* (Paris: Seuil, 2007), 164.

22. Devra Davis, *The Secret History of the War on Cancer* (New York: Basic Books, 2007), xii.

23. Gérin et al., *Environnement et santé publique*, 90.

24. To be precise, it should be added that the OR is followed by two numbers in parentheses which, as in all statistics, indicate the confidence interval within which the results are located.

25. Barbier and Farrachi, *La Société cancérigène*, 163–64.

5: Pesticides and Cancer: Consistent Studies

1. Recall that the herbicides 2,4,5-T and 2,4-D, the two components of Agent Orange, are in this category (chlorophenols). 2,4-D is still one of the pesticides in most widespread use around the world.

2. M. Alavanja, J.A. Hoppin, and F. Kamel, "Health Effects of Chronic Pesticide Exposure: Cancer and Neurotoxicity," *Annual Review of Public Health* 25 (2004): 155–97.

3. Coated seeds have been treated with pesticides by the manufacturer before delivery.

4. Emphasis added.

5. The Avignon TASS transferred Bony's case to the regional occupational disease committee in Montpellier, which was supposed to render an opinion in February 2011.

6. David Michaels, *Doubt Is Their Product: How Industry's Assault on Science Threatens Your Health* (New York: Oxford University Press, 2008), 61.

7. I suggest that readers consult this invaluable tool which in late 2010 had more than 20 million references.

8. Alavanja et al., "Health Effects of Chronic Pesticide Exposure."

9. Margaret Sanborn, Donald Cole, Kathleen Kerr, Cathy Vakil, Luz Helena Sanin, and Kate Bassil, *Systematic Review of Pesticide Human Health Effects* (Toronto: Ontario College of Family Physicians, 2004).

10. Lennart Hardell and Mikael Eriksson, "A Case-Control Study of Non-Hodgkin Lymphoma and Exposure to Pesticides," *Cancer* 85, no. 6 (March 15, 1999): 1353–60.

11. Shelia Hoar Zahm et al., "A Case-Control Study of Non-Hodgkin's Lymphoma and the Herbicide 2,4-dichlorophenoxyacetic Acid (2,4-D) in Eastern Nebraska," *Epidemiology* 1, no. 5 (September 1990): 349–56. For this study, 201 disease sufferers were compared to 725 healthy subjects.

12. E. Hansen, H. Hasle, and F.A. Lander, "A Cohort Study on Cancer Incidence Among Danish Gardeners," *American Journal of Industrial Medicine* 21, no. 5 (1992): 651–60.

13. Julie Agopian et al., "Agricultural Pesticide Exposure and the Molecular Connection to Lymphomagenesis," *Journal of Experimental Medicine* 206, no. 7 (July 6, 2009): 1473–83.

14. Aaron Blair, Shelia Hoar Zahm, Neil E. Pearce, Ellen F. Heineman, Joseph F. Fraumeni Jr., "Clues to Cancer Etiology from Studies of Farmers," *Scandinavian Journal of Work, Environment & Health* 18, no. 4 (1992): 209–15; Aaron Blair and Hoar Zahm, "Agricultural Exposures and Cancer," *Environmental Health Perspectives* 103, suppl. 8 (November 1995): 205–8; Aaron Blair and Laura Freeman, "Epidemiologic Studies in Agricultural Populations: Observations and Future Directions," *Journal of Agromedicine* 14, no. 2 (2009): 125–31.

15. John Acquavella et al., "Cancer Among Farmers: A Meta-Analysis," *Annals of Epidemiology* 8, no. 1 (January 1998): 64–74.

16. Samuel Milham, "Letter," *Annals of Epidemiology* 9, no. 1 (January 1999): 71–72; Milham is the author of "Leukemia and Multiple Myeloma in Farmers," *American Journal of Epidemiology* 94, no. 4 (1971): 307–10.

17. Linda Brown, Aaron Blair, Robert Gibson, George D. Everett, and Kenneth P. Cantor, "Pesticide Exposures and Other Agricultural Risk Factors for Leukemia Among Men in Iowa and Minnesota," *Cancer Research* 50, no. 20 (October 15, 1990): 6585–91.

18. Alavanja et al., "Health Effects of Chronic Pesticide Exposure"; Sadik Khuder, "Meta-Analyses of Multiple Myeloma and Farming," *American Journal of Industrial Medicine* 32, no. 5 (November 1997): 510–16.

19. Isabelle Baldi and Pierre Lebailly, "Cancers et pesticides," *La Revue du praticien* 57, suppl. (June 15, 2007): 50–54.

20. Dorothée Provost et al., "Brain Tumors and Exposure to Pesticides: A Case Control Study in Southwestern France," *Occupational and Environmental Medicine* 64, no. 8 (2007): 509–14.

21. Vines cover only 10 percent of cultivated land in France but use 80 percent of the fungicides and 46 percent of the insecticides used nationally (Bernard Delemotte et al., "Le risqué pesticide en agriculture," *Archives des maladies professionnelles* 48, no. 6 [1987]: 467–75).

22. Jean-François Viel et al., "Brain Cancer Mortality Among French Farmers: The Vineyard Pesticide Hypothesis," *Archives of Environmental Health* 53, no. 1 (1998):

65–70; Jean-François Viel, "Étude des associations géographiques entre mortalité par cancers en milieu agricole et exposition aux pesticides," doctoral thesis, Faculté de médecine Paris-Sud, 1992.

23. André Fougeroux, "Les produits phytosanitaires: Évaluation des surfaces et des tonnages par type de traitement in 1988," *Défense des végétaux* 259 (1989): 3–8. Fougeroux is now manager of biodiversity at Syngenta, a Swiss multinational specializing in pesticides and transgenic seeds.

24. In order of area under cultivation: straw cereals, corn, vines, sunflowers, rapeseed, protein peas, beets, potatoes, apple trees, flax, and pear trees.

25. P. Kristensen, A. Andersen, L.M. Irgens, A.S. Bye, and L. Sundheim, "Cancer in Offspring of Parents Engaged in Agricultural Activities in Norway: Incidence and Risk Factors in the Farm Environment," *International Journal of Cancer* 65, no. 1 (1996): 39–50.

26. Professional pesticide applicators are employees who work for firms specializing in crop spraying or the treatment of storage sites such as grain silos.

27. Michael Alavanja et al., "Use of Agricultural Pesticides and Prostate Cancer in the Agricultural Health Study Cohort," *American Journal of Epidemiology* 157, no. 9 (2003): 800–814.

28. Agricultural Health Study, aghealth.nci.nih.gov.

29. Michael Alavanja et al., "Cancer Incidence in the Agricultural Health Study," *Scandinavian Journal of Work and Environmental Health* 31, suppl. 1 (2005): 39–45.

30. Prohibited by the 1987 Montreal protocol because of its effect on the stratospheric ozone layer, methyl bromide was until 2005 one of the most widely used insecticides in the world. It was used to disinfect agricultural soils (especially in the cultivation of hothouse tomatoes), grain fumigation, the protection of stored crops, and the cleaning of silos and mills. In 2005, France obtained the right to use 194 tons on the grounds that, for certain uses, there was no alternative.

31. Geneviève Van Maele-Fabry and Jean-Louis Willems, "Prostate Cancer Among Pesticide Applicators: A Meta-Analysis," *International Archives of Occupational and Environmental Health* 77, no. 8 (2004): 559–70. It should be noted that the ORs found in the twenty-two studies selected ranged from 0.63 to 2.77.

32. Bas-Rhin, Calvados, Côte-d'Or, Doubs, Gironde, Haut-Rhin, Isère, Loire-Atlantique, Manche, Somme, Tarn, et Vendée.

33. Emphasis in original.

34. Baldi and Lebailly, "Cancers et pesticides," 50–54.

6: The Unstoppable Rise of Pesticides and Neurodegenerative Diseases

1. Fabrice Nicolino and François Veillerette, *Pesticides, révélations sur un scandale français* (Paris: Fayard, 2007), 56.

2. "Le Gaucho retenu tueur official des abeilles. 450 000 ruches ont disparu depuis 1996," *Libération*, October 9, 2000.

3. It should be noted that political tendencies had no effect on the case: the inaction of the two ministers of agriculture concerned—the socialist Jean Glavany (October 1998–February 2002) and the RPR (Gaullist political party) Hervé Gaymard (May 2002–November 2004)—was strictly identical.

4. For more details on Catherine Geslain-Lanéelle's career, see Nicolino and Veillerette, *Pesticides, révélations sur un scandale français*, 60.

5. It is estimated that between 1995 and 2003, French honey production fell from 32,000 to 16,500 tons. Simultaneously, another insecticide, just as toxic, BASF's Regent, was also decimating bees. It was prohibited in turn in 2005.

6. M.C. Alavanja, J.A. Hoppin, and F. Kamel, "Health Effects of Chronic Pesticide Exposure: Cancer and Neurotoxicity," *Annual Review of Public Health* 25 (2004): 155–97. See Chapter 5.

7. Freya Kamel et al., "Pesticide Exposure and Self-Reported Parkinson's Disease in the Agricultural Health Study," *American Journal of Epidemiology* 165, no. 4 (2006): 364–74.

8. It was the French physician Jean-Martin Charcot (1825–1893) who would name the disease.

9. Cited in Paul Blanc, *How Everyday Products Make People Sick: Toxins at Home and in the Workplace* (Berkeley: University of California Press, 2007), 243.

10. Ibid.

11. Louis Casamajor, "An Unusual Form of Mineral Poisoning Affecting the Nervous System: Manganese," *Journal of the American Medical Association* 60 (1913): 646–49 (cited by Blanc, *How Everyday Products Make People Sick*, 250).

12. Hugo Mella, "The Experimental Production of Basal Ganglion Symptomatology in Macacus Rhesus," *Archives of Neurology and Psychiatry* 11 (1924): 405–17 (cited by Blanc, *How Everyday Products Make People Sick*, 251).

13. H.B. Ferraz, P.H. Bertolucci, J.S. Pereira, J.G.C, Lima, and L.A. Andrade, "Chronic Exposure to the Fungicide Maneb May Produce Symptoms and Signs of CSN Manganese Intoxication," *Neurology* 38 (1988): 550–52.

14. G. Meco, V. Bonifati, N. Vanacore, and E. Fabrizio, "Parkinsonism After Chronic Exposure to the Fungicide Maneb (Manganese-ethylene-bis-dithiocarbamate)," *Scandinavian Journal of Work Environment and Health* 20 (1994): 301–5.

15. William Langston, "The Aetiology of Parkinson's Disease with Emphasis on the MPTP Story," *Neurology* 47 (1996): 153–60.

16. It was precisely due to the use of rotenone and paraquat that a Parisian gardener, who worked for a large horticultural company for thirty-four years, was granted occupational disease status in 2009. The gardener had developed Parkinson's disease at the age of forty-eight. Maria Gonzales from the University Hospital Center of Strasbourg, who took part in the expert committee called upon by the CRRMP of Paris, explained this in an interview with *Hygiène, Sécurité, Environnement* (Hygiene, Safety, Environment) on June 19, 2009.

17. A report published in January 2011 by Future Generations and Pesticides Action Network Europe revealed that in Europe, submissions for dispensations to use prohibited pesticides had increased by 500 percent between 2007 and 2010. The European directive on pesticides (91/414) contains an article, 8.4, that permits obtaining a "dispensation of twenty days" that gives a member state the possibility of using prohibited pesticides "in the case of unforeseeable danger." In Europe, we thus went from 59 dispensations in 2007 to 321 in 2010, 74 of which were for France (Future Generations and Pesticides Action Network Europe, "La question des derogations accordéesdans le cadre de la legislation européennesur les pesticides [The issue of dispensations given in the context of European legislation on pesticides]," January 26, 2011).

18. "Maïs: le désherbage en prélevéeestrecommandé," *Le Syndicat agricole*, April 19, 2007.

19. WHO/United Nations Environment Programme, *Sound Management of Pesticides and Diagnosis and Treatment of Pesticides Poisoning: A Resource Tool* (2006), 92.

20. Isabelle Baldi et al., "Neuropsychologic Effects of Long-Term Exposure to Pesticides: Results from the French Phytoner Study," *Environmental Health Perspective* 109, no. 8 (August 2001): 839–44.

21. Isabelle Baldi et al., "Neurodegenerative Diseases and Exposure to Pesticides in the Elderly," *American Journal of Epidemiology* 1, no. 5 (March 2003): 409–14.

22. Caroline Tanner et al., "Occupation and Risk of Parkinsonism. A Multicenter Case-Control Study," *Archives of Neurology* 66, no. 9 (2009): 1106–13. The study was carried out on five hundred patients compared to an equivalent control group.

23. In 2008, sixteen illnesses were on this list, including various cancers (respiratory tract, prostate, soft tissue sarcoma, leukemia, and non-Hodgkin's lymphoma), but also type-2 diabetes and peripheral neuropathy.

24. William Langston, whom I cited for his study on MPTP (see note 15), the contaminant in synthetic heroin, works for the Parkinson's Institute in Sunnyvale.

25. Alexis Elbaz et al., "CYP2D6 Polymorphism, Pesticide Exposure and Parkinson's Disease," *Annals of Neurology* 55 (March 2004): 430–34. The Prix Épidaure was created by *Le Quotidien du médecin* to encourage research in medicine and ecology.

26. Martine Perez, "Parkinson: le role des pesticides reconnu [Parkinson's: Pesticides' role recognized]," *Le Figaro*, September 27, 2006.

27. Alexis Elbaz and Frédéric Moisan, "Professional Exposure to Pesticides and Parkinson's Disease," *Annals of Neurology* 66 (October 2009): 494–504.

28. Sadie Costello, Myles Cockburn, Jeff Bronstein, Xinbo Zhang, and Beate Ritz, "Parkinson's Disease and Residential Exposure to Maneb and Paraquat from Agricultural Applications in the Central Valley of California," *American Journal of Epidemiology* 169, no. 8 (April 15, 2009): 919–26.

29. David Pimentel, "Amounts of Pesticides Reaching Target Pests: Environmental Impacts and Ethics," *Journal of Agricultural and Environmental Ethics* 8 (1995): 17–29.

30. Hayo van der Werf, "Assessing the Impact of Pesticides on the Environment," *Agriculture, Ecosystems and Environment*, no. 60 (1996): 81–96.

31. Ibid. For more information, see Dwight E. Glotfelty, Alan W. Taylor, Benjamin C. Turner, and William H. Zoller, "Volatization of Surface-Applied Pesticides from Fallow Soil," *Journal of Agriculture and Food Chemistry* 32 (1984): 638–43; and Dennis Gregor and William Gummer, "Evidence of Atmospheric Transport and Deposition of Organochlorine Pesticides and Polychlorinated Biphenyls in Canadian Arctic Snow," *Environmental Science and Technology* 23 (1989): 561–65.

32. David Pimentel, "Amounts of Pesticides Reaching Target Pests: Environmental Impacts and Ethics," *Journal of Agricultural and Environmental Ethics* 8 (1995).

33. Beate Ritz, "Pesticide Exposure Raises Risk of Parkinson's Disease," www .niehs.nih.gov

34. Robert Repetto and Sanjay S. Baliga, *Pesticides and the Immune System: The Public Health Risks* (Washington, DC: World Resources Institute, 1996).

35. Telephone interview with Robert Repetto, June 11, 2009.

36. Repetto and Baliga, *Pesticides and the Immune System*, 22–35.

37. Michael Fournier et al., "Limited Immunotoxic Potential of Technical Formulation of the Herbicide Atrazine (AAtrex) in Mice," *Toxicology Letters* 60 (1992): 263–74.

38. J.G. Vos, E.I. Krajnc, P.K. Beekhof, and M.J. van Logten, "Methods for Testing Immune Effects of Toxic Chemicals: Evaluation of the Immunotoxicity of Various Pesticides in the Rat," in *Pesticide Chemistry, Human Welfare and the Environment: Proceedings of the 5th International Congress of Pesticide Chemistry*, ed. Junshi Miyamoto (Oxford: Pergamon Press, 1983), 497–504.

39. A. Walsh and William E. Ribelin, "The Pathology of Pesticide Poisoning," in *The Pathology of Fishes*, ed. William E. Ribelin and George Migaki (Madison: University of Wisconsin Press, 1975), 515–57.

40. S. De Guise, D. Martineau, P. Béland, and M. Fournier, "Possible Mechanisms of Action of Environmental Contaminants on St. Lawrence Beluga Whales (*Delphinapterusleucas*)," *Environmental Health Perspectives* 103, suppl. 4 (May 1995): 73–77.

41. Marlise Simons, "Dead Mediterranean Dolphins Give Nations Pause," *New York Times*, February 2, 1992.

42. Alex Aguilar, "The Striped Dolphin Epizootic in the Mediterranean Sea," *Ambio* 22 (December 1993): 524–28.

43. Rik de Swart, "Impaired Immunity in Harbor Seals (*Phocavitulina*) Exposed to Bioaccumulated Environmental Contaminants: Review of a Long-Term Feeding Study," *Environmental Health Perspectives* 104, no. 4 (August 1996): 823–28.

44. Arthur I. Holleb, Diane J. Fink, and Gerald Patrick Murphy, "Principles of Tumor Immunology," in *The America Cancer Society Textbook of Clinical Oncology* (Atlanta, GA: The Society, 1991), 71–79.

45. K. Abrams, D.J. Hogan, and H.I. Maibach, "Pesticide-Related Dermatoses in Agricultural Workers," *Occupational Medicine. State of the Art Reviews* 6, no. 3 (July–September 1991): 463–92.

46. "Unknown two centuries ago, atopic syndrome, the predisposition to develop an allergy, affects more than 15% of the population worldwide today and, in all likelihood, 20–30% in industrialized countries," notes Mohamed Laaidi ("Synergie entre pollens et polluantschimiques de l'air: les risquescroisés [Synergy Between Pollen and Chemical Pollutants in the Air: Intertwined Risks]," *Environnement, Risques& Santé* 1, no. 1 (March–April 2002): 42–49.

47. WHO/UNEP, *Sound Management of Pesticides*, 94.

48. John Acquavella et al., "A Critique of the World Resources Institutes Report 'Pesticides and the Immune System: The Public Health Risks,'" *Environmental Health Perspectives* 106 (February 1998): 51–54.

7: The Sinister Side of Progress

1. Author interview with Peter Infante, Washington, DC, October 16, 2009. Among his studies, see Peter Infante and Gwen K. Pohl, "Living in a Chemical World: Actions and Reactions to Industrial Carcinogens," *Teratogenesis, Carcinogenesis and Mutagenesis* 8, no. 4 (1988): 225–49. The authors write: "The synthesis of chemical substances has resulted in technological benefits for society and has also caused increased chemical exposures and risks of related cancers."

2. Peter Infante directed, at OSHA, the Office of Carcinogen Identification and Classification from 1978 to 1983, then the Office of Standards Review from 1983 to 2002.

3. In 1992, Albert Arnold Gore, a Democrat, was elected vice president of the United States, under William "Bill" Clinton (1992–2000). An unlucky presidential candidate who ran against George W. Bush in November 2000, he became the spokesperson for the fight against global warming, thanks to his film *An Inconvenient Truth* (2006).

4. Letter from Albert Gore Jr. to Thorne G. Auchter, July 1, 1981. The letter is accessible on the NIH website: www.profiles.nlm.nih.gov.

5. Geneviève Barbier and Armand Farrachi, *La Société cancérigène* (Paris: Seuil, 2007).

6. Ibid., 16.

7. Neoplasia is the formation of tumors, either benign or malignant (cancerous).

8. Jean Guilaine, ed., *La Préhistoire française. Civilisations néolithiques et protohistoriques*, vol. 2 (Paris: Éditions du CNRS, 1976).

9. John Newby and Vyvyan Howard, "Environmental Influences in Cancer Aetiology," *Journal of Nutritional & Environmental Medicine* (2006): 1–59.

10. For Hippocrates, the body was composed of four "humors": blood, phlegm (in the brain), yellow bile (in the gallbladder) and black bile (in the spleen).

11. Ibid., 9.

12. Vilhjalmur Stefansson, *Cancer: Disease of Civilization? An Anthropological and Historical Study* (New York: Hill & Wang, 1960); see also Zac Goldsmith, "Cancer: A Disease of Industrialization," *The Ecologist*, no. 28 (March–April 1998): 93–99.

13. John Lyman Bulkley, "Cancer Among Primitive Tribes," *Cancer* 4 (1927): 289–95 (cited by Stefansson, *Cancer*).

14. Goldsmith, "Cancer," 95.

15. Weston A. Price, "Report of an Interview with Dr. Joseph Herman Romig: Nutrition and Physical Degeneration," 1939 (cited by Stefansson, *Cancer*).

16. Alexander Berglas, "Cancer: Nature, Cause and Cure," Institut Pasteur, Paris, 1957 (cited by Stefansson, *Cancer*).

17. Frederick Hoffman, "Cancer and Civilization," speech to Belgian National Cancer Congress at Brussels, 1923 (cited by Stefansson, *Cancer*).

18. Albert Schweitzer, *À l'orée de la forêt vierge* (Lausanne: La Concorde, 1923) (cited by Barbier and Farrachi, *La Société cancérigène*, 18; English translation by Ch. Th. Campion, *On the Edge of the Primeval Forest* [London: Whitefriars Press, 1924]).

19. R. de Bovis, "L'augmentation de la fréquence des cancers. Sa predominance dans les villes et sa predilection pour le sexe feminine sont-elles réelles ou apparentes?" *La Semaine médicale*, September 1902 (cited by Barbier and Farrachi, *La Société cancérigène*, 19).

20. Giuseppe Tallarico, *La Vie des aliments* (Paris: Denoël, 1947), 249.

21. Pierre Darmon, "Le mythe de la civilisation cancérogène (1890–1970)," *Communications*, no. 57 (1993): 73.

22. Ibid., 71.

23. In 1906, in France, infectious pathologies made up 19 percent of the causes of mortality, with tuberculosis and diphtheria at the top of the list; today they make up 1.8 percent of mortality causes, and cancer, 27 percent.

24. Roger Williams, "The Continued Increase of Cancer with Remarks as to Its Causations," *British Medical Journal* (1896): 244 (cited by Darmon, "Le mythe de la civilisation cancérogène," 71).

25. Situated in southern Sweden, Follingsbro was known for its foundries, and then its steel industry, both of which started in the eighteenth century.

26. Darmon, "Le mythe de la civilisation cancérogène."

27. Ibid., p. 73.

28. This work was translated into English in 1912 by an engineer working in American mines, Herbert Hoover (with his wife, Lou), who would become the thirty-first president of the United States (1929–33).

29. Bernardo Ramazzini, *Diseases of Workers* (New York: Hafner Publishing Company, 1964), 248. Emphasis mine.

30. Paul Blanc, *How Everyday Products Make People Sick: Toxins at Home and in the Workplace* (Berkeley: University of California Press, 2007), 31.

31. Karl Marx, *Das Kapital: Book One*, trans. Samuel Moor and Edward Aveling (Marx/Engels Internet Archive [marxists.org], 1999).

32. Kerrie Schoffer and John O'Sullivan, "Charles Dickens: The Man, Medicine and Movement Disorders," *Journal of Clinical Neuroscience* 13, no. 9 (2006): 898–901.

33. Alex Wilde, "Charles Dickens Could Spot the Shakes," *ABC Science*, www .abc.net.au/science/articles/2006/10/19/1763460.htm, October 19, 2006.

34. Percivall Pott, *The Chirurgical Works of Percivall Pott*, vol. 5 (London: Hawes Clark and Collins, 1775), 50–54 (cited by Blanc, *How Everyday Products Make People Sick*, 228).

35. Henry Butlin, "On Cancer of the Scrotum in Chimney-Sweeps and Others: Three Lectures Delivered at the Royal College of Surgeons of England," British Medical Association, 1892 (cited by Blanc, *How Everyday Products Make People Sick*, 228).

36. Creosote is an oil extracted from wood or coal tar. In light of its carcinogenic properties, the European Union banned the sale of wood treated with creosote in 2001.

37. Hugh Campbell Ross and John Westray Cropper, "The Problem of the Gasworks Pitch Industry and Cancer," in *The John Howard McFadden Researches* (London: John Murray, 1912).

38. Blanc, *How Everyday Products Make People Sick*, 229.

39. Ibid., 132.

40. A carbon disulfide-based pesticide, metam sodium, is still widely used today for the disinfection of soil (before strawberry cultivation), as a germination inhibitor, or for treating grain.

41. Auguste Delpech, "Accidents que développe chez les ouvriers en caoutchouc l'inhalation du sulfure de carbone en vapeur," *L'Union médicale* 10, no. 60 (May 31, 1856) (cited by Blanc, *How Everyday Products Make People Sick*, 142).

42. Auguste Delpech, "Accidents produits par l'inhalation du sulfure de carbone en vapeur: expériences sur les animaux," *Gazette hebdomadaire de médecine et de chirurgie*, May 30, 1856, 384–85 (cited by Blanc, *How Everyday Products Make People Sick*).

43. Auguste Delpech, "Industrie du caoutchouc soufflé: recherches sur l'intoxication spéciale que détermine le sulfure de carbone," *Annales d'hygiène publique et de médecine légale* 19 (1863): 65–183 (cited by Blanc, *How Everyday Products Make People Sick*, 143).

44. "Unhealthy Trades," *London Times*, September 26, 1863.

45. Jean-Martin Charcot, "Leçon de mardi à La Salpêtrière: Policlinique 1888–1889, notes de cours de MM. Blin, Charcot, Henri Colin," *Le Progrès médical*, 1889, 43–53 (cited by Blanc, *How Everyday Products Make People Sick*, 146).

46. The term "gassed" was invented by the British rubber industry, as the *Oxford English Dictionary* explains, which cites an article from the *Liverpool Daily* in 1889.

47. Thomas Oliver, "Indiarubber: Dangers Incidental to the Use of Bisulphide of Carbon and Naphtha," in *Dangerous Trades* (London: J. Murray, 1902), 470–74 (cited by Blanc, *How Everyday Products Make People Sick*, 151).

48. Paul Blanc reports that in the 1930s, women working in "rayon" workshops in Belgium were transported by a special train to avoid contact with the other personnel, due to their "licentious behavior" (Blanc, *How Everyday Products Make People Sick*, 159).

49. Ibid., 168.

50. Isaac Berenblum, "Cancer Research in Historical Perspective: An Autobiographical Essay," *Cancer Research*, January 1977, 1–7.

51. Devra Davis, *The Secret History of the War on Cancer* (New York: Basic Books, 2007), 18.

52. "International Cancer Congress," *Nature* 137 (March 14, 1936): 426.

53. Davis, *Secret History of the War on Cancer*, 19–21.

54. William Cramer, "The Importance of Statistical Investigations in the Campaign Against Cancer," *Report of the Second International Congress of Scientific and Social Campaign Against Cancer*, Brussels, 1936 (cited by Davis, *Secret History of the War on Cancer*, 21).

55. Davis, *Secret History of the War on Cancer*, 23.

56. In 1925, Switzerland and Germany included bladder cancer in their tables of occupational diseases linked to benzidine and BNA. In France, table 15 *ter* (15 c) concerning "proliferative lesions of the bladder brought on by aromatic amines and their salts" only dates from 1995.

57. International Labour Office, "Cancer of the Bladder Among Workers in Aniline Factories," Studies and Reports Series F, no. 1, Geneva, 1921.

58. David Michaels, "When Science Isn't Enough: Wilhelm Hueper, Robert A. M. Case and the Limits of Scientific Evidence in Preventing Occupational Bladder Cancer," *International Journal of Occupational and Environmental Health* 1 (1995): 278–88.

59. Edgar E. Evans, "Causative Agents and Protective Measures in the Anilin Tumor of the Bladder," *Journal of Urology* 38 (1936): 212–15.

60. Wilhelm Hueper, unpublished autobiography, National Library of Medicine, Washington, DC (cited by David Michaels, *Doubt Is Their Product: How Industry's Assault on Science Threatens Your Health* [New York: Oxford University Press, 2008], 21).

61. W.C. Hueper, F.H. Wiley, and H.D. Wolfe, "Experimental Production of Bladder Tumours in Dogs by Administration of Beta-naphtylamine," *Journal of Industrial Hygiene and Toxicology* 20 (1938): 46–84.

62. In 1982, DuPont removed "through *chemistry*" from its slogan, to eventually end up with "Better living through the miracles of science."

63. Hueper, unpublished autobiography (cited by Michaels, "When Science Isn't Enough," 283).

64. Michaels, *Doubt Is Their Product*, 24.

65. Ibid., 19–20. This letter can be consulted on David Michaels's website: www .defendingscience.org/upload/Evans_1947.pdf.

66. E. Ward, A. Carpenter, S. Markowitz, D. Roberts, and W. Halperin, "Excess Number of Bladder Cancers in Workers Exposed to Ortho-toluidine and Aniline," *Journal of the National Cancer Institute* 3 (1991): 501–6.

67. Michaels, "When Science Isn't Enough," 286.

8: Industry Lays Down the Law

1. Devra Davis, *The Secret History of the War on Cancer* (New York: Basic Books, 2007), 78.

2. Gerald Markowitz and David Rosner, *Deceit and Denial: The Deadly Politics of Industrial Pollution* (Berkeley: University of California Press, 2002), 15.

3. Ibid., 137.

4. William Kovarik, "Ethyl-Leaded Gasoline: How a Classic Occupational Disease Became an International Public Health Disaster," *International Journal of Occupational and Environmental Health* (October–December 2005), 384–439.

5. Markowitz and Rosner, *Deceit and Denial*. Chapter 2 is devoted to the "house of butterflies," 12–25.

6. Gerald Markowitz and David Rosner, "A Gift of God? The Public Health Controversy over Leaded Gasoline in the 1920s," *American Journal of Public Health* 75 (1985): 344–51.

7. Kovarik, "Ethyl-Leaded Gasoline," 384.

8. "Bar Ethyl Gasoline as 5th Victim Dies," *New York Times*, October 31, 1924.

9. Leaded gasoline was not definitively banned in the United States until 1986, and in Europe not until 2000.

10. "Chicago Issues Ban on Leaded Gasoline," *New York Times*, September 8, 1984.

11. "Bar Ethyl Gasoline as 5th Victim Dies."

12. "Use of Ethylated Gasoline Barred Pending Inquiry," *The World*, October 31, 1924.

13. "No Reason for Abandonment," *New York Times*, November 28, 1924.

14. Kehoe Papers, University of Cincinnati (cited by Davis, *Secret History of the War on Cancer*, 81).

15. Davis, *Secret History of the War on Cancer*, 81.

16. Ibid., 94.

17. René Allendy, *Paracelse. Le médecin maudit* (Paris: Dervy-Livres, 1987).

18. Paracelsus, "Liber paragraphorum," *Sämtliche Werke* (Éditions K. Sudhoff), vol. 4, 1–4.

19. Andrée Mathieu, "Le 500e anniversaire de Paracelse," *L'Agora* 1, no. 4 (December 1993–January 1994).

20. An inveterate rebel, Paracelsus did not write in Latin, but in German. For German speakers, the original phrase is: "Alle Ding sind Gift, und nichts ohne Gift; allein die Dosis macht, das ein Ding kein Gift ist," the literal translation of which is: "All things are poison, and nothing is without poison; only the dose allows something not to be poisonous."

21. Michel Gérin et al., *Environnement et santé publique: Fondements et pratiques* (Montreal: Edisem, 2003), 120. It is suspected that the poisons used by the unfortunate king—who was eventually killed by a mercenary—were too old and had lost their potency.

22. Kovarik, "Ethyl-Leaded Gasoline," 391.

23. Statement of Robert Kehoe, June 8, 1966, *Hearings before a Subcommittee on Air and Water Pollution of the Committee on Public Works* (Washington, DC: GPO, 1966), 222 (cited by Kovarik, "Ethyl-Leaded Gasoline").

24. Markowitz and Rosner, *Deceit and Denial*, 110.

25. Kovarik, "Ethyl-Leaded Gasoline," 391.

26. Wilhelm Hueper, unpublished autobiography, 222–23 (cited by Davis, *Secret History of the War on Cancer*, 98).

27. I attended and filmed this conference.

28. The book has since been published: Devra Davis, *Disconnect: The Truth About Cell Phone Radiation, What the Industry Has Done to Hide It, and How to Protect Your Family* (New York; Dutton Adult, 2010).

29. This experience nourished her first book, *When Smoke Ran Like Water: Tales of Environmental Deception and the Battle Against Pollution* (New York: Basic Books, 2002).

30. Author interview with Devra Davis, Pittsburgh, October 15, 2009.

31. See in particular Gérard Dubois, *Le Rideau de fumée: Les méthodes secrètes de l'industrie du tabac* (Paris: Seuil, 2003).

32. John Hill, *Cautions Against the Immoderate Use of Snuff* (London, 1761), 27–38.

33. Étienne Frédéric Bouisson, *Tribut à la chirurgie* (Paris: Baillière, 1858–61), vol. 1, 259–303.

34. Angel Honorio Roffo, "Der Tabak als Krebserzeugendes Agens," *Deutsche Medizinische Wochenschrift* 63 (1937): 1267–71.

35. The Buenos Aires researcher published his work in journals coming out of Germany, the only country interested in tobacco at the time, because the prevalence of cancer there was the highest in the world (59 percent for stomach cancer and 23 percent for lung cancer).

36. Franz Hermann Müller, "Tabakmissbrauch und Lungencarcinom," *Zeitschrift für Krebsforschung* 49 (1939): 57–85. By "very heavy smoker," Franz Müller meant someone who smokes "ten to fifteen cigars, plus thirty five cigarettes or fifty grams of pipe tobacco" daily.

37. Robert N. Proctor, *The Nazi War on Cancer* (Princeton, NJ: Princeton University Press, 2000), 174; see also Robert N. Proctor, "The Nazi War on Tobacco:

Ideology, Evidence and Possible Cancer Consequences," *Bulletin of the History of Medicine* 71, no. 3 (1997): 435–88.

38. Eberhard Schairer and Erich Schöniger, "Lungenkrebs und Tabakverbrauch," *Zeitschrift für Krebsforschung* 54 (1943): 261–69. The results of this study were re-evaluated in 1995 with more modern statistical tools; the conclusion was that the probability they were due to chance was one in 10 million (George Davey et al., "Smoking and Death," *British Medical Journal* 310 [1995]: 396).

39. Anecdote told by Richard Doll to Robert Proctor in 1997 (Proctor, *Nazi War on Cancer*, 46).

40. Richard Doll and Bradford Hill, "Smoking and Carcinoma of the Lung," *British Medical Journal* 2 (September 30, 1950): 739–48.

41. Davis, *Secret History of the War on Cancer*, 147.

42. Cuyler Hammond and Daniel Horn, "The Relationship Between Human Smoking Habits and Death Rates: A Follow-Up Study of 187,766 Men," *Journal of the American Medical Association*, August 7, 1954, 1316–28. The other studies are Ernest Wynder and Evarts Graham, "Tobacco Smoking as a Possible Etiologic Factor in Bronchiogenic Carcinoma," *Journal of the American Medical Association* 143 (1950): 329–36; Robert Schrek, Lyle A. Baker, George P. Ballard, and Sidney Dolgoff, "Tobacco Smoking as an Etiologic Factor in Disease. I. Cancer," *Cancer Research* 10 (1950): 49–58; Morton L. Levin, Hyman Goldstein, and Paul R. Gerhardt, "Cancer and Tobacco Smoking: A Preliminary Report," *Journal of the American Medical Association* 143 (1950): 336–38; Ernest L. Wynder, Evarts A. Graham, and Adele B. Croninger, "Experimental Production of Carcinoma with Cigarette Tar," *Cancer Research* 13 (1953): 855–64.

43. *Time* magazine, March 22, 1937.

44. *US News and World Report*, July 2, 1954.

45. Brown & Williamson Tobacco Corp., "Smoking and Health Proposal," Brown & Williamson document no. 68056, 1969, 1778–86, legacy.library.ucsf.edu/tid/nvs/40f00 (emphasis in original).

46. "A Frank Statement to Cigarette Smokers," sent to over fifty major newspapers by the TIRC, January 4, 1954. The statement's text can be consulted at www.tobacco.org.

47. Robert N. Proctor, "Tobacco and Health. Expert Witness Report Filed on Behalf of Plaintiffs in the United States of America, Plaintiff, v. Philip Morris, Inc., et al., Defendants," Civil Action no. 99-CV-02496 (GK) (federal case), *Journal of Philosophy, Science & Law* 4 (March 2004).

48. "Project Truth: The Smoking/Health Controversy: A View from the Other Side (Prepared for the *Courier-Journal* and *Louisville Times*)," February 8, 1971 (Brown & Williamson Tobacco Corp. document, cited by David Michaels, *Doubt Is Their Product: How Industry's Assault on Science Threatens Your Health* [New York: Oxford University Press, 2008], 3).

49. *Le Nouvel Observateur*, February 24, 1975 (cited by Gérard Dubois, *Le Rideau de fumée: Les méthodes secrètes de l'industrie du tabac* [Paris: Seuil, 2003], 290).

50. See in particular Nadia Collot's film *Tabac: la conspiration* (2006).

51. Evarts Graham, "Remarks on the Aetiology of Bronchogenic Carcinoma," *The Lancet* 263, no. 6826 (June 26, 1954): 1305–8.

52. Christie Todd Whitman, "Effective Policy Making: The Role of Good Science," remarks at the National Academy of Science's symposium on nutrient overenrichment of coastal waters, October 13, 2000 (cited by Michaels, *Doubt Is Their Product*, x).

53. Cited by Elisa Ong and Stanton Glantz, "Constructing 'Sound Science' and 'Good Epidemiology': Tobacco, Lawyers and Public Relations Firms," *American Journal of Public Health* 91, no. 11 (November 2011): 1749–57 (emphasis added). This document, as well as all those I cite in this section, are accessible on a website launched by Philip Morris following the guilty verdict handed down by the court: www.pmdocs.com/Disclaimer.aspx.

54. Imperial Chemical Industries was bought by AzkoNobel in 2008.

55. Ong and Glantz, "Constructing 'Sound Science.'"

56. André Cicolella and Dorothée Benoît Browaeys, *Alertes santé: Experts et citoyens face aux intérêts privés* (Paris: Fayard, 2005), 301.

57. Ibid., 299.

58. Michaels, *Doubt Is Their Product*, 57.

9: Mercenaries of Science

1. Author interview with Peter Infante, Washington, DC, October 16, 2009.

2. Marie-Monique Robin, *The World According to Monsanto: Pollution, Corruption, and the Control of Our Food Supply*, trans. George Holoch (New York: The New Press, 2010), 46.

3. David Michaels, *Doubt Is Their Product: How Industry's Assault on Science Threatens Your Health* (New York: Oxford University Press, 2008), 60.

4. Ibid., 66.

5. Ibid., 69–70. Emphasis in original.

6. Author interview with Devra Davis, Pittsburgh, October 15, 2009.

7. William Ruckelshaus, "Risk in a Free Society," *Environmental Law Reporter* 14 (1984): 10190 (cited by Michaels, *Doubt Is Their Product*, 69).

8. "Chronic Exposure to Benzene," *Journal of Industrial Hygiene and Toxicology*, October 1939, 321–77.

9. Paul Blanc, *How Everyday Products Make People Sick: Toxins at Home and in the Workplace* (Berkeley: University of California Press, 2007), 62.

10. Benzol is a mix of three hydrocarbons: benzene, toluene, and xylene.

11. Ibid., 67.

12. American Petroleum Institute, "API Toxicological Review: Benzene," New York, 1948 (cited in Michaels, *Doubt Is Their Product*, 70). I recommend reading this document, which can be consulted on David Michaels's website, www.defending science.org. Emphasis in original.

13. Peter Infante, "The Past Suppression of Industry Knowledge of the Toxicity of Benzene to Humans and Potential Bias in Future Benzene Research," *International Journal of Occupational and Environmental Health* 12 (2006): 268–72.

14. I will come back to Dow Chemical later; the company is one of the current leaders in the pesticide market. Along with Monsanto, Dow was one of the main producers of Agent Orange.

15. Dante Picciano, "Cytogenic Study of Workers Exposed to Benzene," *Environmental Research* 19 (1979): 33–38.

16. P.F. Infante, R.A. Rinsky, J.K. Wagoner, and R.J. Young, "Leukemia in Benzene Workers," *The Lancet* 2 (1977): 76–78.

17. Industrial Union Department v. American Petroleum Institute, July 2, 1980, 448 US 607 (accessible at www.publichealthlaw.net).

18. We will see at the end of this book that the European regulation REACH aims precisely to reverse the responsibility of proof, which is clearly a positive goal.

19. Devra Davis, *The Secret History of the War on Cancer* (New York: Basic Books, 2007), 385.

20. Robert Rinsky et al., "Benzene and Leukemia: An Epidemiologic Risk Assessment," *New England Journal of Medicine* 316, no. 17 (1987): 1044–50. Peter Infante's team determined four exposure levels (per day of work): less than 1 ppm, from 1 to 5 ppm, from 5 to 10 ppm, and more than 10 ppm. There were sixty times more leukemia cases in the last level than in the first.

21. OSHA, "Occupational Exposure to Benzene: Final Rule," *Federal Register* 52 (1987): 34460–578.

22. Peter Infante, "Benzene: Epidemiologic Observations of Leukemia by Cell Type and Adverse Health Effects Associated with Low-Level Exposure," *Environmental Health Perspectives* 52 (October 1983): 75–82.

23. Michaels, *Doubt Is Their Product*, 47.

24. Exponent, 2003 Annual Report, Form 10K SEC filing, June 26, 2005.

25. Susanna Rankin Bohme, John Zorabedian, and David Egilman, "Maximizing Profit and Endangering Health: Corporate Strategies to Avoid Litigation and Regulation," *International Journal of Occupational and Environmental Health* 11 (2005): 338–48.

26. Ibid.

27. Gerald Markowitz and David Rosner, *Deceit and Denial: The Deadly Politics of Industrial Pollution* (Berkeley: University of California Press, 2002). Chapter 6 is entitled "Evidence of Illegal Conspiracy by Industry," 168–94.

28. Bohme et al., "Maximizing Profit and Endangering Health."

29. See Robin, *World According to Monsanto*, 30–35, where I tell the story of Times Beach, a little town in Missouri that was evacuated and then razed in 1983, due to contamination by polychlorinated biphenyls (PCBs) and dioxin, produced by Monsanto. As for Love Canal, located not far from Niagara Falls in the state of New York, it was evacuated in 1978 after the discovery of 21,000 tons of toxic products buried near the Hooker Chemicals factory.

30. Hexavalent chromium or "chromium VI" is produced by the oxidization of chromium. Exposure to this highly toxic substance can cause stomach, lung, liver, and kidney cancer.

31. J.D. Zhang and X.L. Li, "Chromium Pollution of Soil and Water in Jinzhou," *Chinese Journal of Preventive Medicine* 2, no. 5 (1987): 262–64.

32. J.D. Zhang and X.L. Li, "Cancer Mortality in a Chinese Population Exposed to Hexavalent Chromium," *Journal of Occupational and Environmental Medicine* 39, no. 4 (1997): 315–19.

33. "Study Tied Pollutant to Cancer; Then Consultants Got Hold of It," *Wall Street Journal*, December 23, 2005.

34. Paul Brandt-Rauf, "Editorial Retraction," *Journal of Occupational and Environmental Medicine* 48, no. 7 (2006): 749.

35. Richard Hayes et al., "Benzene and the Dose-Related Incidence of Hematologic Neoplasm in China," *Journal of the National Cancer Institute* 89, no. 14 (1997): 1065–71.

36. Pamela Williams and Dennis Paustenbach, "Reconstruction of Benzene Exposure for the Pliofilm Cohort (1936–1976) Using Monte Carlo Techniques," *Journal of Toxicology and Environmental Health* 66, no. 8 (2003): 677–81.

37. Michaels, *Doubt Is Their Product*, 46.

38. Bohme et al., "Maximizing Profit and Endangering Health."

39. Markowitz and Rosner, *Deceit and Denial*.

40. Qinq Lan et al., "Hematotoxicity in Workers Exposed to Low Levels of Benzene," *Science* 306 (December 3, 2004): 1774–76.

41. Benzene Health Research Consortium, "The Shanghai Health Study," PowerPoint presentation, February 1, 2003 (cited by Lorraine Twerdok and Patrick Beatty, "Proposed Studies on the Risk of Benzene-Induced Diseases in China: Costs and Funding"; document can be consulted on David Michaels's website, www.defendingscience.org).

42. Craig Parker, "Memorandum to Manager of Toxicology and Product Safety (Marathon Oil). Subject: International Leveraged Research Proposal," 2000 (document can be consulted on David Michaels's website, www.defendingscience.org).

43. Bohme et al., "Maximizing Profit and Endangering Health."

44. Arnold Relman, "Dealing with Conflicts of Interest," *New England Journal of Medicine* 310 (1984): 1182–83.

45. International Committee of Medical Journal Editors, "Uniform Requirements for Manuscripts Submitted to Biomedical Journals. Ethical Considerations in the Conduct and Reporting of Research: Conflicts of Interest," 2001 (see Frank Davidoff et al., "Sponsorship, Authorship and Accountability," *The Lancet* 358 [September 15, 2001]: 854–56).

46. Created in 1971, the CSPI is an American nongovernmental organization specialized in consumer defense and assistance, which conducts independent scientific studies in the field of health and nutrition.

47. Scientific articles often include several author names that comprise the team that conducted the study. The convention is that the two main authors are the first

and last cited. "Omissions" concerned six articles published in the *Journal of the American Medical Association* (out of fifty-three), three in *Environmental Health Perspectives* (out of thirty-five), two in *Toxicology and Applied Pharmacology* (out of thirty-three), and two in the *New England Journal of Medicine* (out of forty-two).

48. Merrill Goozner, *Unrevealed: Non-Disclosure of Conflicts of Interest in Four Leading Medical and Scientific Journals*, Integrity in Science: Project of the Center of Science in the Public Interest, July 12, 2004.

49. Ibid.

50. "Tough Talking Journal Editor Faces Accusations of Leniency," *New York Times*, August 1, 2006.

51. Catherine D. DeAngelis, Phil B. Fontanarosa, and Annette Flanagin, "Reporting Financial Conflicts of Interest and Relationships Between Investigators and Research Sponsors," *Journal of the American Medical Association* 286 (2001): 89–91.

52. Catherine DeAngelis, "The Influence of Money on Medical Science," *Journal of the American Medical Association* 296 (2006): 996–98.

53. Phil Fontanarosa, Annette Flanagin, and Catherine DeAngelis, "Reporting Conflicts of Interest, Financial Aspects of Research and the Role of Sponsors in Funded Studies," *Journal of the American Medical Association* 294, no. 1 (2005): 110–11.

54. DeAngelis, "Influence of Money on Medical Science."

55. David Michaels, "Science and Government: Disclosure in Regulatory Science," *Science* 302, no. 5653 (December 19, 2003): 2073.

56. Justin Bekelman, Yan Li, and Cary Gross, "Scope and Impact of Financial Conflicts of Interest in Biomedical Research: A Systematic Review," *Journal of the American Medical Association* 289 (2003): 454–65.

57. Astrid James, "*The Lancet*'s Policy on Conflicts of Interest," *The Lancet* 363 (2004): 2–3.

58. Wendy Wagner and Thomas McGarity, "Regulatory Reinforcement of Journal Conflict of Interest Disclosures: How Could Disclosure of Interests Work Better in Medicine, Epidemiology and Public Health?," *Journal of Epidemiology and Community Health* 6 (2009): 606–7.

59. Michaels, "Science and Government."

60. Robin, *World According to Monsanto*, 324–25.

61. Michaels, *Doubt Is Their Product*, 256–57.

10: Institutional Lies

1. President's Cancer Panel, *Reducing Environmental Cancer Risk: What We Can Do Now. 2008–2009 Annual Report*, U.S. Department of Health and Human Services, National Institutes of Health, National Cancer Institute, April 2010.

2. *Les Causes du cancer en France*, report published by the Académie national de médecine (National Academy of Medicine), the Académie nationale des sci-

ences (the National Academy of Sciences)/Institut de France, the International Agency for Research on Cancer (WHO-Lyon), the Fédération nationale des centres de lutte contre le cancer (National Federation of Centers for the Fight Against Cancer), with the cooperation of the Institut national du cancer (National Cancer Institute) and the Institut national de veille sanitaire (National Institute of Health Monitoring), 2007. The abridged French version is 48 pages long and the unabridged version in English is 275 pages long. The excerpts I use come from the French version.

3. Ibid., p. 4. Emphasis added.

4. Ibid., p. 6.

5. On the UIPP website, click on "Infos pesticides" (pesticides info), then "santé et pesticides" (health and pesticides) and "produits pharmaceutiques et cancers" (pharmaceutical products and cancer).

6. *Les Causes du cancer en France*, 42. Emphasis added.

7. Author interview with Richard Clapp, Boston, October 29, 2009

8. André Cicolella and Dorothée Benoît Browaeys, *Alertes santé: Experts et citoyens face aux intérêts privés* (Paris: Fayard, 2005), 155.

9. Académie national des sciences/Comité des applications de l'académie des sciences, *Dioxin and Its Analogues* (London: Technique & Documentation, 1994).

10. Author's interview with André Picot, Paris, June 2, 2009.

11. "Circulaire du 30 mai 1997 relative aux dioxines et furanes" (Memorandum of May 30, 1997, relating to dioxins and furans) addressed by the Ministry of the Environment to departmental prefects.

12. International Agency for Research on Cancer (IARC), *Polychlorinated Dibenzo-para-Dioxins and Polychlorinated Dibenzofurans,* IARC Monographs on the Evaluation of Carcinogenic Risks to Humans, vol. 69, July 1997.

13. See notably Roger Lenglet, *L'Affaire de l'amiante* (Paris: La Découverte, 1996).

14. Frédéric Denhez, *Les Pollutions invisibles: Quelles sont les vraies catastrophes écologiques?* (Paris: Delachaux et Niestlé, 2006), 220.

15. Activities at Danmarie-les-Lys stopped in 1993, and at Descartes in 1996. Soon after, former workers of these two factories suffering from mesothelioma filed a lawsuit against Saint-Gobain, supported by the National Association of Defense of Asbestos Victims (Association nationale de défense des victimes de l'amiante, ANDEVA).

16. Gérard Dériot and Jean-Pierre Godefroy, *Le Drame de l'amiante en France: comprendre, mieux réparer, en tirer des leçons pour l'avenir,* Informational report no. 37, Senate, Paris, October 26, 2005.

17. Étienne Fournier, who directed the Fernand-Widal Anti-Poison Center in Paris, was one of those who sponsored the creation of the CPA.

18. Étienne Fournier, "Amiante et protection de la population exposée à l'inhalation de fibres d'amiante dans les bâtiments publics et privés," *Bulletin de l'Académie nationale de médecine*, 180, no. 4–16 (April 30, 1996).

19. At the time, this report was strongly criticized by a scholar named Claude Allègre, who declared: "It is worthless. This report is not scientifically sound" (*Le Point*, October 16, 1997).

20. INSERM, *Effets sur la santé des principaux types d'exposition à l'amiante*, La Documentation française, Paris, January 1997.

21. Joseph LaDou, "The Asbestos Cancer Epidemic," *Environmental Health Perspectives* 112, no. 3 (March 2004): 285–90. It is estimated that 30 million tons of asbestos were used over the course of the twentieth century.

22. *Les Causes du cancer en France*, 24.

23. All of this information comes from a document that can be consulted on IARC's website: "Sécurité et prévention. Risques liés à la prévention des produits cancérogènes. Liste réactualisée des produits génotoxiques classes par le CIRC" (last updated in August 2010).

24. Author interview with Vincent Cogliano, Lyon, February 10, 2010.

25. Ibid. In December 2010 I learned that Vincent Cogliano had returned to his original job at the Environmental Protection Agency.

26. Paolo Bofetta directed the environmental cancer epidemiology section from 1995 to 2003, and then the genetics and epidemiology group.

27. Paolo Bofetta et al., "The Causes of Cancer in France," *Annals of Oncology* 20, no. 3 (March 2009): 550–55.

28. Author interview with Christopher Wild, Lyon, February 10, 2010.

29. *Les Causes du cancer en France*, 47.

30. "Time to Strengthen Public Confidence at IARC," *The Lancet* 371, no. 9623 (May 3, 2008): 1478.

31. "Transparency at IARC," *The Lancet* 361, no. 9353 (January 18, 2003): 189.

32. Lorenzo Tomatis, "The IARC Monographs Program: Changing Attitudes Towards Public Health," *International Journal of Occupational and Environmental Health* 8, no. 2 (April–June 2002): 144–52. Lorenzo Tomatis passed away in 2007.

33. "Letter to Dr. Gro Harlem Brundtland, Director General WHO," February 25, 2002, published in the *International Journal of Occupational and Environmental Health* 8, no. 3 (July–September 2002): 271–73.

34. James Huff et al., "Multiple-Site Carcinogenicity of Benzene in Fischer 344 Rats and B6C3F1 Mice," *Environmental Health Perspectives* 82 (1989): 125–63; James Huff, "National Toxicology Program. NTP Toxicology and Carcinogenesis Studies of Benzene (Case No. 71-43-2) in F344/N Rats and B6C3F1 Mice (Gavage Studies)," National Toxicology Program, *Technical Report Series* 289 (1986): 1–277.

35. Author interview with James Huff, Research Triangle Park, October 27, 2009.

36. Dan Ferber, "NIEHS Toxicologist Receives a 'Gag Order,'" *Science* 297 (August 9, 2002): 215.

37. Ibid.

38. Ibid.

39. Ibid.

40. In 2002, the American Public Health Association, which is made up of 55,000 members, gave James Huff the David P. Rall Award for Advocacy in Public Health, which recognizes those who have "made an outstanding contribution to public health through science-based advocacy."

41. Author interview with James Huff, Research Triangle Park, October 27, 2009.

42. James Huff, "IARC Monographs, Industry Influence, and Upgrading, Downgrading, and Under-Grading Chemicals: A Personal Point of View," *International Journal of Occupational and Environmental Health* 8, no. 3 (July–September 2002): 249–70.

43. Author interview with Vincent Cogliano, Lyon, February 10, 2010.

44. Cicolella and Browaeys, *Alertes santé*, 203.

45. Sawdust is also suspected of causing cancer in the nasal cavity and sinus, which is recognized in France as an occupational disease (table 47).

46. James Huff, "IARC and the DEHP Quagmire," *International Journal of Occupational and Environmental Health* 9, no. 4 (October–December 2003): 402–4; National Toxicology Program, "Carcinogenesis Bioassay of Di (2-ethylhexyl) Phthalate (Case No. 117-81-7) in F344 Rats and B6C3F1 Mice (Feed Studies)," NTP TR 217, Research Triangle Park, 1982; W.M. Kluwe, J.K. Haseman, J.F. Douglas, and J.E. Huff, "The Carcinogenicity of Dietary Di-2-ethylhexyl Phthalate (DEHP) in Fischer 344 Rats and B6C3F1 Mice," *Journal of Toxicology and Environmental Health* 10 (1983): 797–815.

47. Raymond M. David, Michael R. Moore, Dean C. Finney, and Derek Guest, "Chronic Toxicity of Di (2-ethylhexyl) Phthalate in Rats," *Toxicological Sciences* 55 (2000): 433–43.

48. Ronald Melnick, "Suppression of Crucial Information in the IARC Evaluation of DEHP," *International Journal of Occupational and Environmental Health* 9 (October–December 2003): 84–85.

49. Cited in Ronald Melnick, James Huff, Charlotte Brody, and Joseph Digangi, "The IARC Evaluation of DEHP Excludes Key Papers Demonstrating Carcinogenic Effects," *International Journal of Occupational and Environmental Health* 9 (October–December 2003): 400–401.

50. Author interview with Devra Davis, Pittsburgh, October 15, 2009.

51. Author interview with Peter Infante, October 16, 2009.

52. David Michaels, *Doubt Is Their Product: How Industry's Assault on Science Threatens Your Health* (New York: Oxford University Press, 2008), 60–61.

53. Author interview with Vincent Cogliano, Lyon, February 10, 2010. For more information on this critical subject, see Ronald Melnick, Kristina Thayer, and John Bucher, "Conflicting Views on Chemical Carcinogenesis Arising from the Design and Evaluation of Rodent Carcinogenicity Studies," *Environmental Health Perspectives* 116, no. 1 (January 2008): 130–35.

11: An Epidemic of Chronic Diseases

1. Devra Davis, *The Secret History of the War on Cancer* (New York: Basic Books, 2007), 262.

2. Ibid., 147.

3. Ibid., 255.

4. Richard Doll and Richard Peto, "The Causes of Cancer: Quantitative Estimates of Avoidable Risks of Cancer in the United States Today," *Journal of the National Cancer Institute* 66, no. 6 (June 1981): 1191–308.

5. Geneviève Barbier and Armand Farrachi, *La Société cancérigène* (Paris: Seuil, 2007), 49.

6. Lucien Abenhaim, *Rapport de la Commission d'orientation sur le cancer* (Paris: La Documentation française, 2003).

7. *Les Causes du cancer en France*, report published by the Académie national de médecine (National Academy of Medicine), the Académie nationale des sciences (the National Academy of Sciences)/Institut de France, the International Agency for Research on Cancer (WHO-Lyon), and the Fédération nationale des centres de lutte contre le cancer (National Federation of Centers for the Fight Against Cancer), with the cooperation of the Institut national du cancer (National Cancer Institute) and the Institut national de veille sanitaire (National Institute of Health Monitoring), 7.

8. Rory O'Neill, Simon Pickvance, and Andrew Watterson, "Burying the Evidence: How Great Britain Is Prolonging the Occupational Cancer Epidemic," *International Journal of Occupational and Environmental Health* 13 (2007): 432–40.

9. André Cicolella, *Le Défi des épidémies modernes: Comment sauver la Sécu en changeant le système de santé* (Paris: La Découverte, 2007), 48.

10. Eva Steliarova-Foucher et al., "Geographical Patterns and Time Trends of Cancer Incidence and Survival Among Children and Adolescents in Europe Since the 1970s (The ACCIS Project): An Epidemiological Study," *The Lancet* 364, no. 9451 (December 11, 2004): 2097–105.

11. This interview was filmed on January 13, 2010, and the transcription is word for word.

12. Author interview with Devra Davis, Pittsburgh, October 15, 2009.

13. Devra Davis and Joel Schwartz, "Trends in Cancer Mortality: US White Males and Females, 1968–1983," *The Lancet* 331, no. 8586 (1988): 633–36.

14. Devra Davis and David Hoel, "Trends in Cancer in Industrial Countries," *Annals of the New York Academy of Sciences* 609 (1990).

15. Davis, *Secret History of the War on Cancer*, 257.

16. Devra Davis, Abraham Lilienfeld, and Allen Gittelsohn, "Increasing Trends in Some Cancers in Older Americans: Fact or Artifact?" *Toxicology and Industrial Health* 2, no. 1 (1986): 127–44.

17. President's Cancer Panel, *Reducing Environmental Cancer Risk. What We Can Do Now. 2008–2009 Annual Report*, U.S. Department of Health and Human Services, National Institutes of Health, National Cancer Institute, April 2010, 4.

18. Philippe Irigaray et al., "Lifestyle-Related Factors and Environmental Agents Causing Cancer: An Overview," *Biomedicine & Pharmacotherapy* 61 (2007): 640–58.

19. See J.L. Botha, F. Bray, R. Sankila, and D.M. Parkin, "Breast Cancer Incidence and Mortality Trends in 16 European Countries," *European Journal of Cancer* 39 (2003): 1718–29.

20. In France, the first cancer registry was created in 1975. In 2010, there were thirteen registries measuring the incidence of all cancers in eleven regions (out of ninety-six), which covers only 13 percent of the population.

21. Dominique Belpomme et al., "The Growing Incidence of Cancer: Role of Lifestyle and Screening Detection (Review)," *International Journal of Oncology* 30, no. 5 (May 2007): 1037–49.

22. J.A. Newby, C.C. Busby, C.V. Howard, and M.J. Platt, "The Cancer Incidence Temporality Index: An Index to Show Temporal Changes in the Age of Onset of Overall and Specific Cancer (England and Wales, 1971–1999)," *Biomedicine & Pharmacotherapy* 61 (2007): 623–30.

23. Cicolella, *Le Défi des épidémies modernes*, 21–22. The risk of stomach cancer was reduced by a factor of five for women and 2.5 for men. This decrease is attributed to the use of the refrigerator, which led to lower consumption of salted and smoked products, which can cause stomach cancer.

24. Belpomme et al., "Growing Incidence of Cancer."

25. Catherine Hill and Agnès Laplanche, "Tabagisme et mortalité: aspects épidémiologiques," *Bulletin épidémiologique hebdomadaire*, no. 22–23 (May 27, 2003).

26. Barbier and Farrachi, *La Société cancérigène*, 38.

27. Abenhaim, *Rapport de la Commission d'orientation sur le cancer.*

28. Barbier and Farrachi, *La Société cancérigène*, 35.

29. Lennart Hardell and Anita Sandstrom, "Case-Control Study: Soft Tissue Sarcomas and Exposure to Phenoxyacetic Acids or Chlorophenols," *British Journal of Cancer* 39 (1979): 711–17; M. Eriksson, L. Hardell, N.O. Berg, T. Möller, and O. Axelson, "Soft Tissue Sarcoma and Exposure to Chemical Substances: A Case Referent Study," *British Journal of Industrial Medicine* 38 (1981): 27–33; L. Hardell, M. Eriksson, P. Lenner, and E. Lundgren, "Malignant Lymphoma and Exposure to Chemicals, Especially Organic Solvents, Chlorophenols and Phenoxy Acids," *British Journal of Cancer* 43 (1981): 169–76; Lennart Hardell and Mikael Eriksson, "The Association Between Soft-Tissue Sarcomas and Exposure to Phenoxyacetic Acids: A New Case Referent Study," *Cancer* 62 (1988): 652–56.

30. *Royal Commission on the Use and Effects of Chemical Agents on Australian Personnel in Vietnam, Final Report*, vols. 1–9 (Canberra: Australian Government Publishing Service, 1985).

31. "Agent Orange: The New Controversy. Brian Martin Looks at the Royal Commission That Acquitted Agent Orange," *Australian Society* 5, no. 11 (November 1986): 25–26.

32. Monsanto Australia Ltd., "Axelson and Hardell. The Odd Men Out. Submission to the Royal Commission on the Use and Effects on Chemical Agents on Australian Personnel in Vietnam," 1985.

33. Cited in Lennart Hardell, Mikael Eriksson, and Olav Axelson, "On the Misinterpretation of Epidemiological Evidence, Relating to Dioxin-Containing Phenoxyacetic Acids, Chlorophenols and Cancer Effects," *New Solutions*, Spring 1994.

34. Chris Beckett, "Illustrations from the Wellcome Library. An Epidemiologist at Work: The Personal Papers of Sir Richard Doll," *Medical History* 46 (2002): 403–21.

35. Marie-Monique Robin, *The World According to Monsanto: Pollution, Corruption, and the Control of the World's Food Supply*, trans. George Holoch (New York: The New Press, 2010), 63.

36. Sarah Boseley, "Renowned Cancer Scientist Was Paid by Chemical Firm for 20 Years," *The Guardian*, December 8, 2006.

37. Cristina Odone, "Richard Doll Was a Hero, Not a Villain," *The Observer*, December 10, 2006.

38. Geoffrey Tweedale, "Hero or Villain?—Sir Richard Doll and Occupational Cancer," *International Journal of Occupational and Environmental Health* 13, no. 2 (April 1, 2007): 233–35.

39. Richard Peto, *The Times*, December 9, 2006.

40. Richard Stott, "Cloud Over Sir Richard," *Sunday Mirror*, December 10, 2006.

41. Julian Peto and Richard Doll, "Passive Smoking," *British Journal of Cancer* 54 (1986): 381–83. Julian Peto is Richard Peto's brother.

42. Elizabeth Fontham et al. on behalf of ACS Cancer and the Environment Subcommittee, "American Cancer Society Perspectives on Environmental Factors and Cancer," *Cancer Journal for Clinicians* 59 (2009): 343–51.

43. Author interview with Michael Thun, Atlanta, October 25, 2009.

44. Hans-Olav Adami was recruited by the famous Dennis Paustenbach at Exponent (see Chapter 9) to minimize the toxicity of dioxin, just when the Environmental Protection Agency was reviewing its regulation (see L. Hardell, M.J. Walker, B. Walhjalt, L.S. Friedman, and E.D. Richter, "Secret Ties to Industry and Conflicting Interests in Cancer Research," *American Journal of Industrial Medicine*, November 13, 2006).

45. Gerald Markowitz and David Rosner, *Deceit and Denial: The Deadly Politics of Industrial Pollution* (Berkeley: University of California Press, 2002), 168.

46. Cited in Robin, *World According to Monsanto*, 9.

47. The identification code for PVC is a triangle with "03" written on the inside.

48. Henry Smyth to T.W. Nale, November 24, 1959 (cited in Markowitz and Rosner, *Deceit and Denial*, 172).

49. Cited in Markowitz and Rosner, *Deceit and Denial*, 173.

50. Letter from Robert Kehoe to R. Emmet Kelly, February 2, 1965, Manufacturing Chemists' Association (MCA) archives (cited in ibid., 174).

51. Letter from R. Emmet Kelly to A.G. Erdman, Pringfield, "PVC Exposure," January 7, 1966, MCA archives (cited in ibid., 174).

52. Letter from Rex Wilson to Dr. J. Newman, "Confidential," January 6, 1966, MCA archives (cited in ibid., 174).

53. Rex H. Wilson, William E. McCormick, Caroll F. Tatum, and John L. Creech, "Occupational Acroosteolysis: Report of 31 Cases," *Journal of the American Medical Association* 201 (1967): 577–81.

54. Letter from Verald Rowe, Biochemical Research Laboratory, to William McCormick, Director, Department of Industrial Hygiene and Toxicology, the B.F. Goodrich Company, May 12, 1959. This document can be consulted online at www.pbs.org/tradesecrets/docs.

55. Pierluigi Viola, "Cancerogenic effect of vinyl chloride," paper presented at the Tenth International Cancer Congress, May 22–29, 1970, Houston, TX; P. Viola, A. Bigotti, and A. Caputo, "Oncogenic Response of Rats, Skin, Lungs and Bones to Vinyl Chloride," *Cancer Research* 31 (May 1971): 516–22.

56. Memorandum from L.B. Crider to William McCormick, Goodrich, "Some New Information on the Relative Toxicity of Vinyl Chloride Monomer," March 24, 1969, MCA archives (cited by Markowitz and Rosner, *Deceit and Denial*, 184). Maltoni's study was published in 1975, despite his sponsors' prohibition on doing so: C. Maltoni and G. Lefemine, "Carcinogenicity Bioassays of Vinyl Chloride: Current Results," *Annals of New York Academy of Sciences* 246 (1975): 195–218.

57. Memorandum from A.C. Siegel (Tenneco Chemicals, Inc.) to G.I. Rozland (Tenneco Chemicals, Inc), "Subject: Vinyl Chloride Technical Task Group Meeting," November 16, 1972 (this document can be consulted on David Michaels's website, www.defendingscience. org).

58. Letter from D.M. Elliott (General Manager, Production, Solvents and Monomers Group, Imperial Chemical Industries Limited, Mond Division) to G.E. Best (MCA), October 30, 1972; "Meeting Minutes: Manufacturing Chemists Association, Vinyl Chloride Research Coordinators," January 30, 1973 (these documents can be consulted on David Michaels's website, www.defendingscience.org).

59. The meetings were led by Theodore Torkelson from Dow Chemical. Represented were Union Carbide, Uniroyal, Ethyl Corporation, Goodrich, Shell Oil Company, Exxon Corporation, Tenneco Chemicals, Diamond Shyrock Corporation, Allied Chemical Corporation, Firestone Plastics Company, Continental Oil Company, and Air Products & Chemicals, Inc.

60. "Meeting Minutes: Manufacturing Chemists' Association, Vinyl Chloride Research Coordinators," May 21, 1973, MCA archives (this document can be consulted on David Michaels's website, www.defendingscience.org).

61. H.L. Kusnetz (Manager of Industrial Hygiene, Head Office, Shell Oil Co.), "Notes on the Meeting of the VC Committee," July 17, 1973, MCA archives (ibid.).

62. R.N. Wheeler (Union Carbide), "Memorandum to Carvajal JL, Dernehl CU, Hanks GJ, Lane KS, Steele AB, Zutty NL. Subject: Vinyl Chloride Research: MCA Report to NIOSH," July 19, 1973, MCA archives (ibid.).

63. John Creech et al., "Angiosarcoma of the Liver Among Polyvinyl Chloride Workers," *Morbidity and Mortality Weekly Report* 23, no. 6 (1974): 49–50.

64. OSHA, "News: OSHA Investigating Goodrich Cancer Fatalities," press release, January 24, 1974 (this document can be consulted on David Michaels's website, www.defendingscience.org).

65. Markus Key, *Deposition in the United States District Court for the Western District of New York, in the Matter of Holly M. Smith v. the Dow Chemical Company; PPG Industries, Inc., and Shell Oil Company v. the Goodyear Tire and Rubber Company*. CA no. 94-CV-0393, September 19, 1995 (ibid.).

66. David Michaels, *Doubt Is Their Product: How Industry's Assault on Science Threatens Your Health* (New York: Oxford University Press, 2008), 36.

67. I invite the reader to consult Hill and Knowlton's website, www.hillandknowl ton.com.

68. Hill and Knowlton, "Recommendations for Public Affairs Program for SPI's Vinyl Chloride Committee. Phase 1: Preparation for OSHA Hearings," June 1974 (this document can be consulted on David Michaels's website, www.defendingscience.org).

69. Paul H. Weaver, "On the Horns of Vinyl Chloride Dilemma," *Fortune*, no. 150, October 1974.

70. "PVC Rolls Out of Jeopardy, into Jubilation," *Chemical Week*, September 5, 1977.

71. Polyvinyl chloride (PVC) [9002-86-2] (vol. 19, suppl. 7, 1987).

72. Richard Doll, "Effects of Exposure to Vinyl Chloride: An Assessment of the Evidence," *Scandinavian Journal of Work and Environmental Health* 14 (1988): 61–78. In 1981, Peter Infante had also conducted a meta-analysis on vinyl chloride in which he came to opposite conclusions as Richard Doll's: Peter Infante, "Observations of the Site-Specific Carcinogenicity of Vinyl Chloride to Humans," *Environmental Health Perspectives* 41 (October 1981): 89–94.

73. Jennifer Beth Sass, Barry Castleman, and David Wallinga, "Vinyl Chloride: A Case Study of Data Suppression and Misrepresentation," *Environmental Health Perspectives* 113, no. 7 (July 2005): 809–12.

74. Richard Doll, "Deposition of William Richard Shaboe Doll, Ross v. Conoco, Inc.," Case no. 90-4837, LA 14th Judicial District Court, London, January 27, 2000.

75. "The Paris Appeal," accessed on the ARTAC website, www.artac.info/fr, on May 4, 2013.

76. ARTAC was created in 1984 by cancer specialist Dominique Belpomme "with a group of researchers, patients, and their families." His work is geared "towards the determination of the causes of cancer origins and prevention" (see Dominique Belpomme, *Avant qu'il ne soit trop tard* [Paris: Fayard, 2007], 21–25).

77. Dominique Belpomme, *Ces maladies créées par l'homme: Comment la dégradation de l'environnement met en péril notre santé* (Paris: Albin Michel, 2004).

78. Barbier and Farrachi, *La Société cancérigène*, 114.

79. Jacques Ferlay, Philippe Autier, Mathieu Boniol, Michael Heanue, and M. Colombet, "Estimates of the Cancer Incidence and Mortality in Europe in 2006," *Annals of Oncology* 3 (March 2007): 581–92.

80. Eva Stelliarova-Foucer et al., "Geographical Patterns and Time Trends of Cancer Incidence and Survival Among Children and Adolescents in Europe Since the 1970s (The ACCIS Project): An Epidemiological Study," *The Lancet* 364, no. 9451 (December 11, 2004): 2097–105.

81. WHO Regional Office for Europe, "Largely Preventable Chronic Diseases Cause 86% of Deaths in Europe," press release EURO/05/06, Copenhagen, September 11, 2006. Emphasis added.

82. AFSSET/INSERM, *Cancers et Environnement: Expertise collective*, October 2008.

83. The incidence rate of prostate cancer rose, every year, 6.3 percent between 1980 and 2005, and the increase was even more significant between 2000 and 2005 (+8.5 percent). For breast cancer, the average annual increase from 1980 to 2005 was 2.4 percent; +6 percent for thyroid cancer, +2.5 percent for testicular cancer, and +1 percent for brain cancer.

84. Suketami Tominaga, "Cancer Incidence in Japanese in Japan, Hawaii, and Western United States," *National Cancer Institute Monograph* 69 (December 1985): 83–92; see also Gertraud Maskarinec, "The Effect of Migration on Cancer Incidence Among Japanese in Hawaii," *Ethnicity & Disease* 14, no. 3 (2004): 431–39.

85. André Cicolella and Dorothée Benoît Browaeys, *Alertes santé: Experts et citoyens face aux intérêts privés* (Paris: Fayard, 2005), 25.

86. Ibid., 23.

87. Paul Lichtenstein et al., "Environmental and Heritable Factors in the Causation of Cancer Analyses of Cohorts of Twins from Sweden, Denmark and Finland," *New English Journal of Medicine* 343, no. 2 (July 13, 2000): 78–85.

88. "Action Against Cancer," European Parliament resolution on the European Commission communication on Action Against Cancer: European Partnership, May 6, 2010.

12: The Colossal Scientific Masquerade Behind Poisons' "Acceptable Daily Intakes"

1. Author interview with Erik Millstone, Brighton, January 12, 2010.

2. Author interview with Herman Fontier, Parma, January 19, 2010. Emphasis added.

3. Bruno Latour, *Science in Action: How to Follow Scientists and Engineers Through Society* (Cambridge, MA: Harvard University Press, 1987). The citations that follow are from 21, 49, 43, and 23.

4. Léopold Molle, "Éloge du professeur René Truhaut," *Revue d'histoire de la pharmacie* 72, no. 262 (1984): 340–48.

5. Toxicokinetics studies what becomes of medications and chemical substances in the body by analyzing the mechanisms of absorption, distribution, metabolism, and excretion.

6. Jean Lallier, *Le Pain et le Vin de l'an 2000*, documentary aired on the French national public television and radio agency L'Office de Radiodiffusion de Télévision Française (ORTF) on December 17, 1964. This film is part of the bonus materials on the DVD of my film *Notre poison quotidien*.

7. René Truhaut, "The Concept of the Acceptable Daily Intake: An Historical Review," *Food Additives and Contaminants* 8, no. 2 (March–April 1991): 151–62.

8. René Truhaut, "25 Years of JECFA Achievements," report presented at the 25th session of the JECFA, March 23–April 1, 1981, Geneva (WHO archives).

9. "Joint FAO/WHO Conference on Food Additives," *World Health Organization Technical Report Series*, no. 107, Geneva, September 19–22, 1955.

10. Truhaut, "Concept of the Acceptable Daily Intake."

11. Ibid.

12. Interview broadcast on the ORTF televised news show on June 3, 1974.

13. Truhaut, "Concept of the Acceptable Daily Intake." Emphasis added.

14. Ibid.

15. Ibid. Emphasis added.

16. Truhaut, "25 years of JECFA Achievements."

17. Ibid.

18. Parathion was banned in Europe in 2003 because of its high toxicity. It is one of the insecticides that joined the list of the "dirty dozen" persistent pollutants to be banished at any cost. Until its prohibition, it had an ADI of 0.004 mg per kg of body weight.

19. Truhaut, "Concept of the Acceptable Daily Intake."

20. Ibid.

21. "The ADI Concept: A Tool for Ensuring Food Safety," ILSI Workshop, Limelette, Belgium, October 18–19, 1990.

22. The complete list of the sixty-eight financing members of the European branch of the ILSI, created in 1986, can be consulted on ILSI Europe's website, www.ilsi.org /Europe. Based in Washington, DC, the ILSI has branches on every continent.

23. www.ilsi.org/Europe.

24. "WHO Shuts Life Sciences Industry Group Out of Setting Health Standards," *Environmental News Service*, February 2, 2006.

25. WHO/FAO, "Carbohydrates in Human Nutrition," FAO Food and Nutrition Paper, no. 66, Rome, 1998.

26. Tobacco Free Initiative, "The Tobacco Industry and Scientific Groups. ILSI: A Case Study," www.who.int, February 2001.

27. Derek Yach and Stella Bialous, "Junking Science to Promote Tobacco," *American Journal of Public Health* 91 (2001): 1745–48.

28. "WHO Shuts Life Sciences Industry Group Out of Setting Health Standards."

29. Environmental Working Group, "EPA Fines Teflon Maker DuPont for Chemical Cover-up," www.ewg.org, Washington, DC, December 14, 2005. See also Amy Cortese, "DuPont, Now in the Frying Pan," *New York Times*, August 8, 2004.

30. Michael Jacobson, "Lifting the Veil of Secrecy from Industry Funding of Nonprofit Health Organizations," *International Journal of Occupational and Environmental Health* 11 (2005): 349–55.

31. Diane Benford, *The Acceptable Daily Intake, a Tool for Ensuring Food Safety*, ILSI Europe Concise Monographs Series, International Life Sciences Institute, 2000.

32. Ibid. Emphasis added.

33. René Truhaut, "Principles of Toxicological Evaluation of Food Additives," Joint FAO/WHO Expert Committee on Food Additives, WHO, Geneva, July 4, 1973. Emphasis added.

34. Author interview with Diane Benford, London, January 11, 2010.

35. House of Representatives, *Problems Plague the EPA Pesticide Registration Activities*, U.S. Congress, House Report 98-1147, 1984.

36. Office of Pesticides and Toxic Substances, *Summary of the IBT Review Program*, EPA, Washington, July 1983.

37. "Data Validation. Memo from K. Locke, Toxicology Branch, to R. Taylor, Registration Branch," EPA, Washington, DC, August 9, 1978.

38. Communications and Public Affairs, "Note to Correspondents," EPA, Washington, DC, March 1, 1991.

39. *New York Times*, March 2, 1991.

40. Benford, *Acceptable Daily Intake*.

41. Truhaut, "Principles of Toxicological Evaluation of Food Additives."

42. Author interview with Ned Groth, Washington, October 17, 2009.

43. Truhaut, "Principles of Toxicological Evaluation of Food Additives." Emphasis is mine.

44. Benford, *Acceptable Daily Intake*.

45. Author interview with Erik Millstone, Brighton, January 12, 2010.

46. Author interview with James Turner, Washington, DC, October 17, 2009.

47. Author interview with Angelika Tritscher, Geneva, September 21, 2009.

48. Author interview with Herman Fontier, Parma, January 19, 2010.

49. Rachel Carson, *Silent Spring* (1962; New York: Houghton Mifflin 2002), 242.

50. Ulrich Beck, *Risk Society: Towards a New Modernity* (London: Sage Publications, 1992), 19. Emphasis in original.

51. Ibid., 49. Emphasis in original.

52. Ibid., 40. Emphasis in original.

53. Ibid., 19.

54. Benford, *Acceptable Daily Intake*. Emphasis is mine.

55. Truhaut, "Principles of Toxicological Evaluation of Food Additives."

56. Council Directive 91/414/EEC, July 15, 1991, concerning the placing of plant protection products on the market, *Official Journal*, no. l 230 (August 19, 1991): 0001–32. Emphasis added.

57. Éliane Patriarca, "Le texte des rapporteurs UMP est révélateur du rétropédalage de la droite sur les objectifs du Grenelle," *Libération*, May 4, 2010.

58. Claude Gatignol and Jean-Claude Étienne, *Pesticides et Santé*, Office parlementaire des choix scientifiques et technologiques, Paris, April 27, 2010.

59. Federal Insecticide, Fungicide, and Rodenticide Act (FIFRA), 2 (bb).

60. Michel Gérin et al., *Environment et santé publique: Fondements et pratiques* (Montreal: Edisem, 2003), 371.

13: The Unsolvable Conundrum of "Maximum Residue Limits"

1. From 1963 to 2010, the JMPR evaluated some 230 pesticides.

2. According to the fact sheet, chlorpyrifos-methyl is considered less toxic than chlorpyrifos, sold under the name Lorsban or Durban.

3. Directorate-General for Health and Consumer Protection, *Review Report for the Active Substance Chlorpyrifos-Methyl*, European Commission, SANCO/3061/99, June 3, 2005. This document is sixty-six pages long!

4. Its NOAEL (no-observed-adverse-effect level) at that time was l mg/kg/day.

5. See the European Union's website, which shows the standards for all pesticides used in Europe: EU Pesticides Database, ec.europa.eu/sanco_pesticides/pub lic/index.cfm#. A list of all agricultural products containing chlorpyrifos-methyl can be found there.

6. Author interview with Bernadette Ossendorp, Geneva, September 22, 2009.

7. Author interview with James Huff, Research Triangle Park, October 27, 2009.

8. Ulrich Beck, *Risk Society: Towards a New Modernity* (London: Sage Publications, 1992), 59.

9. Ibid., 64, 68. Emphasis in original.

10. Ibid., 65, 68.

11. Directorate-General for Health and Consumer Protection, *Review Report for the Active Substance Chlorpyrifos-Methyl*.

12. R. Tresdale, "Residues of Chlorpyrifos-methyl in Tomatoes at Harvest and Processed Fractions (Canned Tomatoes, Juice and Puree) Following Multiple Applications of RELDAN 22 (EF-1066), Italy 1999," R99-106/GHE-P-8661, 2000, Dow GLP (unpublished).

13. A. Doran and A.B. Clements, "Residues of Chlorpyrifos-methyl in Wine Grapes at Harvest Following Two Applications of EF-1066 (RELDAN 22) or GF-71, Southern Europe 2000," (N137) 19952/GHE-P-9441, 2002, Dow GLP (unpublished).

14. Author interview with Angelo Moretto, Geneva, September 21, 2009.

15. Author interview with Erik Millstone, Brighton, January 12, 2010.

16. Joint FAO/WHO Meeting on Pesticide Residues 2009, "List of Substances Scheduled for Evaluation and Request for Data. Meeting Geneva, 16–25 September 2009," October 2008.

17. See Thomas Zeltner, David A. Kessler, Anke Martiny, and Fazel Randera, *Tobacco Companies' Strategies to Undermine Tobacco Control Activities at the World Health Organization*, Report of the Committee of Experts on Tobacco Industry Documents, WHO, July 2000. See also Sheldon Krimsky, "The Funding Effect in Science and Its Implications for the Judiciary," *Journal of Law and Policy*, December 16, 2005.

18. Author interview with Angelika Tritscher, Geneva, September 21, 2009.

19. All of the citations in this section come from e-mails I very carefully saved.

20. BASF, Bayer CropScience, Dow AgroSciences, DuPont, FMC, Monsanto, Sumitomo, Syngenta.

21. Author telephone interview with Jean-Charles Bocquet, February 11, 2010.

22. E-mail sent by Sue Breach on February 24, 2010, without specifying the author of the "written response."

23. For more details, see Deborah Cohen and Philip Carter, "WHO and the Pandemic Flu 'Conspiracies,'" *British Medical Journal*, June 3, 2010.

24. Author interview with Ned Groth, Washington, DC, October 17, 2009.

25. Erik Millstone, Eric Brunner, and Ian White, "Plagiarism or Protecting Public Health?," *Nature* 371 (October 20, 1994): 647–48.

26. Erik Millstone, "Science in Trade Disputes Related to Potential Risks: Comparative Case Studies," European Commission, Joint Research Centre Institute for Prospective Technological Studies, Eur21301/EN, August 2004; Erik Millstone et al., "Risk-Assessment Policies: Differences Across Jurisdictions," European Commission, Joint Research Centre Institute for Prospective Technological Studies, January 2008.

27. FAO/WHO, "Principles and Methods for the Risk Assessment of Chemicals in Food," *Environmental Health Criteria*, no. 240 (2009). Emphasis is mine.

28. René Truhaut, "Principles of Toxicological Evaluation of Food Additives," Joint FAO/WHO Expert Committee on Food Additives, WHO, Geneva, July 4, 1973.

29. "Reasoned Opinion of EFSA Prepared by the Pesticides Unit (PRAPeR) on MRLs of Concern for the Active Substance Procymidone (Revised Risk Assessment)," *EFSA Scientific Report*, no. 227 (January 21, 2009): 1–26. Emphasis added.

30. Hypospadias is a congenital malformation in boys, which manifests as an opening of the urethra on the underside of the penis.

31. The epididymis is a small tube attached to the testicle that conserves and transports sperm.

32. Author interview with Angelo Moretto, Geneva, September 21, 2009.

33. Beck, *Risk Society*, xxx. Emphasis in original.

34. Author interview with Herman Fontier, Parma, January 19, 2010.

35. Emphasis added.

36. Eurobarometer, "Risk Issues. Executive Summary on Food Safety," February 2006.

37. *Official Journal of the European Communities*, no. L 225/263 (August 21, 2001).

38. An active substance can produce many formulations of different pesticides.

39. Author interview with Manfred Krautter, Hamburg, October 5, 2009.

40. Lars Neumeister, "Die unsicheren Pestizidhöchstmengen in der EU. Überprüfung der harmonisierten EU-Höchstmengen hinsichtlich ihres potenziellen akuten und chronischen Gesundheitsrisikos," Greenpeace and GLOBAL 2000, Friends of the Earth/Austria, March 2008.

41. Author interview with Herman Fontier, Parma, January 19, 2010.

42. "2007 Annual Report on Pesticide Residues," *EFSA Scientific Report (2009)*, no. 305 (June 10, 2009).

43. For example, the percentage of baby food jars that exceeded MRLs varied from 0 percent to 9.09 percent according to country.

44. Author interview with Eberhard Schüle, Stuttgard, October 6, 2009.

45. Author interview with Herman Fontier, Parma, January 19, 2010.

14: Aspartame and Regulation: How Industry Is Pulling the Strings

1. Edgar Monsanto Queeny, *The Spirit of Enterprise* (New York: Charles Scribner's Sons, 1943).

2. D.R. Lucas and J.P. Newhouse, "The Toxic Effect of Sodium L-glutamate on the Inner Layers of the Retina," *AMA Archives of Ophthalmology* 58, no. 2 (August 1957): 193–201.

3. Dale Purves et al., *Neuroscience* (Brussels: De Boeck, 2005), 145.

4. John Olney, "Brain Lesions, Obesity, and Other Disturbances in Mice Treated with Monosodium Glutamate," *Science* 164, no. 880 (May 1969): 719–21; J.W. Olney, L.G. Sharpe, and R.D. Feigin, "Glutamate-Induced Brain Damage in Infant Primates," *Journal of Neuropathology and Experimental Neurology* 31, no. 3 (July 1972): 464–88; John Olney, "Excitotoxins in Foods," *Neurotoxicology* 15, no. 3 (1994): 535–44.

5. Author interview with John Olney, New Orleans, October 20, 2009.

6. MSG is the sodium salt of glutamic acid and is responsible for "Chinese Restaurant Syndrome," which can trigger headaches, nausea, aches and rashes in the minutes or hours after ingestion. MSG is used by the food processing industry to

amplify salty flavors and stimulate the appetence of preparations while reducing their spice contents.

7. "Council Directive of 21 December 1988 on the Approximation of the Laws of the Member States Concerning Food Additives Authorized for Use in Foodstuffs Intended for Human Consumption (89/107/EEC)," *Official Journal*, no. L 040 (February 11, 1989): 0027–33. Emphasis added.

8. For example, flavoring one ton of ice cream with natural vanilla costs $1,020, versus only $5 with ethylvanillin, an artificial chemical flavoring (see Charles Wart, *L'Envers des etiquettes: Choisir son alimentation* [Brussels: Éditions Amyris, 2005]).

9. "Council Directive No 95/2/EC of 20 February 1995 on Food Additives Other Than Colours and Sweeteners," February 20, 1995, *Official Journal of the European Union*, no. L 61 (March 18, 1995).

10. "The Early Show, Artificial Sweeteners, New Sugar Substitute," BBC, September 28, 1982.

11. Saccharin was banned in Canada in 1977 because it was suspected of causing cancer (notably of the bladder). The International Agency for Research on Cancer reclassified it in 1987 as group 2B, "possible carcinogenic to humans," then in 1999 as group 3, "not classifiable." It remains legal in the rest of the world, with an ADI of 5 mg/kg.

12. Pat Thomas, "Bestselling Sweetener," *The Ecologist*, September 2005, 35–51.

13. John Henkel, "Sugar Substitutes: Americans Opt for Sweetness and Lite," *FDA Consumer Magazine*, November–December 1999.

14. See the website of "Mission Possible," the association created by Betty Martini: www.dorway.com.

15. Ulrich Beck, *Risk Society: Towards a New Modernity* (London: Sage Publications, 1992), 55.

16. R.E. Ranney, J.A. Oppermann, E. Muldoon, and F.G. McMahon, "Comparative Metabolism of Aspartame in Experimental Animals and Humans," *Journal of Toxicology and Environmental Health* 2 (1976): 441–51.

17. Methanol is a highly toxic substance that can cause blindness or even death if accidentally ingested. In case of poisoning, the best antidote is ethanol.

18. Herbert Helling, "Food and Drug Sweetener Strategy. Memorandum Confidential-Trade Secret Information to Dr. Buzard, Dr. Onien, Dr. Jenkins, Dr. Moe, and Mr. O'Bleness," December 28, 1970.

19. John Olney, "Brain Damage in Infant Mice Following Oral Intake of Glutamate, Aspartate or Cysteine," *Nature* 227, no. 5258 (August 8, 1970): 609–11; Bruce Schainker and John Olney, "Glutamate-Type Hypothalamic-Pituitary Syndrome in Mice Treated with Aspartate or Cysteate in Infancy," *Journal of Neural Transmission* 35 (1974): 207–15; J.W. Olney, J. Labruyere, and T. de Gubareff, "Brain Damage in Mice from Voluntary Ingestion of Glutamate and Aspartate," *Neurobehavioral Toxicology and Teratology* 2 (1980): 125–29.

20. James Turner and Ralph Nader, *The Chemical Feast: The Ralph Nader Study Group Report on Food Protection and the Food and Drug Administration* (London: Penguin, 1970). Ralph Nader is an attorney renowned for defending consumer rights, and who has also run for president on four occasions, twice as the Green Party nominee.

21. Note that studies on this product did not have the same effects on both sides of the Atlantic: cyclamate (E 952) is still authorized in Europe in nonalcoholic beverages, desserts, and candies, with an ADI of 7 mg/kg, whereas the ADI set by JECFA is 11 mg/kg.

22. Author interview with James Turner, Washington, October 17, 2009.

23. Letters from Adrian Gross to Senator Howard M. Metzenbaum, October 30 and November 3, 1987 (available at www.dorway.com).

24. Committee on Labor and Public Health, "Record of Hearings of April 8–9 and July 10, 1976, Held by Sen. Edward Kennedy, Chairman, Subcommittee on Administrative Practice and Procedure, Committee on the Judiciary, and Chairman, Subcommittee on Health," 3–4.

25. Food and Drug Administration, *Bressler Report*, August 1, 1977.

26. Author interview with John Olney, New Orleans, October 20, 2009.

27. Andy Pasztor and Joe Davidson, "Two Ex-U.S. Prosecutors' Roles in Case Against Searle Are Questioned in Probe," *Wall Street Journal*, February 7, 1986.

28. Phenylketonuria is a genetic disorder that prevents the metabolism of phenylalanine. Screening for the disease is obligatory in many countries, like France, because when untreated, it causes brain disorders and mental retardation.

29. John Olney, "Aspartame Board of Inquiry. Prepared Statement," University School of Medicine St. Louis, Missouri, September 30, 1980.

30. Ibid.

31. Department of Health and Human Services, "Aspartame: Decision of the Public Board of Inquiry," Food and Drug Administration, docket no 75F-0355, September 30, 1980.

32. "Medical Professor at Pennsylvania State Is Nominated to Head Food and Drug Agency," *New York Times*, April 3, 1981.

33. Florence Graves, "How Safe Is Your Diet Soft Drink?," *Common Cause Magazine*, July–August 1984.

34. Letter from Adrian Gross to Senator Howard M. Metzenbaum, November 3, 1987. Emphasis added.

35. World Health Organization (WHO), *Evaluation of Certain Food Additives: Some Food Colours, Thickening Agents, Smoke Condensates, and Certain Other Substances (Nineteenth Report of the Joint FAO/WHO Expert Committee on Food Additives),*" WHO Technical Report Series, no. 576 (1975).

36. WHO, *Evaluation of Certain Food Additives (Twentieth Report of the Joint FAO/WHO Expert Committee on Food Additives)*, WHO Technical Report Series, no. 599 (1976).

37. WHO, *Evaluation of Certain Food Additives (Twenty-First Report of the Joint FAO/WHO Expert Committee on Food Additives)*, WHO Technical Report Series, no. 617 (1978).

38. WHO, *Evaluation of Certain Food Additives (Twenty-Fourth Report of the Joint FAO/WHO Expert Committee on Food Additives)*, WHO Technical Report Series, no. 653 (1980).

39. H. Ishii, T. Koshimizu, S. Usami, and T. Fujimoto, "Toxicity of Aspartame and Its Diketopiperazine for Wistar Rats by Dietary Administration for 104 Weeks," *Toxicology* 21, no. 2 (1981): 91–94.

40. WHO, *Evaluation of Certain Food Additives (Twenty-Fifth Report of the Joint FAO/WHO Expert Committee on Food Additives)*, WHO Technical Report Series, no. 669 (1981).

41. Author interview with Hugues Kenigswald, Parma, January 19, 2010.

15: The Dangers of Aspartame and the Silence of Public Authorities

1. Emphasis added.

2. According to the *Chicago Tribune*, Monsanto bought Searle for $2.7 billion. The sale earned the Searle family $1 billion and Donald Rumsfeld $12 million ("Winter Comes for a Beltway Lion; Rumsfeld Rose and Fell with His Conviction Intact," *Chicago Tribune*, November 12, 2006).

3. "Proceedings and Debates of the 99th Congress, First Session," *Congressional Record* 131 (May 7, 1985).

4. *Hearing Before the Committee on Labor and Human Resources United States Senate, One Hundredth Congress. Examining the Health and Safety Concerns of NutraSweet (Aspartame)*, November 3, 1987.

5. For more information about the "revolving doors" of the aspartame industry, see Gregory Gordon, "NutraSweet: Questions Swirl," United Press International Investigative Report, October 12, 1987.

6. "FDA Handling of Research on NutraSweet Is Defended," *New York Times*, July 18, 1987.

7. In the U.S. Air Force newsletter *Flying Safety* (May 1992), Colonel Roy Poole warns pilots against the dangers of aspartame: "Vertigo, epilepsy, sudden memory loss and progressive vision loss."

8. Richard Wurtman and Timothy Maher, "Possible Neurologic Effects of Aspartame, a Widely Used Food Additive," *Environmental Health Perspectives* 75 (November 1987): 53–57; Richard Wurtman, "Neurological Changes Following High Dose Aspartame with Dietary Carbohydrates," *New England Journal of Medicine* 309, no. 7 (1983): 429–30.

9. Richard Wurtman, "Aspartame: Possible Effects on Seizures Susceptibility," *The Lancet* 2, no. 8463 (1985): 1060.

10. The letter was published in Gordon's "NutraSweet."

11. The blood–brain barrier, also called the "hematoencephalic barrier," protects the brain from pathogenic agents circulating in the blood.

12. Ibid.

13. Jacqueline Verrett and Jean Carper, *Eating May Be Hazardous to Your Health* (New York: Simon & Schuster, 1974), 19–21.

14. Citrus Red 2 (E 121) has been banned in Europe since 1977. It is classified as "possibly carcinogenic to humans" (group 2B) by the International Agency for Research on Cancer (IARC). It is still authorized in the United States, uniquely to color orange skins. If you buy oranges from Florida, it is recommended that you wash your hands after peeling them.

15. Phocomelia presents as atrophy of the limbs. It is characteristic of children who were exposed in the womb to thalidomide, a drug prescribed to pregnant women in the 1950s and 1960s to combat nausea.

16. Verrett and Carper, *Eating May Be Hazardous to Your Health*, 42, 48.

17. Author interview with David Hattan, Washington, DC, October 19, 2009

18. The document showed that 38.3 percent of complaints were linked to the consumption of caffeinated beverages, 21.7 percent to that of sweeteners, and 4 percent to that of chewing gum.

19. Hyman J. Roberts, *Aspartame (NutraSweet), Is it Safe?* (Philadelphia: Charles Press, 1990), 4.

20. Hyman J. Roberts, "Reactions Attributed to Aspartame-Containing Products: 551 Cases," *Journal of Applied Nutrition* 40, no. 2 (1988): 85–94.

21. Hyman J. Roberts, *Aspartame Disease: An Ignored Epidemic* (West Palm Beach, FL: Sunshine Sentinel Press, 2001).

22. Letter from John Olney to Howard Metzenbaum, December 8, 1987.

23. Richard Wurtman, *Dietary Phenylalanine and Brain Function* (Boston: Birkhauser, 1988).

24. Ralph Walton, Robert Hudak, and Ruth Green-Waite, "Adverse Reactions to Aspartame: Double-Blind Challenge in Patients from Vulnerable Population," *Biological Psychiatry* 34, no. 1 (July 1993): 13–17.

25. Author interview with Ralph Walton, New York, October 30, 2009.

26. J.W. Olney, N.B. Farber, E. Spitznagel, and L.N. Robins, "Increasing Brain Tumor Rates: Is There a Link to Aspartame?" *Journal of Neuropathology and Experimental Neurology* 55, no. 11 (1996): 1115–23.

27. David Michaels, *Doubt Is Their Product: How Industry's Assault on Science Threatens Your Health* (New York: Oxford University Press, 2008), 143.

28. Paula Rochon et al., "A Study of Manufacturer-Supported Trials of Nonsteroidal Anti-Inflammatory Drugs in the Treatment of Arthritis," *Archives of Internal Medicine* 154, no. 2 (1994): 157–63. See also Sheldon Krimsky, "The Funding Effect in Science and Its Implications for the Judiciary," *Journal of Law Policy* 13, no. 1 (2005): 46–68.

29. Henry Thomas Stelfox, Grace Chua, Keith O'Rourke, and Allan S. Detsky, "Conflict of Interest in the Debate Over Calcium-Channel Antagonists," *New England Journal of Medicine* 338, no. 2 (1998): 101–6.

30. J.E. Bekelman, Y. Li, and C.P. Gross, "Scope and Impact of Financial Conflicts of Interest in Biomedical Research," *Journal of the American Medical Association* 289 (2003): 454–65; Valerio Gennaro and Lorenzo Tomatis, "Business Bias: How Epidemiologic Studies May Underestimate or Fail to Detect Increased Risks of Cancer and Other Diseases," *International Journal of Occupational and Environmental Health* 11 (2005): 356–59.

31. Bruno Latour, *Science in Action: How to Follow Scientists and Engineers Through Society* (Cambridge, MA: Harvard University Press, 1987), 38.

32. Harriett Butchko and Frank Kotsonis, "Acceptable Daily Intake vs Actual Intake: The Aspartame Example," *Journal of the American College of Nutrition* 10, no. 3 (1991): 258–66.

33. Lewis Stegink and Jack Filer, "Repeated Ingestion of Aspartame-Sweetened Beverage: Effect on Plasma Amino Acid Concentrations in Normal Adults," *Metabolism* 37, no. 3 (March 1988): 246–51.

34. Richard Smith, "Peer Review: Reform or Revolution?," *British Medical Journal* 315, no. 7111 (1997): 759–60. See also Richard Smith, "Medical Journals Are an Extension of the Marketing Arm of Pharmaceutical Companies," *PLoS Medicine* 2, no. 5 (2005): 138.

35. The NTP is under the management of the NIEHS, but an executive committee, which includes representatives from every American regulatory agency, including OSHA, the Environmental Protection Agency (EPA), and the FDA, determines its research subjects.

36. Author interview with James Huff, Research Triangle Park, October 27, 2009.

37. Cited by Greg Gordon, "FDA Resisted Proposals to Test Aspartame for Years," *Star Tribune*, November 22, 1996.

38. NTP, *Toxicology Studies of Aspartame (Case No. 22839-47-0) in Genetically Modified (FVB Tg.AC Hemizygous) and B6.129-Cdkn2atm1Rdp (N2) Deficient Mice and Carcinogenicity Studies of Aspartame in Genetically Modified [B6.129-Trp53tm1Brd (N5) Haploinsufficient] Mice (Feed Studies)*, October 2005.

39. There are two types of carcinogenic agents: genotoxic ones, which act directly on genes by initiating the first stage of the carcinogenesis process via genetic mutations; and nongenotoxic agents, which do not act directly on genes, but participate in the carcinogenesis process (promotion or progression stage) while favoring the proliferation of mutated or "initiated" cells (see the "Occupational Cancers" section at www.cancer-environnement.fr).

40. The following comment was appended to the NTP study: "Because this is a new model, there is uncertainty whether the study possessed sufficient sensitivity to detect a carcinogenic effect."

41. Cesare Maltoni, "The Collegium Ramazzini and the Primacy of Scientific Truth," *European Journal of Oncology* 5, suppl. 2 (2000): 151–52.

42. M. Soffritti, F. Belpoggi, F. Minardi, L. Bua, and C. Maltoni, "Mega-Experiments to Identify and Assess Diffuse Carcinogenic Risks," *Annals of the New York Academy of Sciences* 895 (December 1999): 34–55.

43. See M. Soffritti, F. Belpoggi, F. Minardi, and C. Maltoni, "History and Major Projects, Life-Span Carcinogenicity Bioassay Design, Chemicals Studied, and Results," *Annals of the New York Academy of Sciences* 982 (2002): 26–45; Cesare Maltoni and Morando Soffritti, "The Scientific and Methodological Bases of Experimental Studies for Detecting and Quantifying Carcinogenic Risks," *Annals of the New York Academy of Sciences* 895 (1999): 10–26.

44. Morando Soffritti et al., "First Experimental Demonstration of the Multipotential Carcinogenic Effects of Aspartame Administered in the Feed to Sprague-Dawley Rats," *Environmental Health Perspectives* 114, no. 3 (March 2006): 379–85; Fiorella Belpoggi et al., "Results of Long-Term Carcinogenicity Bioassay on Sprague-Dawley Rats Exposed to Aspartame Administered in Feed," *Annals New York Academy of Sciences* 1076 (2006): 559–77.

45. Center for Food Safety and Applied Nutrition, "FDA Statement on European Aspartame Study," April 20, 2007.

46. "Opinion of the Scientific Panel on Food Additives, Flavorings, Processing Aids and Materials in Contact with Food (AFC) Related to a New Long-Term Carcinogenicity Study on Aspartame," EFSA-Q-2005-122, May 3, 2006.

47. M. Soffritti, F. Belpoggi, E. Tibaldi, D.D. Esposti, and M. Lauriola, "Life-Span Exposure to Low Doses of Aspartame Beginning During Prenatal Life Increases Cancer Effects in Rats," *Environmental Health Perspectives* 115 (2007): 1293–97.

48. "Updated Opinion on a Request from the European Commission Related to the 2nd ERF Carcinogenicity Study on Aspartame, Taking into Consideration Study Data Submitted by the Ramazzini Foundation in February 2009," EFSA-Q-2009-00474, March 19, 2009. Emphasis added.

49. "EFSA Re-Confirms the Safety of Aspartame and Dismisses Claims Made by the Ramazzini Institute," April 23, 2009. At the time of writing, I have learned that the Ramazzini Institute has published a new study conducted on gravid mice, which shows that aspartame induces liver and lung cancers among the males (Morando Soffritti et al., "Aspartame Administered in Feed, Beginning Prenatally Through Life-Span, Induces Cancers of the Liver and Lung in Male Swiss Mice," *American Journal of Industrial Medicine* 53, no. 12 [December 2010]: 1197–206).

50. See William Reymond, "Coca-Cola serait-il bon pour la santé?," *Bakchich*, April 19–20, 2008.

51. "Les boissons light? C'est le sucré . . . sans sucres," *La Dépêche*, September 29, 2009; "Souvent accusé, le faux sucre est blanchi," www.Libération.fr, September 14, 2009.

52. Author interview with Catherine Geslain-Lanéelle, Parma, January 19, 2010.

16: "Men in Peril": Is the Human Species in Danger?

1. I borrowed this title from the excellent documentary by Sylvie Gilman and Thierry de Lestrade, which was broadcast on the European television channel Arte on November 25, 2008; its creators were the first in France to reveal some of the astonishing observations mentioned in this chapter.

2. Gérard Bapt (Parti socialiste, PS) is president of the environmental health group in the French National Assembly; Bérengère Poletti (Union pour un Mouvement Populaire, UMP) is president of a group that monitors the national environmental health plan.

3. Author interview with Ana Soto and Carlos Sonnenschein, Tufts University, Boston, October 28, 2009.

4. Nonylphenol is part of a family of synthetic chemical products called "alkylphenols." Its global production amounts to 600,000 tons per year.

5. A.M. Soto, H. Justicia, J.W. Wray, and C. Sonnenschein, "P-Nonylphenol: An Estrogenic Xenobiotic Released from 'Modified' Polystyrene," *Environmental Health Perspectives* 92 (May 1991): 167–73.

6. Author interview with Ana Soto and Carlos Sonnenschein, Tufts University, Boston, October 28, 2009.

7. Rachel Carson, *Silent Spring* (1962; New York: Houghton Mifflin 2002), 207.

8. "Rachel Carson Talks About Effects of Pesticides on Children and Future Generations," www.bbcmotiongallery.com, January 1, 1963.

9. Theo Colborn, Dianne Dumanoski, and John Peterson Myers, *Our Stolen Future: Are We Threatening Our Fertility, Intelligence, and Survival? A Scientific Detective Story* (New York: Plume, 1996).

10. Ibid., 145.

11. Ibid., 106.

12. Ibid., 91.

13. Eric Dewailly, Albert Nantel, Jean-P. Weber, and François Meyer, "High Levels of PCBs in Breast Milk of Inuit Women from Arctic Quebec," *Bulletin of Environmental Contamination and Toxicology* 43, no. 5 (November 1989): 641–46.

14. Joseph L. Jacobson, Sandra W. Jacobson, Pamela M. Schwartz, Greta G. Fein, and Jeffrey K. Dowler, "Prenatal Exposure to an Environmental Toxin: A Test of the Multiple Effects Model," *Developmental Psychology* 20, no. 4 (July 1984): 523–32.

15. Joseph Jacobson and Sandra Jacobson, "Intellectual Impairment in Children Exposed to Polychlorinated Biphenyls *in Utero*," *New England Journal of Medicine* 335 (September 12, 1996): 783–89.

16. Author interview with Theo Colborn, Paonia, December 10, 2009.

17. Among the many studies published by Louis Guillette, I recommend the following: Louis Guillette et al., "Developmental Abnormalities of the Gonad and Abnormal Sex Hormone Concentrations in Juvenile Alligators from Contaminated and

Control Lakes in Florida," *Environmental Health Perspectives* 102, no. 8 (August 1994): 680–88.

18. Howard Bern et al. "Statement from the Work Session on Chemically-Induced Alterations in Sexual Development: The Wildlife/Human Connection," in *Chemically-Induced Alterations in Sexual and Functional Development: The Wildlife/Human Connection*, ed. Theo Colborn and Coralie Clement (Princeton, NJ: Princeton Scientific Publishing Co., 1992), 1–8.

19. André Cicolella and Dorothée Benoît Browaeys, *Alertes Santé: Experts et citoyens face aux intérêts privés* (Paris: Fayard, 2005), 231.

20. Bernard Jégou, Pierre Jouannet, and Alfred Spira, *La Fertilité est-elle en danger?* (Paris: La Découverte, 2009), 54.

21. Ibid., 147.

22. Colborn et al., *Our Stolen Future*, 82.

23. Jégou et al., *La Fertilité est-elle en danger?*, 60.

24. E. Carlsen, A. Giwercman, N. Keiding, and N. Skakkebaek, "Evidence for Decreasing Quality of Semen During Past 50 Years," *British Medical Journal* 305, no. 6854 (September 12,1992): 609–13.

25. J. Auger, J.M. Kunstmann, F. Czyglik, and P. Jouannet, "Decline in Semen Quality Among Fertile Men in Paris During the Last 20 Years," *New England Journal of Medicine* 332 (1995): 281–85.

26. Jégou et al., *La Fertilité est-elle en danger?*, 61.

27. Shanna Swan, "The Question of Declining Sperm Density Revisited: An Analysis of 101 Studies Published 1934–1996," *Environmental Health Perspectives* 108, no. 10 (October 2000): 961–66.

28. Jégou et al., *La Fertilité est-elle en danger?*, 71–74.

29. Richard Sharpe and Niels Skakkebaek, "Are Oestrogens Involved in Falling Sperm Counts and Disorders of the Male Reproductive Tract?," *The Lancet* 29, no. 341 (May 29, 1993): 1392–95.

30. N.E. Skakkebaek, E. Rajpert-De Meyts, and K.M. Main, "Testicular Dysgenesis Syndrome: An Increasingly Common Developmental Disorder with Environmental Aspects," *Human Reproduction* 16, no. 5 (May 2001): 972–78.

31. Katharina Main et al., "Human Breast Milk Contamination with Phthalates and Alterations of Endogenous Reproductive Hormones in Infants Three Months of Age," *Environmental Health Perspectives* 114, no. 2 (February 2006): 270–76. Numerous studies have shown this link, such as Shanna Swan et al., "Decrease in Anogenital Distance Among Male Infants with Prenatal Phthalate Exposure," *Environmental Health Perspectives* 113, no. 8 (August 2005): 1056–61.

32. "Alerte aux poêles à frire," www.Libération.fr, September 30, 2009. DuPont de Nemours, holder of the Teflon trademark since 1954, announced that it would stop using PFOA by 2015.

33. Ulla Nordström Joensen et al., "Do Perfluoroalkyl Compounds Impair Human Semen Quality?," *Environmental Health Perspectives* 117, no. 6 (June 2009): 923–27.

34. Dawn Forsythe let me photocopy one hundred or so documents from her personal archives, including those she cites in this interview.

35. The authors of this "draft" are Dave Fischer (Bayer), Richard Balcomb (American Cyanamid), C. Holmes (BASF), T. Hall (Sandoz), K. Reinert and V. Kramer (Rohm & Haas), Ellen Mihaich (Rhône-Poulenc), R. McAllister, and J. McCarthy (ACPA).

36. This refers to the Food Quality Protection Act (FQPA) and to amendments made to the Safe Drinking Water Act (SDWA) in 1996. In 2010, the Obama administration asked the EPA to accelerate the program. On its website, the EPA wrote (in 1998) that its problem was "the lack of endocrine disruptor effects-related data on the vast majority of [87,000] chemicals and their breakdown products," which would allow for an evaluation of risks associated with the endocrine system (*EDSTAC Final Report*, www.epa.gov, August 1998, chap. 4).

17: Distilbene: The "Perfect Model"?

1. NIEHS News, "Women's Health Research at NIEHS," *Environmental Health Perspectives* 101, no. 2 (June 1993).

2. E.C. Dodds, L. Goldberg, W. Lawson, and R. Robinson, "Œstrogenic Activity of Certain Synthetic Compounds," *Nature* 141 (February 1938): 247–48.

3. Howard Burlington and Verlus Frank Linderman, "Effect of DDT on Testes and Secondary Sex Characters of White Leghorn Cockerels," *Proceedings of the Society for Experimental Biology and Medicine* 74, no. 1 (May 1950): 48–51.

4. As written by Kehoe in a report prepared for a U.S. and British Intelligence Committee in January 1947 (cited by Devra Davis, *The Secret History of the War on Cancer* [New York: Basic Books, 2007], 91).

5. Ibid., 90.

6. A. Parkes, E.C. Dodds, and R.L. Noble, "Interruption of Early Pregnancy by Means of Orally Active Oestrogens," *British Medical Journal* 2, no. 4053 (September 10, 1938): 557–59.

7. Sidney John Folley et al., "Induction of Abortion in the Cow by Injection with Stilboestrol Diproporniate," *The Lancet* 2 (1939).

8. Antoine Lacassagne, "Apparition d'adénocarcinomes mammaires chez des souris mâles traitées par une substance œstrogène synthétique," *Comptes rendus des séances de la Société de biologie* 129 (1938): 641–43.

9. R.R. Greene, M.W. Burrill, and A.C. Ivy, "Experimental Intersexuality: The Paradoxical Effects of Estrogens on the Sexual Development of the Female Rat," *Anatomical Record* 74, no. 4 (August 1939): 429–38.

10. "Estrogen Therapy: A Warning," *Journal of the American Medical Association* 113, no. 26 (December 23, 1939): 2323–24.

11. As cited by Jacqueline Verrett and Jean Carper, *Eating May Be Hazardous to*

Your Health (New York: Simon & Schuster, 1974), 146. Use of distilbene in poultry and cattle would be banned in 1959 and 1980, respectively.

12. Olive Smith and George Smith, "Diethylstilbestrol in the Prevention and Treatment of Complications of Pregnancy," *American Journal of Obstetrics and Gynecology* 56, no. 5 (1948): 821–34; Olive Smith and George Smith, "The Influence of Diethylstilbestrol on the Progress and Outcome of Pregnancy as Based on a Comparison of Treated with Untreated Primigravidas," *American Journal of Obstetrics and Gynecology* 58, no. 5 (1949): 994–1009.

13. Susan E. Bell, *DES Daughters: Embodied Knowledge and the Transformation of Women's Health Politics* (Philadelphia: Temple University Press, 2009), 16.

14. Eclampsia is a dangerous pregnancy complication characterized by seizures.

15. James Ferguson, "Effect of Stilbestrol on Pregnancy Compared to the Effect of a Placebo," *American Journal of Obstetrics and Gynecology* 65, no. 3 (March 1953): 592–601.

16. W.J. Dieckmann, M.E. Davis, L.M. Rynkiewicz, and R.E. Pottinger, "Does the Administration of Diethylstilbestrol During Pregnancy Have Therapeutic Value?," *American Journal of Obstetrics and Gynecology* 66, no. 5 (November 1953): 1062–81.

17. Y. Brackbill and H.W. Berendes, "Dangers of Diethylstilbestrol: Review of a 1953 Paper," *The Lancet* 2, no. 8088 (1978): 520.

18. On November 16, 2012, French health authorities admitted that Mediator, a drug produced by Servier Laboratories and indicated for diabetic and overweight patients (but without any efficacy), largely used as an appetite suppressant, caused at least five hundred deaths and thousands of hospitalizations due to heart valve problems between 1976 and November 2009.

19. Pat Cody, *DES Voices: From Anger to Action* (Jupiter, FL: DES Action, 2008), 13.

20. William Gardner, "Experimental Induction of Uterine Cervical and Vaginal Cancer in Mice," *Cancer Research* 19, no. 2 (February 1959): 170–76.

21. A.M. Bongiovanni, A.M. Di George, and M.M. Grumbach, "Masculinization of the Female Infant Associated with Estrogenic Therapy Alone During Gestation: Four Cases," *Journal of Clinical Endocrinology and Metabolism* 19 (August 1959): 1004.

22. Norman M. Kaplan, "Male Pseudohermaphroditism: Report of a Case, with Observations on Pathogenesis," *New England Journal of Medicine* 261 (1959): 641.

23. Theo Colborn, Dianne Dumanoski, and John Peterson Myers, *Our Stolen Future: Are We Threatening Our Fertility, Intelligence, and Survival? A Scientific Detective Story* (New York: Plume, 1996), 49.

24. Ibid., 50. Emphasis added.

25. Ibid.

26. "The Full Story of the Drug Thalidomide," *Life* magazine, August 10, 1962.

27. "Rachel Carson Talks About Effects of Pesticides on Children and Future Generations," www.bbcmotiongallery.com, January 1, 1963.

28. Colborn et al., *Our Stolen Future*, 50.

29. Arthur Herbst, Howard Ulfelder, and David Poskanzer, "Adenocarcinoma of the Vagina: Association of Maternal Stilbestrol Therapy with Tumor Appearance in Young Women," *New England Journal of Medicine* 284, no. 15 (April 22, 1971): 878–81.

30. Verrett and Carper, *Eating May Be Hazardous to Your Health*, 142.

31. In France, distilbene was not banned for pregnant women until 1977 (see Véronique Mahé, *Distilbène: Des mots sur un scandale* [Paris: Albin Michel, 2010]). It is estimated that in France, DES was prescribed to approximately 200,000 pregnant women who gave birth to 160,000 children.

32. Note that in epidemiology, a cluster refers to a remarkable aggregation of a given disease in a given place or population.

33. Colborn et al., *Our Stolen Future*, 53.

34. Bell, *DES Daughters*, 1.

35. Cody, *DES Voices*, 4.

36. Ibid., 43.

37. Ibid., 93.

38. Ibid., 85.

39. Ibid., 90.

40. Ibid., 97.

41. Ibid., 96.

42. Bell, *DES Daughters*, 23.

43. Ibid., 27.

44. Sheldon Krimsky, *Hormonal Chaos: The Scientific and Social Origins of the Environmental Hypothesis* (Baltimore: Johns Hopkins University Press, 2002), 2. Emphasis added.

45. Ibid., 11.

46. Bell, *DES Daughters*, 27.

47. It is impossible to cite here all the studies conducted by John McLachlan on DES. I will only reference two: Retha Newbold and John McLachlan, "Vaginal Adenosis and Adenocarcinoma in Mice Exposed Prenatally or Neonatally to Diethylstilbestrol," *Cancer Research* 42, no. 5 (May 1982): 2003–11; John McLachlan and Retha Newbold, "Reproductive Tract Lesions in Male Mice Exposed Prenatally to Diethylstilbestrol," *Science* 190, no. 4218 (December 5, 1975): 991–92.

48. Author interview with John McLachlan, New Orleans, October 22, 2009.

49. Newbold and McLachlan, "Vaginal Adenosis and Adenocarcinoma."

50. Retha Newbold, "Cellular and Molecular Effects of Developmental Exposure to Diethylstilbestrol: Implications for Other Environmental Estrogens," *Environmental Health Perspectives* 103 (October 1995): 83–87.

51. Retha Newbold et al., "Increased Tumors but Uncompromised Fertility in the Female Descendants of Mice Exposed Developmentally to Diethylstilbestrol," *Carcinogenesis* 19, no. 9 (September 1998): 655–63.

52. Retha Newbold et al., "Proliferative Lesions and Reproductive Tract Tumors in Male Descendants of Mice Exposed Developmentally to Diethylstilbestrol," *Carcinogenesis* 21, no. 7 (2000): 1355–63.

53. R.R. Newbold, E. Padilla-Banks, and W.N. Jefferson, "Adverse Effects of the Model Environmental Estrogen Diethylstilbestrol Are Transmitted to Subsequent Generations," *Endocrinology* 147, suppl. 6 (June 2006): 11–17; Retha Newbold, "Lessons Learned from Perinatal Exposure to Diethylstilbestrol," *Toxicology and Applied Pharmacology* 199, no. 2 (September 1, 2004): 142–50.

54. Among them, a study conducted by the Netherlands Cancer Institute: Helen Klip et al., "Hypospadias in Sons of Women Exposed to Diethylstilbestrol *in Utero*: A Cohort Study," *The Lancet* 359, no. 9312 (March 30, 2002): 1101–7.

55. Felix Grün and Bruce Blumberg, "Environmental Obesogens: Organotins and Endocrine Disruption via Nuclear Receptor Signaling," *Endocrinology* 47, no. 6 (2006): 50–55.

56. R.R. Newbold, E. Padilla-Banks, W.N. Jefferson, and J.J. Heindel, "Effects of Endocrine Disruptors on Obesity," *International Journal of Andrology* 31, no. 2 (April 2008): 201–8; R.R. Newbold, E. Padilla-Banks, R.J. Snyder, T.M. Phillips, and W.N. Jefferson, "Developmental Exposure to Endocrine Disruptors and the Obesity Epidemic," *Reproductive Toxicology* 23, no. 3 (April–May 2007): 290–96.

18: The Case of Bisphenol A: A Pandora's Box

1. www.bisphenol-a.org.

2. The BPA identification code is a "7" in the middle of a triangle. But it can also be found in products labeled with a "3" or a "6."

3. A. Krishnan, P. Stathis, S.F. Permuth, L. Tokes, and D. Feldman, "Bisphenol-A: An Estrogenic Substance Is Released from Polycarbonate Flasks During Autoclaving," *Endocrinology* 132, no. 6 (June 1993): 2279–86.

4. Cited by Theo Colborn, Dianne Dumanoski, and John Peterson Myers, *Our Stolen Future: Are We Threatening Our Fertility, Intelligence, and Survival? A Scientific Detective Story* (New York: Plume, 1996), 131.

5. Liza Gross, "The Toxic Origins of Disease," *PLoS Biology* 5, no. 7 (June 26, 2007): 193.

6. Patricia Hunt et al., "Bisphenol A Exposure Causes Meiotic Aneuploidy in the Female Mouse," *Current Biology* 13, no. 7 (April 2003): 546–53; Martha Susiarjo, Terry J. Hassold, Edward Freeman, and Patricia Hunt, "Bisphenol A Exposure *in Utero* Disrupts Early Oogenesis in the Mouse," *PLoS Genetics* 3, no. 1 (January 12, 2007): 5.

7. Nena Baker, *The Body Toxic: How the Hazardous Chemistry of Everyday Things Threatens Our Health and Well-Being* (New York: North Point Press, 2008), 151.

8. Elizabeth Grossman, "Two Words: Bad Plastic," *Salon.com*, August 2, 2007.

9. Caroline Markey, Enrique Luque, Monica Muñoz De Toro, Carlos Sonnenschein, and Ana Soto, "*In Utero* Exposure to Bisphenol A Alters the Development and Tissue Organization of the Mouse Mammary Gland," *Biology of Reproduction* 65, no. 4 (October 1, 2001): 1215–23.

10. Colborn et al., *Our Stolen Future*, 30.

11. Ibid., 31.

12. Frederick Vom Saal and Franklin Bronson, "Sexual Characteristics of Adult Female Mice Are Correlated with Their Blood Testosterone Levels During Prenatal Development," *Science* 208, no. 4444 (May 9, 1980): 597–99 (cited in Colborn et al., *Our Stolen Future*, 34).

13. Colborn et al., *Our Stolen Future*, 39.

14. F.S. vom Saal, M.M. Clark, B.G. Gale, L.C. Drickamer, and J.G. Vandenbergh, "The Intra-Uterine Position (IUP) Phenomenon," in *Encyclopedia of Reproduction*, vol. 2, ed. Ernst Knobil and Jimmy Neill (New York: Academic Press, 1999), 893–900; "Science Watch: Prenatal Womb Position and Supermasculinity," *New York Times*, March 31, 1992.

15. Colborn et al., *Our Stolen Future*, 35.

16. Ibid., 38.

17. Ibid., 39.

18. Ibid., 42.

19. Bernard Jégou, Pierre Jouannet, and Alfred Spira, *La Fertilité est-elle en danger?* (Paris: La Découverte, 2009), 10–12.

20. Centers for Disease Control and Prevention, *Fourth National Report on Human Exposure to Environmental Chemicals*, www.cdc.gov, Atlanta, 2009. In Chapter 19, I will return to this report, which explains the "chemical body burden" of thousands of American citizens.

21. Antonia Calafat, "Exposure of the U.S. Population to Bisphenol A and 4-Tertiary-Octylphenol: 2003–2004," *Environmental Health Perspectives* 116 (2008): 39–44.

22. Antonia Calafat et al., "Exposure to Bisphenol A and Other Phenols in Neonatal Intensive Care Unit Premature Infants," *Environmental Health Perspectives* 117, no. 4 (April 2009): 639–44.

23. In 2010, the AFSSA became part of the French Agency for Food, Environmental and Occupational Health and Safety (Agence nationale de sécurité sanitaire de l'alimentation, de l'environnement et du travail, ANSES).

24. Author interview with Frederick vom Saal, New Orleans, October 22, 2009.

25. Frederick Vom Saal et al., "Prostate Enlargement in Mice Due to Foetal Exposure to Low Doses of Estradiol or Diethylstilbestrol and Opposite Effects at Low Doses," *Proceedings of the National Academy of Sciences of the USA* 94, no. 5 (March 1997): 2056–61.

26. Susan Nagel, Frederick Vom Saal, et al., "Relative Binding Affinity-Serum Modified Access (RBA-SMA) Assay Predicts the Relative *in Vivo* Bioactivity of the Xenoestrogens Bisphenol A and Octylphenol," *Environmental Health Perspectives* 105, no. 1 (January 1997): 70–76.

27. Liza Gross, "The Toxic Origins of Disease," *PLoS Biology* 5, no. 7 (2007): 193.

28. Frederick Vom Saal et al., "A Physiologically Based Approach to the Study of Bisphenol A and Other Estrogenic Chemicals on the Size of Reproductive Organs, Daily Sperm Production, and Behavior," *Toxicology and Industrial Health* 14, no. 1–2 (January–April 1998): 239–60.

29. Gross, "Toxic Origins of Disease."

30. Ibid.

31. Ibid.

32. The "anogenital distance" separates the anus from the genital organs. It is normally twice as long in males as in females, and any variation can indicate a congenital defect in the male reproductive organs (see Chapter 19).

33. Channda Gupta, "Reproductive Malformation of the Male Offspring Following Maternal Exposure to Estrogenic Chemicals," *Proceedings of the Society for Experimental Biology and Medicine* 224 (1999): 61–68. Shortly after the publication of Channda Gupta's study, the same journal published an editorial underlining that Frederick vom Saal's initial results had been confirmed: Daniel Sheehan, "Activity of Environmentally Relevant Low Doses of Endocrine Disruptors and the Bisphenol A Controversy: Initial Results Confirmed," *Proceedings of the Society for Experimental Biology and Medicine* 224, no. 2 (2000): 57–60.

34. Gross, "Toxic Origins of Disease."

35. Barbara Elswick, Frederick Miller, and Frank Welsch, "Comments to the Editor Concerning the Paper Entitled 'Reproductive Malformation of the Male Offspring Following Maternal Exposure to Estrogenic Chemicals' by C. Gupta," *Proceedings of the Society for Experimental Biology and Medicine* 226 (2001): 74–75.

36. Channda Gupta, "Response to the Letter by B. Elswick *et alii* from the Chemical Industry Institute of Toxicology," *Proceedings of the Society for Experimental Biology and Medicine* 226 (2001): 76–77.

37. See notably Derek Yach and Stella Aguinaga Bialous, "Tobacco, Lawyers and Public Health, Junking Science to Promote Tobacco," *American Journal of Public Health* 91, no. 11 (November 2001): 1745–48.

38. Cited by Cindy Skrzycki, "Nominee's Business Ties Criticized," *Washington Post*, May 15, 2001.

39. "George M. Gray," www.sourcewatch.org.

40. Lorenz Rhomberg, "Needless Fear Drives Proposed Plastics Ban," *San Francisco Chronicle*, January 17, 2006.

41. George Gray et al., "Weight of the Evidence Evaluation of Low-Dose Reproductive and Developmental Effects of Bisphenol A," *Human and Ecological Risk Assessment* 10 (October 2004): 875–921.

42. Frederick vom Saal and Claude Hughes, "An Extensive New Literature Concerning Low-Dose Effects of Bisphenol A Shows the Need for a New Risk Assessment," *Environmental Health Perspectives* 113 (August 2005): 926–33.

43. See John Peterson Myers and Frederick vom Saal, "Should Public Health Standards for Endocrine-Disrupting Compounds Be Based Upon 16th Century Dogma or Modern Endocrinology?," *San Francisco Medicine* 81, no. 1 (2008): 30–31.

44. Evanthia Diamanti-Kandarakis et al., "Endocrine-Disrupting Chemicals: An Endocrine Society Scientific Statement," *Endocrine Reviews* 30, no. 4 (June 2009): 293–342.

45. "Opinion of the Scientific Panel on Food Additives, Flavourings, Processing Aids and Materials in Contact with Food on a Request from the Commission Related to 2,2-bis (4-Hydroxyphenyl) Propane (Bisphenol A)," Question no. EFSA-Q-2005-100, November 29, 2006.

46. Rochelle Tyl et al., "Three-Generation Reproductive Toxicity Study of Dietary Bisphenol A in CD Sprague-Dawley Rats," *Toxicological Sciences* 68 (2002): 121–46.

47. When it conducted its evaluation, the EFSA had only a preliminary report of Rochelle Tyl's study ("Draft Final Report"), which was published in 2008: Rochelle Tyl et al., "Two-Generation Reproductive Toxicity Evaluation of Bisphenol A in CD-1 (Swiss Mice)," *Toxicological Sciences* 104, no. 2 (2008): 362–84.

48. John Peterson Myers et al., "Why Public Health Agencies Cannot Depend on Good Laboratory Practices as a Criterion for Selecting Data: The Case of Bisphenol A," *Environmental Health Perspectives* 117, no. 3 (March 2009): 309–15. Its authors include Ana Soto, Carlos Sonnenschein, Louis Guillette, Theo Colborn, and John McLachlan.

49. Meg Missinger and Susanne Rust, "Consortium Rejects FDA Claim of BPA's Safety: Scientists Say 2 Studies Used by U.S. Agency Overlooked Dangers," *Journal Sentinel*, April 11, 2009.

50. Myers et al., "Why Public Health Agencies Cannot Depend on Good Laboratory Practices."

51. As reported by John Peterson Myers, co-author of *Our Stolen Future*, who attended the hearing (John Peterson Myers, "The Missed Electric Moment," *Environmental Health News*, September 18, 2008).

52. Missinger and Rust, "Consortium Rejects FDA Claim of BPA's Safety."

53. "Opinion of the Scientific Panel."

54. Myers et al., "Why Public Health Agencies Cannot Depend on Good Laboratory Practices."

55. Ibid.

56. Frederick Vom Saal et al., "Chapel Hill Bisphenol A Expert Panel Consensus Statement: Integration of Mechanisms, Effects in Animals and Potential to Impact Human Health at Current Levels of Exposure," *Reproductive Toxicology* 24 (2007): 131–38. The signatories included Ana Soto, Carlos Sonnenschein, Retha Newbold, John Peterson Myers, Louis Guillette, and John McLachlan.

57. National Toxicology Program, "NTP–CERHR Monograph on the Potential Human Reproductive and Developmental Effects of Bisphenol A," September 2008. Emphasis in original.

58. Health Canada, "Draft Screening Assessment for Phenol, 4,4'-(1-Methylethylidene) Bis- (80-05-7)," April 2008.

59. Xu-Liang Cao, "Levels of Bisphenol A in Canned Liquid Infant Formula Products in Canada and Dietary Intake Estimates," *Journal of Agricultural and Food Chemistry* 56, no. 17 (2008): 7919–24; Xu-Liang Cao and Jeannette Corriveau, "Migration of Bisphenol A from Polycarbonate Baby and Water Bottles into Water Under Severe Conditions," *Journal of Agricultural and Food Chemistry* 56, no. 15 (2008): 6378–81. A third study indicated the same migration phenomenon in carbonated beverage cans: Xu-Liang Cao et al., "Levels of Bisphenol A in Canned Soft Drink Products in Canadian Markets," *Journal of Agricultural and Food Chemistry* 57, no. 4 (2009): 1307–11.

60. On May 17, 2010, the French Parliament would finally adopt a law banning the sale of polycarbonate baby bottles containing BPA.

61. Once BPA penetrates the body, it decomposes into "free bisphenol" and into two main metabolites: BPA-glucuronide and BPA-sulfate.

62. "Toxicokinetics of Bisphenol A: Scientific Opinion of the Panel on Food Additives, Flavourings, Processing Aids and Materials in Contact with Food (AFC)," Question no. EFSA- Q-2008-382, July 9, 2008.

63. AFSSA, "Avis de l'Agence française de sécurité sanitaire des aliments relatif au bisphénol A dans les biberons en polycarbonate susceptibles d'être chauffés au four à micro-ondes. Saisine no 2008-SA-0141," October 24, 2008.

64. The Plastics Portal, www.plasticseurope.org.

65. In the opinion submitted in September 2010 on the neurological effects of BPA, the EFSA specified, for the first time, that "a minority opinion is expressed by a Panel member," seeing fit to further specify: "EFSA considers it important that scientists are able to express an opinion which diverges from an adopted opinion, that is called a minority opinion." Nonetheless, the experts "concluded that no new study could be identified, which would call for a revision of the current ADI of 0.05 mg/kg b.w./day."

19: The Cocktail Effect

1. Syngenta was created in 2000 through a merger between AstraZeneca and Novartis. For that matter, Novartis was itself created in 1996 from a merger between Sandoz Agro and Ciba-Geigy (see Chapter 16).

2. In February 2011, the French Council of State repealed the market authorization for Cruiser; relying on European directive 91/214, it demanded data proving the substance's safety in the long term.

3. WuQiang Fan et al., "Atrazine-Induced Aromatase Expression Is SF-1 Dependent: Implications for Endocrine Disruption in Wildlife and Reproductive Cancers in Humans," *Environmental Health Perspectives* 115 (May 2007): 720–27.

4. Decision 2004/141/EC from February 12, 2004.

5. "Pesticide Atrazine Can Turn Male Frogs into Females," *Science Daily*, March 1, 2010.

6. Nena Baker, *The Body Toxic: How the Hazardous Chemistry of Everyday Things Threatens Our Health and Well-Being* (New York, North Point Press, 2008), 67.

7. World Wide Fund for Nature (WWF), *Gestion des eaux en France et politique agricole: un long scandale d'État*, June 15, 2010. The two departments the most heavily contaminated with atrazine (and nitrates) in France are l'Eure-et-Loir and la Seine-et-Marne.

8. "Regulators Plan to Study Risks of Atrazine," *New York Times*, October 7, 2009.

9. Baker, *Body Toxic*, 67.

10. The U.S. Geological Survey is a public agency created in 1879 to monitor the evolution of ecosystems and the environment (water quality, earthquakes, hurricanes, etc.). In Illinois, fourteen water providers collectively filed a class action lawsuit in 2004 against Syngenta, demanding $350 million to clean water resources heavily contaminated by atrazine. In 2010, seventeen water providers from six Midwestern states were preparing another class action lawsuit (Rex Dalton, "E-mails Spark Ethics Row: Spat Over Health Effects of Atrazine Escalates," *Nature* 446, no. 918 [August 18, 2010]).

11. A. Donna, P.G. Betta, F. Robutti, and D. Bellingeri, "Carcinogenicity Testing of Atrazine: Preliminary Report on a 13-Month Study on Male Swiss Albino Mice Treated by Intraperitoneal Administration," *Giornale italiano di medicina del lavoro* 8, no. 3–4 (May–July 1986): 119–21; A. Donna et al., "Preliminary Experimental Contribution to the Study of Possible Carcinogenic Activity of Two Herbicides Containing Atrazine-Simazine and Trifuralin as Active Principles," *Pathologica* 73, no. 1027 (September–October 1981): 707–21.

12. A. Pinter et al., "Long-Term Carcinogenicity Bioassay of the Herbicide Atrazine in F344 Rats," *Neoplasma* 37, no. 5 (1990): 533–44.

13. *Occupational Exposures in Insecticide Application and Some Pesticides*, IARC Monographs on the Evaluation of Carcinogenic Risks to Humans, vol. 53, WHO/IARC, 1991.

14. Lawrence Wetzel, "Chronic Effects of Atrazine on Estrus and Mammary Tumor Formation in Female Sprague-Dawley and Fischer 344 Rats," *Journal of Toxicology and Environmental Health* 43, no. 2 (1994): 169–82; James Stevens, "Hypothesis for Mammary Tumorigenesis in Sprague-Dawley Rats Exposed to Certain Triazine Herbicides," *Journal of Toxicology and Environmental Health* 43, no. 2 (1994): 139–53; J. Charles Eldridge, "Factors Affecting Mammary Tumor Incidence in Chlorotriazine-Treated Female Rats: Hormonal Properties, Dosage, and Animal Strain," *Environmental Health Perspectives* 102, suppl. 1 (December 1994): 29–36.

15. M.A. Kettles, S.R. Browning, T.S. Prince, and S.W. Hostman, "Triazine Exposure and Breast Cancer Incidence: An Ecologic Study of Kentucky Counties," *Environmental Health Perspectives* 105, no. 11 (1997): 1222–27.

16. *Some Chemicals That Cause Tumours of the Kidney or Urinary Bladder in Rodents and Some Other Substances*, IARC Monographs on the Evaluation of Carcinogenic Risks to Humans, vol. 73, WHO/IARC, 1999. During my February 2010 meeting with Vincent Cogliano, head of the IARC Monographs Program, he informed me that atrazine was on the list of priority products to be reevaluated.

17. Paul Maclennan, "Cancer Incidence Among Triazine Herbicide Manufacturing Workers," *Journal of Occupational and Environmental Medicine* 44, no. 11 (November 2002): 1048–58. Note that two years later, Exponent scientists (see Chapter 9) published another study in the same publication, showing there was no link between factory exposure to atrazine and prostate cancer. Patrick Hessel et al., "A Nested Case-Control Study of Prostate Cancer and Atrazine Exposure," *Journal of Occupational and Environmental Medicine* 46, no. 4 (2004): 379–85.

18. Tyrone Hayes et al., "Hermaphroditic, Demasculinized Frogs After Exposure to the Herbicide Atrazine at Low Ecologically Relevant Doses," *Proceedings of the National Academy of Sciences USA* 99 (2002): 5476–80; Tyrone Hayes et al., "Feminization of Male Frogs in the Wild," *Nature* 419 (2002): 895–96; Tyrone Hayes et al., "Atrazine-Induced Hermaphroditism at 0.1 ppb in American Leopard Frogs (*Rana pipiens*): Laboratory and Field Evidence," *Environmental Health Perspectives* 111 (2002): 568–75.

19. That same year (2002), Tyrone Hayes received "Berkeley's Distinguished Teaching Award," which rewards professors at the renowned university for the quality of their teaching.

20. William Brand, "Research on the Effects of a Weedkiller on Frogs Pits Hip Berkeley Professor Against Agribusiness Conglomerate," *Oakland Tribune*, July 21, 2002.

21. EPA, "Potential for Atrazine to Affect Amphibian Gonadal Development," October 2007 (Docket ID: EPA-HQ-OPP-2007-0498).

22. Tyrone Hayes, "There Is No Denying This: Defusing the Confusion About Atrazine," *BioScience* 5, no. 12 (2004): 1138–49.

23. Robert Gilliom et al., "The Quality of Our Nation's Waters: Pesticides in the Nation's Streams and Ground Water, 1992–2001," U.S. Geological Survey, March 2006.

24. Tyrone Hayes, "Pesticide Mixtures, Endocrine Disruption and Amphibian Declines: Are We Underestimating the Impact?" *Environmental Health Perspectives* 114, no. 1 (April 2006): 40–50. In this study, Hayes notes that, since 1980, 32 percent of frog species have disappeared and 43 percent are in decline.

25. Department of Health and Human Services, *Fourth National Report on Human Exposure to Environmental Chemicals* (Atlanta: Centers for Disease Control and Prevention, 2009).

26. Cited by Baker, *Body Toxic*, 25.

27. In California, Governor Arnold Schwarzenegger signed a decree in 2006 to implement a biomonitoring program initially aimed at pregnant women.

28. "La chimie ronge le sang des députés européens," *Libération*, April 22, 2004.

29. "Une cobaye verte et inquiète," *Libération*, April 22, 2004.

30. WWF/Greenpeace, "A Present for Life, Hazardous Chemicals in Umbilical Cord Blood," September 2005.

31. In 2010, the Environmental Working Group of Washington published a similar study on the umbilical cords of ten minority newborns in Michigan, Florida, California, and Wisconsin. A total of 232 chemical products were detected (Environmental Working Group, "Pollution in People: Cord Blood Contaminants in Minority Newborns," 2010).

32. Robin Whyatt and Dana Barr, "Measurement of Organophosphate Metabolites in Postpartum Meconium as a Potential Biomarker of Prenatal Exposure: A Validation Study," *Environmental Health Perspectives* 109, no. 4 (2001): 417–20.

33. Robin Whyatt et al., "Contemporary-Use Pesticides in Personal Air Samples During Pregnancy and Blood Samples at Delivery Among Urban Minority Mothers and Newborns," *Environmental Health Perspectives* 111 (2003): 749–56. The infants were monitored throughout their childhood to measure the eventual effects of pesticides on their neurocognitive development.

34. Cécile Chevrier et al., "Biomarqueurs urinaires d'exposition aux pesticides des femmes enceintes de la cohorte Pélagie réalisée en Bretagne, France (2002–2006)," *Bulletin épidémiologique hebdomadaire*, special edition, June 16, 2009, 23–28.

35. John Adgate et al., "Measurement of Children's Exposure to Pesticides: Analysis of Urinary Metabolite Levels in a Probability-Based Sample," *Environmental Health Perspectives* 109 (2001): 583–90. Similar results were obtained in Iowa: Brian Curwin et al., "Urinary Pesticide Concentrations Among Children, Mothers and Fathers Living in Farm and Non-Farm Households in Iowa," *Annals of Occupational Hygiene* 51, no. 1 (2007): 53–65.

36. Pesticide Action Network North America, "Chemical Trespass: Pesticides in Our Bodies and Corporate Accountability," May 2004.

37. "Enquête sur les substances chimiques présentes dans notre alimentation," Future Generations with the Health and Environmental Alliance, the Environmental Health Network (RES) and WWF France, 2010.

38. "Une association alerte sur les substances chimiques contenues dans les repas des enfants," www.LeMonde.fr, December 1, 2010.

39. "Des résidus chimiques dans l'assiette des enfants," *Le Monde*, December 1, 2010.

40. Ulla Hass et al., "Combined Exposure to Anti-androgens Exacerbates Disruption of Sexual Differentiation in the Rat," *Environmental Health Perspectives* 115, suppl. 1 (December 2007): 122–28; Stine Broeng Metzdorff et al., "Dysgenesis and Histological Changes of Genitals and Perturbations of Gene Expression in Male Rats after *in Utero* Exposure to Antiandrogen Mixtures," *Toxicological Science* 98, no. 1 (July 2007): 87–98.

41. Sofie Christiansen et al., "Synergistic Disruption of External Male Sex Organ Development by a Mixture of Four Antiandrogens," *Environmental Health Perspectives* 117, no. 12 (December 2009): 1839–46.

42. Ulrich Beck, *Risk Society: Towards a New Modernity* (London: Sage Publications, 1992), 67. Emphasis in original.

43. Andreas Kortenkamp, "Breast Cancer and Exposure to Hormonally Active Chemicals: An Appraisal of the Scientific Evidence," Health & Environment Alliance, www.envhealth.org, April 2008.

44. The incidence of breast cancer in North America, Europe, and Australia ranges between 75 and 92 per 100,000 (after age adjustment), versus 20 per 100,000 in Asia and Africa.

45. See, namely, Warren Porter, James Jaeger, and Ian Carlson, "Endocrine, Immune and Behavioral Effects of Aldicarb (Carbamate), Atrazine (Triazine) and Nitrate (Fertilizer) Mixtures at Groundwater Concentrations," *Toxicology and Industrial Health* 15, no 1–2 (1999): 133–50.

46. Jesus Ibarluzea et al., "Breast Cancer Risk and the Combined Effect of Environmental Oestrogens," *Cancer Causes and Control* 15 (2004): 591–600.

47. A. Kortenkamp, M. Faust, M. Scholze, and T. Backhaus, "Low-Level Exposure to Multiple Chemicals: Reason for Human Health Concerns?," *Environmental Health Perspectives* 115, suppl. 1 (December 2007): 106–14.

48. REACH is the acronym of the European regulation "Registration, Evaluation, Authorization and Restriction of Chemicals," which entered into force on June 1, 2007.

49. Philippe Grandjean and Philippe Landrigan, "Developmental Neurotoxicity of Industrial Chemicals: A Silent Pandemic," *The Lancet,* November 8, 2006; Philippe Landrigan, "What Causes Autism? Exploring the Environmental Contribution," *Current Opinion in Pediatrics* 22, no. 2 (April 2010): 219–25; Philippe Landrigan et al., "Environmental Origins of Neurodegenerative Disease in Later Life," *Environmental Health Perspectives* 113, no. 9 (September 2005): 1230–33.

50. Beck, *Risk Society,* 69.

51. Mark Blainey et al., *The Benefits of Strict Cut-Off Criteria on Human Health in Relation to the Proposal for a Regulation Concerning Plant Protection Products,* Comité de l'environnement, de la santé publique et de la sécurité alimentaire du Parlement européen, October 2008, IP/A/ENVI/ST/2008-18.

52. Francesca Valent et al., "Burden of Disease Attributable to Selected Environmental Factors and Injury Among Children and Adolescents in Europe," *The Lancet* 363 (2004): 2032–39.

53. Vincent Garry, "Pesticides and Children," *Toxicology and Applied Pharmacology* 198 (2004): 152–63.

54. Deborah Rice and Stan Barone, "Critical Periods of Vulnerability for the Developing Nervous System: Evidence from Humans and Animal Models," *Environmental Health Perspectives* 108, suppl. 3 (2000): 511–33.

55. Patricia M. Rodier, "Developing Brain as a Target of Toxicity," *Environmental Health Perspectives* 103, suppl. 6 (September 1995): 73–76.

56. Gary Ginsberg, Dale Hattis, and Babasaheb Sonawane, "Incorporating Pharmacokinetic Difference Between Children and Adults in Assessing Children's Risk to Environmental Toxicants," *Toxicology and Applied Pharmacology* 198 (2004): 164–83.

57. Cynthia F. Bearer, "How Are Children Different from Adults?," *Environmental Health Perspectives* 103, suppl. 6 (1995): 7–12.

58. M. Lackmann, K.H. Schaller, and J. Angerel, "Organochlorine Compounds in Breastfed vs Bottle-Fed Infants: Preliminary Results at Six Weeks of Age," *Science of the Total Environment* 329 (2004): 289–93; G. Solomon and P. Weiss, "Chemical Contaminants in Breast Milk: Time Trends and Regional Variability," *Environmental Health Perspectives* 110, no. 6 (2002): 339–47.

59. Author interview with Jacqueline Clavel, Villejuif, France, January 6, 2010.

60. Jérémie Rudant et al., "Household Exposure to Pesticides and Risk of Childhood Hematopoietic Malignancies: The ESCALE Study (SFCE)," *Environmental Health Perspectives* 115, no. 12 (December 2007): 1787–93.

61. D.T. Wigle, M.C. Turner, and D. Krewski, "A Systematic Review and Meta-analysis of Childhood Leukemia and Parental Occupational Pesticide Exposure," *Environmental Health Perspectives* 117, no. 5 (May 2009): 1505–13. One of the reference studies is that of C. Infante-Rivard, D. Labuda, M. Krajinovic, and D. Sinnett, "Risk of Childhood Leukemia Associated with Exposure to Pesticides and with Gene Polymorphisms," *Epidemiology* 10 (September 1999): 481–87.

62. Marie-Monique Robin, *The World According to Monsanto: Pollution, Corruption, and the Control of the World's Food Supply*, trans. George Holoch (New York: The New Press, 2010), 64–68.

63. V.F. Garry, D. Schreinemachers, M.E. Harkins, and J. Griffith, "Pesticide Appliers, Biocides, and Birth Defects in Rural Minnesota," *Environmental Health Perspectives* 104, no. 4 (1996): 394–99; Vincent Garry et al., "Birth Defects, Season of Conception, and Sex of Children Born to Pesticide Applicators Living in the Red River Valley of Minnesota, USA," *Environmental Health Perspectives* 110, suppl. 3 (2002): 441–49; V.F. Garry, S.E. Holland, L.L. Erickson, and B.L. Burroughs, "Male Reproductive Hormones and Thyroid Function in Pesticide Applicators in the Red River Valley of Minnesota," *Journal of Toxicology and Environmental Health* 66 (2003): 965–86.

64. Numerous studies show the link between maternal exposure to pesticides and birth defects, but see especially Ana M. García, Tony Fletcher, Fernando G. Benavides, and Enrique Orts, "Parental Agricultural Work and Selected Congenital Malformations," *American Journal of Epidemiology* 149 (1999): 64–74; and P. Kristensen, L.M. Irgens, A. Andersen, A.S. Bye, and L. Sundheim, "Birth Defects Among Offspring of Norwegian Farmers, 1967–1991," *Epidemiology* 8, no. 5 (1997): 537–54.

Conclusion: A Paradigm Shift

1. Stéphane Foucart, "Pourquoi on vit moins vieux aux États-Unis," *Le Monde*, January 27, 2011.

2. Catherine Vincent, "Une personne sur dix est obèse dans le monde," *Le Monde*, February 7, 2011.

3. Pierre Weill, *Tous gros demain?* (Paris: Plon, 2007), 21.

4. David Servan-Schreiber, *Anti-Cancer: Prevent and Fight Through Our Natural Defenses* (New York: Viking Penguin, 2009), 67.

5. See Richard Béliveau and Denis Gingras, *Les Aliments contre le cancer* (Paris: Solar, 2005).

6. Jed Fahey, "Broccoli Sprouts: An Exceptionally Rich Source of Inducers of Enzymes That Protect Against Chemical Carcinogens," *Proceedings of the National Academy of Sciences USA* 94, no. 19 (September 1997): 10367–72.

7. Denis Gingras and Richard Béliveau, "Induction of Medulloblastoma Cell Apoptosis by Sulforaphane, a Dietary Anticarcinogen from Brassica Vegetables," *Cancer Letters* 203, no. 1 (January 2004): 35–43.

8. Michel Demeule and Richard Béliveau, "Diallyl Disulfide, a Chemopreventive Agent in Garlic, Induces Multidrug Resistance-Associated Protein 2 Expression," *Biochemical and Biophysical Research Communications* 324, no. 2 (November 2004): 937–45.

9. Lyne Labrecque and Richard Béliveau, "Combined Inhibition of PDGF and VEGF Receptors by Ellagic Acid, a Dietary-Derived Phenolic Compound," *Carcinogenesis* 26, no. 4 (April 2004).

10. Borhane Annabi et al., "Radiation Induced-Tubulogenesis in Endothelial Cells Is Antagonized by the Antiangiogenic Properties of Green Tea Polyphenol (-) Epigallocatechin-3-Gallate," *Cancer Biology & Therapy* 2, no. 6 (November–December 2003): 642–49; Anthony Pilorget and Richard Béliveau, "Medulloblastoma Cell Invasion Is Inhibited by Green Tea Epigallocatechin-3-Gallate," *Journal of Cellular Biochemistry* 90, no. 4 (November 2003): 745–55.

11. John Weisburger, "Chemopreventive Effects of Cocoa Polyphenols on Chronic Diseases," *Experimental Biology and Medicine* 226, no. 10 (November 2001): 891–97.

12. Meishiang Jang et al., "Cancer Chemopreventive Activity of Resveratrol, a Natural Product Derived from Grapes," *Nature* 25, no. 2 (1999): 65–77.

13. B.B. Aggarwal, A. Kumar, and A.C. Bharti, "Anticancer Potential of Curcumin: Preclinical and Clinical Studies," *Anticancer Research* 23 (2003): 363–98; Bharat Aggarwal, "Prostate Cancer and Curcumin Add Spice to Your Life," *Cancer Biology & Therapy* 7, no. 9 (September 2008): 1436–440; S. Aggarwal et al., "Curcumin (Diferuloylmethane) Down-Regulates Expression of Cell Proliferation and Antiapoptotic and Metastatic Gene Products Through Suppression of IκBα Kinase and Akt Activation," *Molecular Pharmacology* 69, no. 1 (2006): 195–206; Ajaikumar Kunnu-Makkara, Preetha Anand, and Bharat Aggarwal, "Curcumin Inhibits Prolif-

eration, Invasion, Angiogenesis and Metastasis of Different Cancers Through Interaction with Multiple Cell Signaling Proteins," *Cancer Letters* 269, no. 2 (October 2008): 199–225.

14. Author interview with Arvind Chaturvedi, Bhubaneswar, India, December 21, 2009.

15. Cynthia L. Curl, Richard A. Fenske, and Kai Elgethun, "Organophosphorus Pesticide Exposure of Urban and Suburban Preschool Children with Organic and Conventional Diets," *Environmental Health Perspectives* 111, no. 3 (2003): 377–82.

16. Chensheng Lu et al., "Organic Diets Significantly Lower Children's Dietary Exposure to Organophosphorus Pesticides," *Environmental Health Perspectives* 114, no. 2 (2006): 260–63.

17. C. Lu, D.B. Barr, M.A. Pearson, and L.A. Waller, "Dietary Intake and Its Contribution to Longitudinal Organophosphorus Pesticide Exposure in Urban/Suburban Children," *Environmental Health Perspectives* 116, no. 4 (April 2008): 537–42.

18. David Egilman and Susanna Rankin Bohme, "Over a Barrel. Corporate Corruption of Science and Its Effects on Workers and the Environment," *International Journal of Occupational and Environmental Health* 11, no. 4 (2005): 331–37.

19. Cited by André Cicolella, *Le Défi des épidémies modernes: Comment sauver la Sécu en changeant le système de santé* (Paris: La Découverte, 2007), 5.

20. "Long duration diseases" refer to chronic illnesses for which treatments (which are often very expensive) are 100 percent covered by the French National Insurance Fund, thanks to the exemption of user fees.

21. Cicolella, *Le Défi des épidémies modernes*, 17.

22. Ibid., 29.

23. Tom Muir and Marc Zegarac, "Societal Costs of Exposure to Toxic Substances: Economic and Health Costs of Four Case Studies That Are Candidates for Environmental Causation," *Environmental Health Perspectives* 109, suppl. 6 (December 2001): 885–903.

24. Mark Blainey, Catherine Ganzleben, Gretta Goldenman, and Iona Pratt, "The Benefits of Strict Cut-Off Criteria on Human Health in Relation to the Proposal for a Regulation Concerning Plant Protection Products," 2008, IP/A/ENVI/ST/2008-18.

25. David Pimentel et al., "Environmental and Economic Costs of Pesticide Use," *Bioscience* 42, no. 10 (1992): 750–60.

26. Jacques Ferlay et al., "Estimates of the Cancer Incidence and Mortality in Europe in 2006," *Annals of Oncology* 18, no. 3 (2007): 581–92.

27. European Commission, *Commission Staff Working Paper*, 2003.

28. Michael Ganz, "The Lifetime Distribution of the Incremental Societal Costs of Autism," *Archives of Pediatrics and Adolescent Medicine* 161, no. 4 (April 2007): 343–49; Krister Järbrink, "The Economic Consequences of Autistic Spectrum Disorder Among Children in a Swedish Municipality," *Autism* 11, no. 5 (September 2007): 453–63.

29. European Commission DG XXIV (consumer health), December 1998.

30. Michel Callon, Pierre Lascoumes, and Yannick Barthes, *Acting in an Uncertain World: An Essay on Technical Democracy*, trans. Graham Burchell (Cambridge, MA: MIT Press, 2000).

31. Ibid., 196.

32. Ibid., 215.

33. The following quotes come from Michel Callon, "Secluded Research," in Callon et al., *Acting in an Uncertain World*, 41.

34. Michel Gérin et al., *Environnement et santé publique: Fondements et pratiques* (Montreal: Edisem, 2003), 79.

35. Jacqueline Verrett and Jean Carper, *Eating May Be Hazardous to Your Health* (New York: Simon & Schuster, 1974), 85.

Index

Marie-Monique Robin is an award-winning journalist and filmmaker. She received the prestigious 1995 Albert-Londres Prize, awarded to investigative journalists in France, and the 2009 Rachel Carson Prize. Robin is also the director and producer of more than thirty documentaries and investigative reports filmed in Latin America, Africa, Europe, and Asia. She is the author of *The World According to Monsanto: Pollution, Corruption, and the Control of Our Food Supply* (The New Press) and *The Photos of the Century: 100 Historic Moments,* among other works. She lives outside Paris.

Allison Schein holds master's degrees in French language and civilization and French–English literary translation from New York University. She lives in New York City.

Lara Vergnaud holds a graduate diploma in specialized English–French translation from the University of Lyon 2 in France and a master's degree in French–English literary translation from New York University. She lives in New York City.

Publishing in the Public Interest

Thank you for reading this book published by The New Press. The New Press is a nonprofit, public interest publisher. New Press books and authors play a crucial role in sparking conversations about the key political and social issues of our day.

We hope you enjoyed this book and that you will stay in touch with The New Press. Here are a few ways to stay up to date with our books, events, and the issues we cover:

- Sign up at www.thenewpress.com/subscribe to receive updates on New Press authors and issues and to be notified about local events
- Like us on Facebook: www.facebook.com/newpressbooks
- Follow us on Twitter: www.twitter.com/thenewpress

Please consider buying New Press books for yourself; for friends and family; or to donate to schools, libraries, community centers, prison libraries, and other organizations involved with the issues our authors write about.

The New Press is a 501(c)(3) nonprofit organization. You can also support our work with a tax-deductible gift by visiting www.thenewpress.com/donate.